THE CONCISE HANDBOOK OF

ANALYTICAL SPECTROSCOPY

Theory, Applications, and Reference Materials

Volume 5

Raman Spectroscopy

THE CONCISE HANDBOOK OF ANALYTICAL SPECTROSCOPY

Theory, Applications, and Reference Materials

Volume 5

Raman Spectroscopy

Jerry Workman

Unity Scientific, USA and National University, USA

World Scientific

NEW JERSEY • LONDON • SINGAPORE • BEIJING • SHANGHAI • HONG KONG • TAIPEI • CHENNAI • TOKYO

Published by

World Scientific Publishing Co. Pte. Ltd.

5 Toh Tuck Link, Singapore 596224

USA office: 27 Warren Street, Suite 401-402, Hackensack, NJ 07601

UK office: 57 Shelton Street, Covent Garden, London WC2H 9HE

Library of Congress Cataloging-in-Publication Data

Workman, Jerry.

The concise handbook of analytical spectroscopy : theory, applications, and reference materials / by Jerry Workman (Unity Scientific, USA & National University, USA).

volumes cm

Includes bibliographical references and index.

ISBN 978-9814508056 (Set)
ISBN 978-9814508087 (v. 1 : hardcover : alk. paper)
ISBN 978-9814508094 (v. 2 : hardcover : alk. paper)
ISBN 978-9814508100 (v. 3 : hardcover : alk. paper)
ISBN 978-9814508117 (v. 4 : hardcover : alk. paper)
ISBN 978-9814508124 (v. 5 : hardcover : alk. paper)

1. Spectrum analysis. I. Title.

QD95.W67 2015

543'.5--dc23

2014042606

British Library Cataloguing-in-Publication Data

A catalogue record for this book is available from the British Library.

Typeset by Stallion Press
Email: enquiries@stallionpress.com

Printed in Singapore

Preface

The Concise Handbook of Analytical Spectroscopy: Theory, Applications, and Reference Materials (In 5 Volumes)

Overview and Description

The concept of improving the use of electromagnetic energy to achieve a variety of qualitative and quantitative spectroscopic measurements on solid and liquid materials has been proliferating at a rapid rate. The use of such technologies to measure chemical composition and appearance; for classification; and to achieve a detailed understanding of material interactions has prompted a dramatic expansion in the use and development of spectroscopic techniques over a variety of academic and commercial fields. This multi-volume handbook is designed specifically as a reference tool for students, commercial development and quality scientists, and researchers or technologists in a variety of measurement endeavors. Each volume covers the theory, instrumentation, sampling methods, experimental design, and data analysis techniques, as well as essential reference tables, figures, and spectra for each spectroscopic region. The detailed practical aspects of applying spectroscopic tools for many of the most exciting and current applications are covered. Featured applications are included as well as common commercial analysis methods (i.e., color, appearance), spectroscopic quantitative and qualitative techniques, and advanced methods.

Outline

Each volume covers a specific region of the electromagnetic spectrum used for electronic and vibrational spectroscopic measurements. The volumes are not a series of essays, or a compilation of isolated papers on specific topics, but represent an integrated whole. The Handbook is compiled as a highly useful reference set for those directly involved in making and understanding spectroscopic measurement techniques. Each volume contains a logical progression of the material as follows.

Organization of Contents for Each Volume

Each volume commences with Front Reference Material for General Spectroscopy, preceding the following chapters.

1. General Introduction and History of Technique (includes useful basic concepts)
2. Theory of Technique (including delineation of common equations and references)
3. Types of Instrumentation (Types and design aspects)
4. Sampling Considerations (Cookbook approach to preparing samples)
5. Spectrum Data Acquisition (Experimental design and methods for data acquisition)
6. Data Analysis and Presentation Methods (The chemometrics and data analysis techniques applicable for each spectral region)
7. Applications and Spectral Examples (Cookbook approaches, results, and extensive literature reviews)
8. Reference Materials, Tables, Figures, and Spectra (The key look-up aspects for each technique)
9. Glossary of Specific Terms: Detailed Index for Reference Material and Text (Each volume includes a detailed table of contents and index that delineates the minutia of using the text and reference material in each volume.)

Each volume contains specifics for the spectral region covered, such as: common and unusual sampling devices; sampling methods (i.e., reagents, solvents, and sampling cells); and reference standard materials. Optics include: filters, etalons, prisms, and lenses; and fiber optics (principles, equipment, and connectors). In addition, reference tables and spectra, glossaries of terms, common equations and calculation methods, cookbook directions, and solved problems are included.

Reference Materials

Fundamental reference sources and early publications are referred to whenever possible as a primary source of data and information. Reference material common to each method is summarized within each volume to provide a more practical set of volumes for bench use. The intention is that each volume will stand somewhat alone for the working analyst using a specific spectroscopic method, such that required referral to other volumes is kept to a minimum and that key information is available regarding the individual spectroscopic technique within each volume.

Terminology and Symbols

Terminology use is included within the chapters and glossary sections for each volume. Note that *spectrophotometer* is the precise term used for each instrument described within the five volumes. However for brevity, the term *spectrometer* is sometimes used less precisely. Also, the symbols used are from different disciplines and thus there are some symbols used for more than one variable or parameter, depending on the common usage and the specific academic discipline where the symbology was formalized. A list of symbols used in this set of volumes may be located using the index and front material of each volume. The terms used for each equation within the volumes are defined where the equation is used.

Dedication

This book is dedicated to my fabulous wife Rebecca and to my family who have patiently sacrificed many hours of their lives for the cause of spectroscopy; and to everyone who spends many hours in laboratories analyzing samples for new knowledge, applied research, quality improvement, and the overall betterment of the lives of people ... And to the One Great Designer who made all this marvelous measurement with light possible, discoverable, and such a pleasure to work with ...

Author Biography

Jerome (Jerry) J. Workman, Jr. is Executive Vice President of Research & Engineering for Unity Scientific; Certified Core Adjunct Professor at National University, CA; and Principal at Biotechnology Business Associates. He was formerly Vice President of Technology Research for Masimo Corporation; Director of Research, Technology & Applications Development for Molecular Spectroscopy & Microanalysis for ThermoFisher Scientific; Senior Research Fellow at Kimberly-Clark Analytical Science & Technology; and Principal Scientist at Perkin-Elmer.

Dr. Workman has played a major role in defining and developing over twenty scientific instrument advancements and applications of novel vibrational spectroscopy technology for laboratory, process and remote monitoring of commercial synthetic and natural products and processes; and medical applications of spectroscopy. Over 55 U.S. and International Patent Applications (since 1998); 17 U.S. and International Patents issued, and multiple trade secrets. He has a total of 475 Technical Publications; and 18 reference books on a broad range of spectroscopy, chemometrics, and data processing techniques. He has received awards from the Eastern Analytical Symposium, ASTM International, and the Coblentz Society, as well as multiple fellowships, technical, and government appointments. He has taught annual courses in spectroscopy, chemometrics, and statistics for the Association of Official Analytical Chemists, the American Chemical Society, the Instrument Society of America, and the Federation of Analytical Chemists and Spectroscopy Societies; and at several universities, government agencies, and corporations.

Patented inventions include: multi-spectral monitoring (data fusion), web-based information technology in spectroscopy, scanning systems for positioning materials in manufacturing technology, invention of small molecule metabolite reporters for monitoring multiple parameters in human metabolism, advanced technology and algorithm development for instrument automation, standardization, and calibration, and for manufacturing monitoring methods for consumer products.

He is a Fellow of the American Institute of Chemists, the American Society for Testing and Materials International, and the Royal Society of Chemistry (U.K.). He is also a Chartered Chemist (CChem) of the Royal Society of Chemistry (England) and a Chartered Scientist of the Science Council (UK). He has served as Chair of the Industrial Advisory Board for the Center for Process Analytical Chemistry (CPAC) at the Univ. of Washington; and The Council for Near-Infrared Spectroscopy; and is past Chairman of ASTM Main Committee E13 on Molecular Spectroscopy and Separation Science. He has served on the U.S. National Academies NRC Panel for assessment of NIST programs.

With a B.A. in natural sciences and an M.A. in biological sciences from Saint Mary's University of Minnesota, Dr. Workman's Ph.D. is in biological chemistry from Columbia Pacific University. He is a graduate of the Columbia Senior Executive Program and also holds Columbia Business School Certificates in Executive Development (CIED) and in Business Excellence (CIBE). He also holds a Certificate in Strategy and Innovation from the M.I.T. Sloan School. He is listed in Who's Who in the World, Who's Who in America, and Who's Who in Science and Engineering.

Contents

Preface v

Author Biography ix

References and Tables xiii

Chapter 1. Introduction to Raman Spectroscopy 1

Chapter 2. Theory of Raman Spectroscopy, Light Interaction with Matter 51

Chapter 3. Raman Instrumentation 69

Chapter 4. Sampling Considerations for Raman Spectroscopy 99

Chapter 5. Data Acquisition for Raman Spectra 119

Chapter 6. Raman Data Analysis and Presentation 125

Chapter 7. Applications and Spectral Examples for Raman Spectroscopy 173

Chapter 8. Raman Reference Material Spectra and Tables 229

Chapter 9. Glossary of Raman Spectroscopy Terms 251

Volume 5 Index 257

Master Index 265

Reference Tables: Comparison of Spectroscopic Techniques, the Electromagnetic Spectrum, Raman Analysis, and Useful Spectral Ranges for Sources, Detectors and Optical Materials

0.1 Physical Constants and Conversion Factors for SI and non-SI Units

Table 0.1 Spectroscopy constants.[1]

Physical constant	Value	Standard uncertainty	Concise form
Avogadro constant. N_A, L	$6.022\ 141\ 29\times10^{23}$ mol^{-1}	$0.000\ 000\ 27\times10^{23}$ mol^{-1}	$6.022\ 141\ 29(27)\times10^{23}$ mol^{-1}
Boltzmann constant, k	$1.380\ 6488\times10^{-23}$ J K^{-1}	$0.000\ 0013\times10^{-23}$ J K^{-1}	$1.380\ 6488(13)\times10^{-23}$ J K^{-1}
Faraday constant, F	96 485.3365 C mol^{-1}	0.0021 C mol^{-1}	96 485.3365(21) C mol^{-1}
Molar gas constant, R	8.314 4621 J mol^{-1} K^{-1}	0.000 0075 J mol^{-1} K^{-1}	8.314 4621(75) J mol^{-1} K^{-1}
Planck constant, h	$6.626\ 069\ 57\times10^{-34}$ J s	$0.000\ 000\ 29\times10^{-34}$ J s	$6.626\ 069\ 57(29)\times10^{-34}$ J s
Speed of light (in vacuum), c	299 792 458 m s^{-1}	(exact)	299 792 458 m s^{-1}
Stefan-Boltzmann constant, ö	$5.670\ 373\times10^{-8}$ W m^{-2} K^{-4}	$0.000\ 021\times10^{-8}$ W m^{-2} K^{-4}	$5.670\ 373(21)\times10^{-8}$ W m^{-2} K^{-4}

Table 0.2 Wavelength conversions.[2]

To convert Column 1 into Column 2 multiply by	Column 1 SI Units	Column 2 non-SI Units	To convert Column 2 into Column 1 multiply by
2.54	Centimeter, cm (10^{-2} m)	Inch, in	0.394
25.4	Millimeter, mm (10^{-3} m)	Inch, in	3.94×10^{-2}
1.0	Micrometer, μm (10^{-6} m)	Micron, μ	1.0
0.1	Nanometer, nm (10^{-9} m)	Angstrom, A (10^{-10} m)	10
$1/(\text{nm} \times 10^7)$	Nanometer, nm (10^{-9} m)	Wavenumber (cm^{-1})	$1/(\text{cm}^{-1} \times 10^7)$
$1/(\text{nm} \times 2.99792458 \times 10^{17})$	Nanometer, nm (10^{-9} m)	Frequency (Hertz, Hz)	$1/(\text{Hz} \times 2.99792458 \times 10^{17})$

Example 1: To convert Nanometer, nm (10^{-9} m) in Column 1 to Angstrom, A (10^{-10} m) in Column 2 to multiply by 0.1; therefore 0.1 nm = 1 Angstrom.

Example 2: To convert Angstrom, A (10^{-10} m) in Column 2 to Nanometer, nm (10^{-9} m) in Column 1 multiply by 10; therefore 10 Angstroms = 1 nm.

Table 0.3 Length conversions.

To convert Column 1 into Column 2 multiply by	Column 1 SI Units	Column 2 non-SI Units	To convert Column 2 into Column 1 multiply by
1.609	Kilometer, km (10^3 m)	Mile, mi	0.6213
0.914	Meter, m	Yard, yd	1.094
0.304	Meter, m	Foot, ft	3.28
1.0	Micrometer, μm (10^{-6} m)	Micron, μ	1.0
1000	Micrometer, μm (10^{-6} m)	Millimeter, mm (10^{-3} m)	0.001
25.4	Millimeter, mm (10^{-3} m)	Inch, in	3.94×10^{-2}
0.1	Nanometer, nm (10^{-9} m)	Angstrom, A	10

Table 0.4 Area conversions.

To convert Column 1 into Column 2 multiply by	Column 1 SI Units	Column 2 non-SI Units	To convert Column 2 into Column 1 multiply by
0.405	Hectare, ha	Acre	2.47
4.05×10^{-3}	Square kilometer, km^2 $(10^3\ m)^2$	Acre	2.47
2.590	Square kilometer, km^2 $(10^3\ m)^2$	Square mile, mi^2	0.386
4.05×10^{3}	Square meter, m^2	Acre	2.47×10^{-4}
9.29×10^{-2}	Square meter, m^2	Square foot, ft^2	10.76
645	Square millimeter, mm^2 $(10^{-3}\ m)^2$	Square inch, in^2	1.55×10^{-3}

Table 0.5 Volume conversions

To convert Column 1 into Column 2 multiply by	Column 1 SI Units	Column 2 non-SI Units	To convert Column 2 into Column 1 multiply by
102.8	Cubic meter, m^3	Acre-inch	9.73×10^{-3}
2.83×10^{-2}	Cubic meter, m^3	Cubic foot. ft^3	35.3
1.64×10^{-5}	Cubic meter, m^3	Cubic inch, in^3	6.10×10^{4}
35.24	Liter, L $(10^{-3}\ m^3)$	Bushel, bu	2.84×10^{-2}
0.946	Liter, L $(10^{-3}\ m^3)$	Quart (liquid), qt	1.057
28.3	Liter, L $(10^{-3}\ m^3)$	Cubic foot, ft^3	3.53×10^{-2}
3.78	Liter, L $(10^{-3}\ m^3)$	Gallon	0.265
2.96×10^{-2}	Liter, L $(10^{-3}\ m^3)$	Ounce (fluid), oz	33.78
0.473	Liter, L $(10^{-3}\ m^3)$	Pint (fluid), pt	2.11

Table 0.6 Mass conversions.

To convert Column 1 into Column 2 multiply by	Column 1 SI Units	Column 2 non-SI Units	To convert Column 2 into Column 1 multiply by
453.6	Gram, g (10^{-3} kg)	Pound, lb	2.2×10^{-3}
28.4	Gram, g (10^{-3} kg)	Ounce (Avoirdupois, avdp), oz	3.52×10^{-2}
0.454	Kilogram, kg	Pound, lb	2.205
100	Kilogram, kg	Quintal (metric), q	0.01
907	Kilogram, kg	Ton (2000 lb), U.S. ton	1.10×10^{-3}
0.907	Megagram, Mg (Tonne)	Ton (U.S.), ton	1.102
0.907	Tonne, t	Ton (U.S.), ton	1.102

Table 0.7 Rate conversions.

To convert Column 1 into Column 2 multiply by	Column 1 SI Units	Column 2 non-SI Units	To convert Column 2 into Column 1 multiply by
0.447	Meter per second, m s^{-1}	Mile per hour	2.237

Table 0.8 Specific surface conversions.

To convert Column 1 into Column 2 multiply by	Column 1 SI Units	Column 2 non-SI Units	To convert Column 2 into Column 1 multiply by
0.1	Square meter per kilogram, m^2 kg^{-1}	Square centimeter per gram, cm^2 g^{-1}	10
0.001	Square meter per kilogram, m^2 kg^{-1}	Square millimeter per gram, mm^2 g^{-1}	1000

Table 0.9 Density conversions.

To convert Column 1 into Column 2 multiply by	Column 1 SI Units	Column 2 non-SI Units	To convert Column 2 into Column 1 multiply by
1.00	Megagram per cubic meter, Mg m^{-3}	Gram per cubic centimeter, g cm^{-3}	1.00

Table 0.10 Pressure conversions.

To convert Column 1 into Column 2 multiply by	Column 1 SI Units	Column 2 non-SI Units	To convert Column 2 into Column 1 multiply by
0.101	Megapascal, MPa (10^6 Pa)	Atmosphere	9.869
0.1	Megapascal, MPa (10^6 Pa)	Bar	10
47.9	Pascal, Pa	Pounds per square foot, lb ft^{-2}	2.09×10^{-2}
6.90×10^3	Pascal, Pa	Pounds per square inch, lb in^{-2}	1.45×10^{-4}

Table 0.11 Temperature conversions.

To convert Column 1 into Column 2 multiply by	Column 1 SI Units	Column 2 non-SI Units	To convert Column 2 into Column 1 multiply by
100(K-273)	Kelvin, K	Celsius, °C	1.00 (°C + 273)
(°R − 491.67) × 5/9	Rankine, °R	Celsius, °C	(°C + 273.15) × 9/5
(9/5 °C) + 32	Celsius, °C	Fahrenheit, °F	5/9 (°F − 32)
°R − 459.67	Rankine, °R	Fahrenheit, °F	°F + 459.67
[K] × 9/5 − 459.67	Kelvin, K	Fahrenheit, °F	(°F + 459.67) × 5/9
[K] × 9/5	Kelvin, K	Rankine, °R	[°R] × 5/9

Table 0.12 Energy, work, quantity of heat conversions.

To convert Column 1 into Column 2 multiply by	Column 1 SI Units	Column 2 non-SI Units	To convert Column 2 into Column 1 multiply by
1.05×10^3	Joule, J	British thermal unit, Btu	9.48×10^{-4}
4.19	Joule, J	Calorie, cal	0.239
10^{-7}	Joule, J	erg	107
1.36	Joule, J	Foot-pound	0.735
4.19×10^4	Joule per square meter, $J\,m^{-2}$	Calorie per square centimeter (Langley)	2.387×10^{-5}
10^{-5}	Newton, N	Dyne	10^5
698	Watt per square meter, $W\,m^{-2}$	Calorie per square centimeter minute (irradiance), cal $cm^{-2}\,m^{-1}$	1.43×10^{-3}

Table 0.13 Plane angle conversions.

To convert Column 1 into Column 2 multiply by	Column 1 SI Units	Column 2 non-SI Units	To convert Column 2 into Column 1 multiply by
$1.75 \times 10^{-2} = \pi/180$	Radian, rad	Degrees (angle), °	57.3
60	Minutes (of arc), arcminute	Degrees (angle), °	$1/60 = 0.01667$
3600	Seconds (of arc), arcsecond	Degrees (angle), °	$1/3600 = 0.0002778$

Table 0.14 Electricity and magnetism conversions.

To convert Column 1 into Column 2 multiply by	Column 1 SI Units	Column 2 non-SI Units	To convert Column 2 into Column 1 multiply by
10	Siemen per meter, S m-1	Millimho per centimeter, mmho cm^{-1}	0.1
10^{-4}	Tesla, T	Gauss, G	10^{4}

Table 0.15 Fluid flow conversions.

To convert Column 1 into Column 2 multiply by	Column 1 SI Units	Column 2 non-SI Units	To convert Column 2 into Column 1 multiply by
102.8	Cubic meter, m^3	Acre-inch, acre-in	9.73×10^{-3}
101.9	Cubic meter per hour, $m^3 h^{-1}$	Cubic foot per second, $ft^3 s^{-1}$	9.81×10^{-3}
0.227	Cubic meter per hour, $m^3 h^{-1}$	U.S. gallon per minute, gal min^{-1}	4.40
0.123	Hectare meter, ha m	Acre-foot, acre-ft	8.11
1.03×10^{-2}	Hectare meter, ha m	Acre-inch, acre-in	97.28
12.33	Hectare centimeter, ha cm	Acre-foot, acre-ft	8.1×10^{-2}

Table 0.16 Concentration conversions.

To convert Column 1 into Column 2 multiply by	Column 1 SI Units	Column 2 non-SI Units	To convert Column 2 into Column 1 multiply by
1	Centimole per kilogram, cmol kg^{-1}	Milliequivalent per 100 grams, meq $100 g^{-1}$	1
0.1	Gram per kilogram, g, kg^{-1}	Percent, %	10
1	Milligrams per kilogram, mg kg^{-1}	Parts per million, ppm	1

Table 0.17 Radioactivity conversions.

To convert Column 1 into Column 2 multiply by	Column 1 SI Units	Column 2 non-SI Units	To convert Column 2 into Column 1 multiply by
3.7×10^{10}	Becquerel, Bq	Curie, Ci	2.7×10^{-11}
37	Becquerel per kilogram, Bq kg^{-1}	Picocurie per gram, pCi g^{-1}	2.7×10^{-2}
0.01	Gray, Gy (Absorbed dose)	Rad, rd	100
0.01	Sievert, Sv (equivalent dose)	Rem (Roentgen equivalent man)	100

0.2 Introduction and Comparison of Spectroscopic Methods[3]

0.2.1 *Ultraviolet (UV) Spectroscopy*

The Ultraviolet region is specified as 190 to 360 nanometers (nm) or 10^{-9} meters. The types of electrons that can be excited by UV/Vis light are few in number: nonbonding electrons, electrons in single bonds, and electrons involved in double bonds. These may be excited to several excited states. The distinction between molecules is that the ability to "jump" to higher states is affected by attached moieties — for example: double bonds, conjugations, and elements such as: oxygen, bromine, and others with pairs of nonbonding electrons. As a consequence, most elements with UV/Vis absorbances have specific wavelengths at which their peak absorbances occur. These peaks may be used to identify a particular molecule. The UV/Vis region is not as "rich" in information as, say, the infrared spectrum, but it reveals enough detail to enable a comparison of a material with a previously identified substance. A common use of this capability is in the Pharmaceutical industry, where UV/Vis detectors are frequently used with HPLC instruments as a final check before a drug product is released for sale to consumers. The various moieties or chromophores associated with ultraviolet absorption include: Nitriles (R-C≡N), 160 nm; Acetylenes (-C≡C-), 170 nm; Alkenes (>C=C<), 175 nm; Alcohols (R-OH), 180 nm and 175–200 nm; Ethers (R-O-R′), 180 nm; Ketones (R-C=O -R′), 180 nm and 280 nm; Amines, primary ($R\text{-}NH_2$), 190 nm and 200–220 nm; Aldehydes (R-C=O-H), 190 nm and 290 nm; Carboxylic acids (R-C=O-OH), 205 nm; Esters (R-C=O-OR′), 205 nm; Amides, primary ($R\text{-}C\text{=}O\text{-}NH_2$), 210 nm; Thiols (R-SH), 210 nm; Nitrites ($R\text{-}NO_2$), 271 nm; and Azo-group (R-N=N-R′), 340 nm.

0.2.2 *Visible (Vis) Spectroscopy*

Visible light (360 to 780 nm) is the light or electromagnetic energy seen by the human eye. The actual light demonstrated by different colors is visible light and spectra in the visible region are subjected to specific mathematical algorithms to determine color coordinates or color space coordinates to enable specific colors and brightness or darkness to be precisely described in mathematical terms. The aspects of color measurement and computation are contained in the text volume on visible spectroscopy and may be located in the index for this set of volumes. The terms "color measurement" and "visible spectroscopy" are the main index topic headings to find detailed information in this set of volumes. Colors are produced by electrons in a pigment moving from one orbital transition to another around the atoms within the molecules of the colored substance. Visible reflectance or transmittance spectra demonstrate the various colors. The color appearance of the human eye to a particular visible spectrum is dependent on: the light source color temperature and emission spectrum, the observer angle of observation, and the object background brightness and color differences. The size, direction, and scattering/absorption characteristics of the object also have an effect on its color appearance. The basic seven colors first described by Isaac Newton, with their corresponding wavelength scale in nanometers, consist of: violet (360–415 nm); indigo (415–444 nm), blue (444–487 nm), green (487–540 nm), yellow (540–590 nm), orange (590–690 nm), and red (690–830 nm). A twelve color system, with corresponding wavelength information is more common in modern terminology, although technically color is defined by a series of color measurement coordinate systems. These color measurement coordinate systems include: XYZ tristimulus values, Yxy color space, L*a*b* color space, L*C*h color space, and Hunter Lab color space. These are described in detail within sections of these volumes (see index and coordinate system names for more information).

There are also molecular absorptions that occur within the visible region and these are described in various chapters within this text and include the notations for the harmonic band or overtone involved: O-H Alkyl alcohol (6ν, no hydrogen bonding), Aromatic C-H Stretch (5ν), O-H Alkyl alcohol (4ν, no hydrogen bonding), Methyl C-H Stretch (5ν), O-H Phenols (4ν, no hydrogen bonding), Methylene C-H Stretch (5ν), O-H Primary Alcohols (4ν), O-H Water (4ν), O-H Secondary Alcohols (4ν), O-H Tertiary Alcohols (4ν), and Alkenes, conjugated RC=C-C=C-R′. The visible spectra include information related to both electronic and molecular interactions of energy with atoms and molecules. Volume 2, Chapter 7 gives more details.

0.2.3 *Near-Infrared (NIR) Spectroscopy*

Near-infrared spectroscopy is used where multicomponent molecular vibrational analysis is required in the presence of interfering substances. The near-infrared spectra consist of overtones and combination bands of the fundamental molecular absorptions found in the mid-infrared region. Near-infrared spectra consist of generally overlapping vibrational bands that are non-specific and poorly resolved. The use of chemometric mathematical data processing can be used to calibrate for qualitative or quantitative analysis despite these apparent spectroscopic limitations. Traditional near-infrared spectroscopy was used in agricultural product analysis for lignin polymers (2270 nm), paraffins and long alkane chain polymers (2310 nm), glucose based polymers such as cellulose (2336 nm), amino acid polymers as proteins (2180 nm), carbohydrates (2100 nm), and moisture (1440 and 1940 nm). The dominant near-infrared spectral features include the methyl C-H stretching vibrations, methylene C-H stretching vibrations, aromatic C-H stretching vibrations, O-H stretching vibrations, methoxy C-H stretching, and carbonyl associated C-H stretching. In addition, N-H from primary amides, secondary amides (both alkyl, and aryl group associations), N-H from primary, secondary, and tertiary amines, and N-H from amine salts predominate near-infrared spectral features of polymers and organic compounds.

0.2.4 *Infrared (IR or Mid-IR) Spectroscopy*

Mid-infrared spectroscopy provides a measurement technique for intense, isolated and reliable absorption bands of fundamental molecular vibrations from polymers and other organic compounds. The spectrometric methodology allows for univariate calibration with higher signal strength (absorptivity) required for solid, liquids or gas phase measurements. Relatively small pathlengths of 0.1 to 1.0 mm are required for hydrocarbon liquids and solids. The technique is generally incompatible with the use of fiber optics, but specialized fiber materials exist. Instrumentation is higher cost than near-infrared spectrophotometers for the most part.

Dominant mid-infrared spectral include the C-H (methyl, methylene, aromatic, methoxy, and carbonyl) fundamental stretching and bending molecular vibrations, O-H (hydroxyl) stretch fundamental vibrations; N-H (amine) stretching, C-F (fluorocarbon) stretching, -C≡N (nitrile) stretching, -C=O (carbonyl) stretch from esters, acetates, and amides; C-Cl stretch from chlorinated hydrocarbons, and -NO_2 from nitro- containing compounds.

0.2.5 *Raman Spectroscopy*

Raman spectroscopy can be used for a variety of measurements on samples that are aqueous in nature or where glass sample holders are present. Carbon dioxide, water, and glass (silica) are weak scatterers and thus there is generally no problems in analyzing samples having these properties. There is typically no sample preparation involved in samples measured using Raman. Raman spectroscopy is complementary to mid infrared spectroscopy in the measurement of fundamental molecular vibrations. Raman measurements are compatible with fiber optics. Raman measurements exhibit high signal-to-noise and a reasonable cost for instrumentation. The dominant Raman spectral features are acetylenic C≡C stretching, olefinic C=C stretch at 1680–1630 cm^{-1}, N=N (azo-) stretching, S-H (thio-) stretching, C=S stretching, C-S stretching, and S-S stretching bands. Raman spectra also contain such molecular vibrational information as CH_2 twist and wagging, carbonyl

C=O stretch associated with esters, acetates, and amides; C-Cl (halogenated hydrocarbons) stretching, and -NO_2 (nitro-/nitrite) stretching. In addition, Raman yields information content of phenyl- containing compounds at 1000 cm^{-1}.

0.3 The Electromagnetic Spectrum

The electromagnetic spectrum consists of many different types of radiation (energy): gamma, x-ray, ultraviolet (UV), visible (Vis), infrared (IR), microwave, and radio waves (Figures 0.1 through 0.3). Each type of radiation occupies its own region of the electromagnetic spectrum, and the major difference between the individual spectral regions is merely the energy involved. This energy is expressed in units of: photon energy (eV), Frequency (Hz), wavenumber (cm^{-1}), or wavelength (nm) and the manner in which these energies interact with matter.

With the possible exception of gamma radiation, most people are familiar with these types of radiation through their experiences in daily living. X-rays penetrate our bodies, allowing physicians to visualize our internal anatomy. UV light is associated with sunburn and tanning. We see colors and objects in the visible spectrum. In the kitchen, we toast our bread with IR radiation, and "zap" our meals with microwaves. We use microwave transmission for cell phone technology and we use radio waves to broadcast sound and images through AM/FM radio and television signals. The spectral region most useful in day-to-day analytical chemistry is the range of wavelengths from just below our visual perception (i.e., Ultraviolet or UV) through the colors we see (i.e., visible or Vis), known collectively as the UV/Vis spectral region. The wavelengths covered in the UV/Vis are measured in nanometers (nm), a unit of length representing one billionth of a meter. The generally accepted ranges for the UV/Vis region are ultraviolet (190–380 nm) and visible (380–750 nm). Some UV/Vis work involves longer wavelengths in the neighboring shortwave near-IR spectral region, from about 750–950 nm.

0.3.1 *Molecular and Electronic Spectroscopy Unit Conversions*

The units used to describe spectra for molecular spectroscopy vary with the academic or engineering discipline describing the spectral regions. For the most part the spectra are described in terms of an optical response specific to energy from a particular spectral region. The responses are in terms of inelastic or elastic collisions of the energy with various materials and are referred to as scattering absorption, absorbance, reflection, transmission, and so on. The various spectral energy for molecular and electronic spectroscopy is described in terms of units of wavelength (e.g., microns (μm) as 10^{-6} meters, or nanometers (nm) as 10^{-9} m, or Angstroms (Å) as 10^{-10} m). Energy is also expressed as frequency (cycles per second or sec.$^{-1}$) termed Hertz (Hz), as Wavenumbers (cm^{-1}), and as electron Volts (eV).

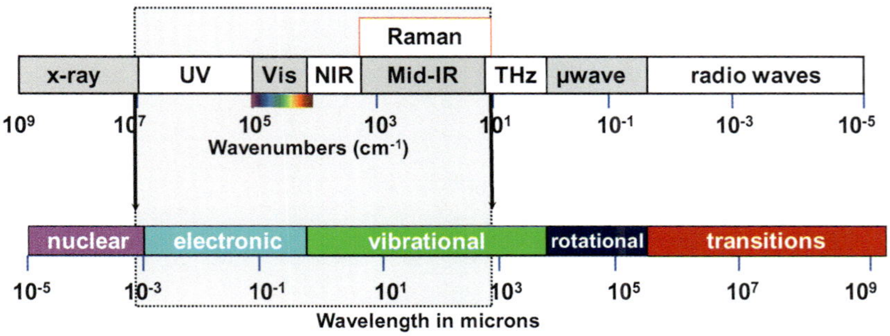

Fig. 0.1 The electromagnetic spectrum highlighting the ultraviolet to infrared/Raman spectral regions. Ultraviolet through Infrared/Raman range is shown in the rectangle. This region is from 190 nanometers to 25,000 nanometers, or 0.19 to 25 microns and represents electronic transitions through vibrational absorptions.

Molecular Electromagnetic Spectrum Regions

Region	Wavelength (nanometers)	Wavenumbers (cm^{-1})	Frequency (Hz)	Energy (eV)
Radio Waves	1e^{9}-1e^{17}	1e^{-2}-1e^{-10}	3.0e^{8}-3.0	1.2e^{-6}-1.2e^{-14}
Microwaves	1e^{6}-1e^{9}	10-0.01	3.0e^{11}-3.0e^{8}	1.2e^{-3}-1.2e^{-6}
Terahertz	2.5x10^{4}-10^{6}	400-10	1.2e^{13}-3.0e^{11}	0.05-0.001
Infrared	2500-25,000	4000-400	1.2e^{14}-1.2e^{13}	0.50-0.05
Raman	2500-25,000	4000-400	1.2e^{14}-1.2e^{13}	0.50-0.05
Near Infrared	780-2500	14,493-3,333	3.8e^{14}-1.2e^{14}	1.6-0.50
Visible	360-780	27,778-12,821	8.3e^{14}-3.8e^{14}	3.4-1.6
Ultraviolet	190-360	52,632-27,778	1.6e^{15}-8.3e^{14}	6.5-3.4
Vacuum UV	10-190	1e^{6}-5.26e^{4}	3.0e^{16}-1.6e^{15}	124-6.5
X-Rays	0.01-10	1e^{9}-1e^{6}	3.0e^{19}-3.0e^{16}	124e^{3}-124
Gamma Rays	<0.01	>1e^{9}	>3e^{19}	>124e^{3}

Fig. 0.2 Comparative Regions of the Electromagnetic Spectrum in wavelength (nanometers), wavenumbers, Frequency (Hertz), and Energy (electron Volts).

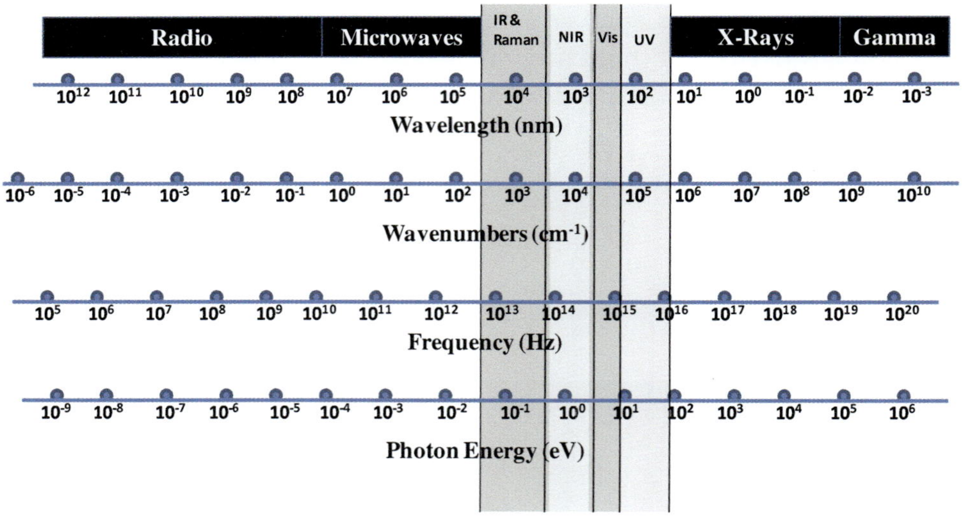

Fig. 0.3 Graphical Illustration of the Comparative Regions of the Electromagnetic Spectrum in wavelength (nanometers), wavenumbers, Frequency (Hertz), and Energy (electron Volts).

The various formulas used for the inter-conversion of units for spectroscopy include the following. Light has both particle and wave properties, quantum theory tells us that the energy of a light "particle" or photon E_p is given by:

$$E_p = h\nu \tag{0.1}$$

$$E_p = hc\bar{\nu} \tag{0.2}$$

where: h = Planck's constant (or 6.6256×10^{-27} erg-sec.); and ν (nu) is the frequency of light, also known as Hertz (Hz) or the number of vibrations per second, in units of sec.$^{-1}$; $\bar{\nu}$ is the wavenumber units as cm^{-1} (i.e., the number of waves per centimeter), c is the velocity of light in a vacuum (or 2.9979×10^{10} cm·sec^{-1}).

$$\bar{\nu} = \frac{\nu}{c \cdot n^{-1}} = \frac{1}{\lambda} \tag{0.3}$$

$$\nu = \frac{c \cdot n^{-1}}{\lambda} \tag{0.4}$$

where: $\bar{\nu}$ is the wavenumber units as cm^{-1} (i.e., the number of waves per centimeter), c is the velocity of light in a vacuum (or 2.9979×10^{10} cm·sec^{-1}), ν (Greek: nu) is the frequency of light, also known as Hertz (Hz) or the number of vibrations per second, in units of (sec.$^{-1}$), and n is the refractive index of the medium the light is passing through (e.g., air = 1.0003), λ (Greek: lambda) is the wavelength in units of cm. Combining the above equations we also note:

$$\bar{\nu} = \frac{hc}{E_p} \Rightarrow \frac{1}{\lambda} = \frac{hc}{E_p} \Rightarrow \lambda = \frac{E_p}{hc} \tag{0.5}$$

0.3.2 *Direct Unit Conversions*

Wavelength (in nanometers) to wavenumbers (cm^{-1}):

$$\bar{\nu}_{cm^{-1}} = \frac{10^7}{\lambda_{nm}} \tag{0.6}$$

Wavenumbers (cm^{-1}) to wavelength (in nanometers):

$$\lambda_{nm} = \frac{10^7}{\bar{\nu}_{cm^{-1}}} \tag{0.7}$$

Wavelength (in nanometers) to frequency (in Hz):

$$\nu_{HZ} = \frac{2.9979 \times 10^{17} nm \cdot \sec^{-1}}{\lambda_{nm}} \tag{0.8}$$

Wavelength (in nanometers) convert to energy (in eV):

$$E_{p_{eV}} = \frac{1240}{\lambda_{nm}} \tag{0.9}$$

0.4 Description of Basic Quantitative Spectroscopic Measurements using Beer's Law[4]

Spectroscopic measurements depend upon the principle that light energy interacting with a material will cause absorption at a specific frequency depending upon the chemical characteristics of that material. The amplitude of the absorption at any particular frequency (or wavelength or wavenumber) is determined by the absorptivity of the molecule being measured and the number of molecules encountered by the beam path of the measuring instrument. It is assumed that a change in spectral response is related to a concentration as described by the Bouguer, Lambert, and Beer relationship, most often termed Beer's law. The Beer's law relationship is described as the absorbance (A, Au, or signal strength) of an analyte being measured using a spectrophotometer is equivalent to the product of the absorptivity (ε) of a specific type of molecular vibration at a given frequency; the concentration (c) of the molecules in the measurement beam; and the pathlength (l) of the sample holder within the measurement beam. This relationship between measured spectral signal and concentration of a molecule is most often expressed as:

$$\mathrm{A} = \varepsilon \mathrm{cl} \tag{0.10}$$

where ε is the molar absorptivity (referred to as molar extinction coefficient by earlier physicists) in units of Liter·Mole^{-1}·cm^{-1}; c is the concentration of molecules in the spectrometer beam in units of Mole·Liter^{-1} (Note: this is a scaled volume fraction unit); and pathlength (l) is the thickness of the sample in units of cm of the measured sample at a specific concentration. The absorptivity for any specific molecule type is calculated by careful measurements of the absorbance of a compound, generally diluted in a suitable organic solvent, and by applying the relationship:

$$\varepsilon = \frac{A}{cl} \tag{0.11}$$

Note that for transmittance (where T = 0.0 to 1.0) and percent transmittance (where %T = 0 to 100.0) spectroscopy, a more complete delineation of the relationships between the various terms is contained in an expression such as:

$$\mathrm{T} = \frac{\mathrm{I}}{\mathrm{I_0}} = 10^{-\varepsilon \mathrm{cl}} \Rightarrow \mathrm{Abs.} = \mathrm{A} = -\log_{10}\left(\frac{\mathrm{I}}{\mathrm{I_0}}\right) = -\log_{10}\mathrm{T} = \varepsilon \mathrm{cl} \tag{0.12}$$

Here the symbols I and $\mathrm{I_0}$ represent the attenuated energy detected after sample interaction and the initial energy incident to the sample, respectively. For reflectance (where %R = 0.0 to 1.0) and percent reflectance (Where R = 0.0 to 100.0) spectroscopy the various relationships are expressed as:

$$R = \frac{I}{I_0} = 10^{-\varepsilon cl} \Rightarrow Abs. = A = -\log_{10}\left(\frac{I}{I_0}\right) = -\log_{10} R = \varepsilon cl \tag{0.13}$$

Noting the relationship exists where the change in intensity (I) of the transmitted or reflected light from a sample is a function of the change in pathlength (l) of the sample as expressed by the absorptivity (ε) of a specific analyte (or molecular substance) and its concentration (c) by:

$$-\frac{\partial I}{\partial l} = \varepsilon cl(\ln 10) \tag{0.14}$$

Modern spectrophotometers utilize these assumptions for making spectroscopic measurement and generally display spectroscopic data as transmission (T), reflection (R), and absorbance A (y-axis or ordinate axis) versus wavelength (nm, microns) or wavenumber (cm^{-1}) (as x-axis, or abscissa axis).

0.5 Spectral Ranges for Sources, Detectors and Optical Materials

Table 0.18 Source emissivity ranges.[3]

Source emissivity ranges (μm)	Start	End
Quartz tungsten halogen	0.22	2.7
Glass tungsten halogen	0.25	2.25
DC deuterium lamp	0.185	3.75
Pulsed xenon arc lamp	0.18	2.5
DC arc lamp	0.20	2.5
Globar (Silicon Carbide)	1	100
Nernst glower	0.3	35
Carbon arc	0.5	100
Mercury lamp	0.3	100
Helium-neon laser (He:Ne)	0.6327	0.6328
Neodymium yttrium aluminum garnet (Nd:YAG) laser	1.0639	1.0640

Table 0.19 Detector sensitivity ranges.[3]

Detector sensitivity ranges (μm)	Start	End
Silicon (Si)	0.30	1.1
Lead Sulfide (PbS)	1.1	3
Indium Arsenide (InAs)	1.7	5.7
Indium Gallium Arsenide (InGaAs), typical	0.90	1.7
Indium Gallium Arsenide (InGaAs), extended	0.65	2.4
Indium Gallium Arsenide (InGaAs), IR extended	0.65	2.7
Germanium (Ge:X)	2	40
Germanium Gold (Ge:Au)	2	9
Germanium Cadmium (Ge:Cd)	2	24
Lead Selenide (PbSe)	1.7	5.5
Germanium Zinc (Ge:Zn)	2	40
Indium Antimonide (InSb)	1.8	6.8
Lead Telluride (PbTe)	1.5	4.5
Deuterated Triglycine Sulfate (DTGS/KBr)	0.83	25
Deuterated Triglycine Sulfate (DTGS/PE)	10	120
Mercury Cadmium Telluride (MCT)	1	17
Triglycine Sulfate (TGS)	10	120
Pyroelectric Lithium Tantalate ($LiTaO_3$) (PLT)	1.5	30
Silicon Photodiode Detectors	0.19	1.1
Photomultipler Tube (PMT), S1 Type Photocathode	0.3	1.2
Photomultipler Tube, Cesium Gallium Arsenide (CsGaAs) Photocathode	0.3	0.93
Photomultipler Tube, Cesium Indium Gallium Arsenide (CsInGaAs) Photocathode	0.3	1
Photomultipler Tube, Cesium Antiminide CsSb Photocathode	0.25	0.75
Photomultipler Tube, Multialkali (Na-K-Sb-Cs) Photocathode	0.3	0.93
Photomultipler Tube, CsI Photocathode	0.2	0.9
Photomultipler Tube, S1 Type Photocathode	0.3	1.2
Photomultipler Tube, Cesium Gallium Arsenide (CsGaAs) Photocathode	0.3	0.93
Photomultipler Tube, Cesium Indium Gallium Arsenide (CsInGaAs) Photocathode	0.3	1
Silicon (Si) CCD	0.2	1.1
Indium Gallium Arsenide (InGaAs) CCD	0.8	1.7
Ultraviolet range	0.19	0.36
Visible Range	0.36	0.78
Near-Infrared Range	0.78	2.5
Infrared	2.5	25
Raman	2.5	25

Table 0.20 Optical material spectral ranges.[3]

Optical material spectral ranges (μm)	Start	End
Methacrylate, Polymer	0.25	1.1
Silica, UV Grade Fused (SiO_2)	0.2	2.5
Silica, Synthetic Fused (SiO_2)	0.23	2.6
Quartz, Crystalline (SiO_2)	0.24	2.6
Quartz, extremely low O-H	0.19	2.7
Glass, Flint (SF 10)	0.38	2.35
Glass, Flint (SF 8)	0.355	2.35
Glass (BK 7)	0.315	2.35
Glass, "Optical Crown"	0.32	2.3
Glass, "Borosilicate Crown"	0.36	2.35
Glass, Pyrex®	0.31	2.5
Glass, Zerodur®Schott	0.4	2.85
Glass, Tempax®	0.36	2.35
Sapphire (Aluminum Oxide) (Al_2O_3) 5 mm is 50 %T at 5.1 microns	0.15	5.1
Sodium Chloride (NaCl)	0.25	16
Suprasil 300	0.19	3.6
Diamond (C)	0.22	4
Silica, Fused as pure SiO_2 (IR Quality)	0.2	4
Quartz, Anhydrous	0.2	2.75
Zirconium Fluoride (ZrF)	0.9	4.76
Chalcogenide (NIR-IR measurements)	2.22	11.1
Cesium Iodide (CsI)	0.3	50
Lead Sulfide (PbS)	1.1	3
Potassium Bromide (KBr)	0.25	26
Potassium Chloride (KCl)	0.25	20
Thallium Bromide-Iodide (KRS-5)	0.6	40
Germanium (Ge)	1.1	30
Arsenic Sulfide (As_2S_3)	0.6	15
Magnesium Fluoride (MgF_2)	0.6	9.5
Zinc Selenide (ZnSe) (IRTRAN-4)	0.6	26
Barium Fluoride (BaF_2)	0.2	13
Zinc Sulfide (ZnS) (Cleartran or IRTRAN-2)	0.6	15
Calcium Fluoride (CaF_2) (IRTRAN-3)	0.2	9
Silver Bromide (AgBr)	0.5	35
Polyethylene, Polymer (High Density)	16	333
Glass, Silica, Fused "Water Free" (SiO_2); 10 mm is 82%T at 3.15 microns	0.15	3.15
Glass, Corning Optical Glasses as HPFS®7978 and 7979 50%T/cm	0.15	4.35

(Continued)

Table 0.20 (*Continued*)

Optical material spectral ranges (μm)	Start	End
Glass, Corning Optical Glasses as HPFS® 7980; 50%T/cm	0.2	2.6
Glass, Corning Optical Glasses as HPFS® 865X and 797X 50%T/cm	0.15	2.6
Glass, ESCO Optical Glasses as S1-UVA and S1-UVB Working Range (Band 1)	0.17	2.2
Glass, ESCO Optical Glasses as S1-UVA and S1-UVB Working Range (Band 2)	2.9	3.6
Glass, ESCO Optical Glasses as I2-IR Working Range	0.25	3.6
Glass, ESCO Optical Glasses as G1 and A1 Working Range	0.27	2.5
Glass, ESCO Optical Glasses as BK7 Working Range	0.35	2
Ultraviolet range	0.19	0.36
Visible Range	0.36	0.78
Near-Infrared Range	0.78	2.5
Infrared	2.5	25
Raman	2.5	25

Table 0.21a UV-Vis gas emission sources and peak positions (in nanometers).[5]

Position	Source	Position	Source	Position	Source
253.652	Hg	452.186	Xe	612.884	Ne
296.728	Hg	462.42	Xe	614.306	Ne
302.15	Hg	466.849	Xe	616.359	Ne
313.155	Hg	469.097	Xe	621.728	Ne
334.148	Hg	469.804	Xe	626.649	Ne
341.79	Ne	473.415	Xe	630.479	Ne
342.391	Ne	479.262	Xe	633.442	Ne
344.77	Ne	480.702	Xe	638.299	Ne
345.076	Ne	482.971	Xe	640.225	Ne
345.419	Ne	484.329	Xe	650.653	Ne
346.052	Ne	491.651	Xe	653.288	Ne
346.658	Ne	492.315	Xe	659.895	Ne
347.257	Ne	503.135	Ne	667.828	Ne
349.806	Ne	503.775	Ne	671.704	Ne
350.121	Ne	508.038	Ne	692.947	Ne
351.519	Ne	511.367	Ne	703.241	Ne
352.047	Ne	511.65	Ne	717.394	Ne
359.353	Ne	540.056	Ne	724.512	Ne
360.017	Ne	546.074	Hg	733.93	Xe
363.366	Ne	556.222	Kr	738.6	Xe
365.015	Hg	557.029	Kr	739.379	Xe
368.573	Ne	576.441	Ne	740.04	Xe
370.122	Ne	576.96	Hg	743.89	Ne
404.656	Hg	579.066	Hg	747.244	Ne
407.783	Hg	582.015	Ne	748.887	Ne
431.958	Kr	585.249	Ne	753.577	Ne
435.833	Hg	587.096	Kr	754.404	Ne
436.264	Kr	588.189	Ne	755.979	Xe
437.612	Kr	594.483	Ne	758.468	Xe
439.997	Kr	597.553	Ne	758.741	Kr
445.392	Kr	602	Ne	760.155	Kr
446.369	Kr	607.433	Ne	764.391	Xe
450.235	Kr	609.616	Ne	768.525	Kr
				769.454	Kr

Table 0.21b NIR gas emission sources and peak positions (in nanometers).[5]

Position	Source	Position	Source	Position	Source
780.265	Xe	1107.887	Ar	1390.748	Ar
785.482	Kr	1117.752	Ne	1409.364	Ar
788.132	Xe	1144.183	Ar	1414.244	Xe
791.343	Kr	1148.811	Ar	1442.679	Kr
796.734	Xe	1152.275	Ne	1473.281	Xe
805.726	Xe	1166.871	Ar	1473.444	Kr
805.95	Kr	1171.949	Ar	1504.65	Ar
806.134	Xe	1181.938	Kr	1517.269	Ar
810.436	Kr	1211.233	Ar	1520.31	Kr
819.006	Kr	1213.974	Ar	1532.934	Ar
823.163	Xe	1220.353	Kr	1537.204	Kr
826.324	Kr	1234.339	Ar	1541.839	Xe
826.652	Xe	1243.932	Ar	1598.949	Ar
829.811	Kr	1248.766	Ar	1605.328	Xe
837.761	Ne	1262.339	Xe	1620.872	Kr
849.536	Ne	1270.228	Ar	1647.29	Xe
866.794	Ar	1273.342	Ar	1656.023	Xe
877.675	Kr	1280.274	Ar	1672.815	Xe
878.375	Ne	1295.666	Ar	1689.676	Kr
881.941	Xe	1300.826	Ar	1694.058	Ar
892.869	Kr	1317.741	Kr	1704.288	Ar
904.545	Xe	1322.811	Ar	1755.35	Kr
912.297	Ar	1322.811	Ar	1763.882	Xe
916.265	Xe	1327.264	Ar	1785.738	Kr
922.45	Ar	1331.321	Ar	1790.45	Xe
935.422	Ar	1336.711	Ar	1800.223	Kr
965.779	Ar	1350.419	Ar	1809.09	Xe
978.45	Ar	1362.266	Ar	1816.733	Kr
979.97	Xe	1363.422	Kr	1832.53	Xe
992.319	Xe	1365.706	Xe	1959.94	Xe
1047.005	Ar	1367.855	Ar	1984.638	Xe
1067.357	Ar	1371.858	Ar	2190.851	Kr
1083.837	Xe	1382.572	Ar		

References

1. NIST: http://physics.nist.gov/cuu/Constants/index.html.
2. Adapted from Near-Infrared Spectroscopy in Agriculture xx-xxiii (Used with Permission). Tables have been changed with different values and conversion factors.
3. J. Workman, *The Academic Press Handbook of Organic Compounds: NIR, IR, Raman, and UV-VIS Spectra Featuring Polymers, and Surfactants*, (3 Volume Set: Vol. 1, Methods and Interpretation; Vol. 2, UV-Vis and NIR Spectra; Vol. 3, IR and Raman Spectra), Academic Press, Boston (2000).
4. J. Workman and L. Weyer, *Practical Guide and Spectral Atlas to Interpretive Near-Infrared Spectroscopy*, 2nd edn (CRC — Taylor and Francis, Boca Raton, FL 2012).
5. Ocean Optics Wavelength Calibration Sources, www.oceanoptics.com. (Used with Permission).

Chapter 1

Introduction to Raman Spectroscopy

Contents

1.1 Introduction 2
  1.1.1 Basic Theory 4
    1.1.1.1 The Electromagnetic Spectrum 4
    1.1.1.2 Molecular and Electronic Spectroscopy Unit Conversion 5
    1.1.1.3 Unit Conversions 6
1.2 Applications of Raman Spectroscopy 7
  1.2.1 Early Work in Raman Spectroscopy 7
1.3 Basic Raman Spectroscopic Instruments (Optical Configurations) 8
  1.3.1 Components of a Raman Spectrometer 11
  1.3.2 The Concepts Behind Raman Spectroscopy Measurements . . 16
  1.3.3 Practical Aspects for Raman Spectral Measurements 22
    1.3.3.1 Basic Applications of Raman Spectroscopy 22
    1.3.3.2 Calibration and Basic Maintenance 24
1.4 Interpretive Raman Spectroscopy 26
  1.4.1 Group Frequencies 26
  1.4.2 Fingerprint Frequencies 29
  1.4.3 Dividing the Raman Spectrum 29
  1.4.4 Coupling of Vibrations 30
  1.4.5 Fermi Resonance (or Second Order Coupling) 30
  1.4.6 Frequencies of Alkanes (Paraffins) 31
    1.4.6.1 Normal hydrocarbons (n-alkanes) 31
    1.4.6.2 Branched Chain Hydrocarbons (iso-, gem-dimethyl, tertiary butyl, and Cyclic Groups) 33
  1.4.7 Group Frequencies of Alkenes (Olefins) 34

1.4.7.1 C-H Stretch of Alkenes 34
1.4.7.2 Olefinic C-H Stretch[23] 35
1.4.7.3 Olefinic C=C Stretch 35
1.4.7.4 Olefinic C-H Bend 37
1.4.8 Molecules with Triple Bonds (-C≡) 38
1.4.8.1 Acetylenes (-C≡ C-) 38
1.4.8.2 Nitriles (R-C≡ N) 38
1.4.9 Cumulated Double Bond Systems (A=B=C) 39
1.4.10 Aromatic Systems 39
1.4.11 Vibrational States in Benzene 41
1.4.12 Carbonyl Compounds (>C=O) 42
1.4.12.1 Special cases of the carbonyl group 43
1.4.13 Amides (R-C=ONH-X) 43
1.4.14 Ethers (C-O-C) 44
1.4.15 Esters (R-CO-OR') 44
1.4.16 Anhydrides (R-C=O-C=O-R) 45
1.4.17 Peroxides (-C-O-O-C-) 45
1.4.18 X-H Functional Groups (O-H and N-H) 45
1.4.19 Hydrogen Bonding (X-H.......Y) 45
1.4.20 The O-H Stretch 46
1.4.21 N-H Group Frequencies and Nitrogen Containing Compounds[36–38] 47
1.4.22 Amines (R-NH-X) 48
References 48

1.1 Introduction

As modern instrumentation for Raman spectrometry (or other forms of molecular spectroscopy) improves, the technique becomes more versatile and easy to use for the analyst. Raman spectrometers are now made portable and for laboratory, process analysis, and field use. Raman spectroscopy is growing more rapidly than many more traditional techniques due to advances in lasers, detectors, and computer processing technologies. Improved designs and increased laser safety has allowed Raman spectrometers to be used for routine laboratory and field analysis at an increasing rate. This section is intended to describe the basic principles and general knowledge related to Raman spectroscopy. This chapter includes an explanation of the electromagnetic spectrum and how radiation interacts with matter to cause the Raman effect. Methods of describing what the Raman spectrum tells us about a sample's composition are examined. Practical information is described including the basic components of Raman spectrometers and useful information such as sample preparation, sampling device selection, and routine maintenance of Raman instrumentation. Finally, a basic glossary of key terms and concepts and a selective bibliography are provided to guide the reader to further discovery on the subject of Raman spectrometry.

Raman spectrometry or spectroscopy provides a measure of molecular inelastic scattering. The Raman technique is a complementary measurement technique to infrared and is used for measuring the molecular structure and composition of pure materials and mixtures. The measure of Raman intensity with respect to Raman shift frequency (in wavenumbers or wavelength) provides a measurement of fundamental molecular vibrations from inorganic and organic compounds, and is particularly useful for measuring synthetic and biopolymer structures.

Raman allows for quantitative or qualitative analysis of solids, liquids or gas phase measurements. The optical pathlength of the sample is determined by the scattering properties of the material, and Raman measurements are compared to reference standards having the same background matrix, except with varying concentrations of the analyte. Raman is excellent for measurements of hydrocarbon liquids and solids, as well as polymers. The technique is generally compatible with the use of fiber optics having low scattering properties achieved with the application of low O-H and low entrained air optical fibers.

Raman spectroscopy can be used for a variety of measurements on samples that are aqueous in nature or where glass sample holders are present. Carbon dioxide, water, and glass (silica) are weak Raman scatterers and thus there is generally no problems in analyzing samples in air and in glass sample vials. There is typically no sample preparation involved for samples measured using the Raman technique.

Raman measurements exhibit high signal-to-noise and a reasonable cost for instrumentation. The dominant Raman spectral features are acetylenic C≡C stretching, olefinic C=C stretch at 1680–1630 cm^{-1}, R-N=N-R′ (azo-) stretching, S-H (thio-) stretching, C=S stretching, C-S stretching, S-S stretching bands, and CH_2 wagging and twisting bands. Raman spectra also contain such molecular vibrational information as CH_2 twist and wagging, carbonyl C=O stretch associated with esters, acetates, and amides, C-Cl (halogenated hydrocarbons) stretching, and $-NO_2$ (nitro-/nitrite) stretching. In addition, Raman yields information content of phenyl- containing compounds at 1000 cm^{-1}. Organic and inorganic materials generally exhibit unique and identifiable Raman spectra. Bands that are strong in infrared but weak in Raman include O-H, N-H, and C-F stretching, as well as C-H out-of-plane bending.

1.1.1 *Basic Theory*

1.1.1.1 *The Electromagnetic Spectrum*

The electromagnetic spectrum consists of many different types of radiation (energy): gamma, x-ray, ultraviolet (UV), visible (Vis), infrared (IR), microwave, and radio waves (see Figures 1.1 and 1.2). Each type of radiation occupies its own region of the electromagnetic spectrum, and the major difference between the individual spectral regions is merely the photon energy involved. This energy is expressed in units of photon energy (eV), Frequency (Hz), wavenumber (cm^{-1}), or wavelength (nm or μm). Each spectral region of electromagnetic radiation interacts differently with matter. In this way the different spectral regions each demonstrate unique spectra for different types of molecules.

Raman spectroscopy measures the specific molecular vibrations and so provides detailed information for the interpretation of the backbone molecular structures as well as attached groups present in a sample. Raman provides information on the

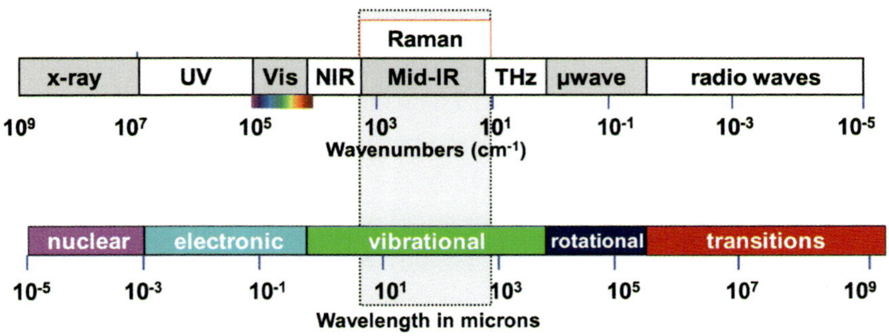

Fig. 1.1 The electromagnetic spectrum, highlighting the Raman region. The Raman spectral region is defined as 4000 through 400 cm^{-1}.

Molecular Electromagnetic Spectrum Regions

Region	Wavelength (nanometers)	Wavenumbers (cm^{-1})	Frequency (Hz)	Energy (eV)
Terahertz	$2.5x10^4$-10^6	400-10	$1.2e^{13}$-$3.0e^{11}$	0.05-0.001
Infrared	2500-25,000	4000-400	$1.2e^{14}$-$1.2e^{13}$	0.50-0.05
Stokes Shift	2857-200,000	3500-50	$1.0e^{14}$-$1.5e^{12}$	0.43-0.01
Raman	2500-25,000	4000-400	$1.2e^{14}$-$1.2e^{13}$	0.50-0.05
Near Infrared	780-2500	14,493-3,333	$3.8e^{14}$-$1.2e^{14}$	1.6-0.50

Fig. 1.2 Raman Spectral Region of the Electromagnetic Spectrum in wavelength (nanometers), wavenumbers (cm^{-1}), Frequency (Hertz), and Energy (electron Volts). The "Stokes shift" region is most typical for modern Raman spectrometers.

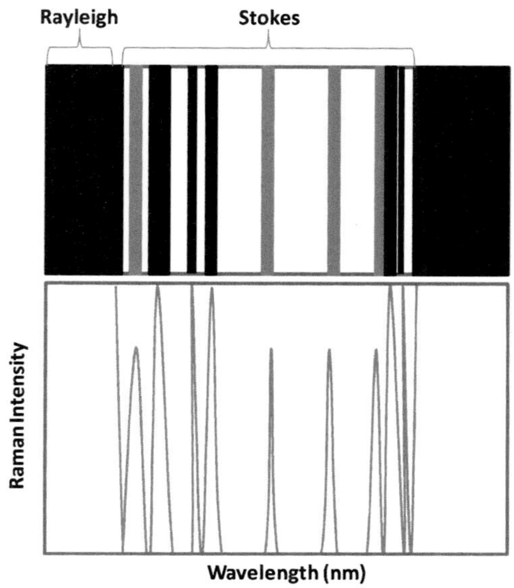

Fig. 1.3 Illustration of original photographic plate (top) with modern Stokes shift spectrum shown as a comparison (bottom).

quantity of molecules under measurement, and the basic chemical and physical form of molecules under measurement (such as polymorphism, crystallinity, backbone structures, saturation, attached groups, etc.). These properties of a material are measured by using scattering techniques and recording the intensity of such scattered energy measurements with respect to Raman shift in units of wavelength, frequency, or wavenumbers. (The most common unit for denoting the position of Raman spectral bands is in + or − cm^{-1} position from the center excitation laser wavenumber position; as illustrated in Figure 1.3.).

1.1.1.2 *Molecular and Electronic Spectroscopy Unit Conversion*

The units used to describe spectra for molecular spectroscopy vary with the academic or engineering discipline describing the particular spectral regions. For the most part the spectra are described in terms of an optical intensity response specific to the interaction of photon energy for a specific frequency. The photon responses are in terms of inelastic or elastic collisions of the energy with the various sample molecules and are referred to in Raman spectroscopy as Rayleigh scattering, Stokes and anti-Stokes scattering, Brillouin scattering, and fluorescence (See Volume 1 for details regarding fluorescence phenomena). The various spectral photon energy for molecular and electronic spectroscopy is described in terms of units of wavelength (e.g., microns (μm) as 10^{-6} meters, or nanometers (nm) as 10^{-9} m, or Angstroms (Å) as 10^{-10} m). Energy is also expressed as frequency (cycles per second or sec.$^{-1}$) termed Hertz (Hz), as Wavenumbers (cm^{-1}), and as electron Volts (eV).

The various formulas used for the inter-conversion of units for spectroscopy include the following. Light has both particle and wave properties, quantum theory tells us that the energy of a light "particle" or photon E_p is given by:

$$E_p = h\nu \tag{1.1}$$

$$E_p = hc\bar{\nu} \tag{1.2}$$

where h = Planck's constant (or 6.6256×10^{-27} erg-sec.; ν (nu) is the frequency of light, also known as Hertz (Hz) or the number of vibrations per second, in units of sec.$^{-1}$; $\bar{\nu}$ is the wavenumber units as cm^{-1} (i.e., the number of waves per centimeter); and c is the velocity of light in a vacuum (or 2.9979×10^{10} cm $\cdot$ sec^{-1}).

$$\bar{\nu} = \frac{\nu}{c \cdot n^{-1}} = \frac{1}{\lambda} \tag{1.3}$$

$$\nu = \frac{c \cdot n^{-1}}{\lambda} \tag{1.4}$$

Where $\bar{\nu}$ is the wavenumber units as cm^{-1} (i.e., the number of waves per centimeter); c is the velocity of light in a vacuum (or 2.9979×10^{10} cm $\cdot$ sec^{-1}); ν (Greek: nu) is the frequency of light, also known as Hertz (Hz) or the number of vibrations per second in units of sec.$^{-1}$; n is the refractive index of the medium the light is passing through (e.g., air = 1.0003); and λ (Greek: lambda) is the wavelength in units of cm. Combining the above equations we also note:

$$\bar{\nu} = \frac{hc}{E_p} \Rightarrow \frac{1}{\lambda} = \frac{hc}{E_p} \Rightarrow \lambda = \frac{E_p}{hc} \tag{1.5}$$

1.1.1.3 *Unit Conversions*

Wavelength (in nanometers) to wavenumbers (cm^{-1}):

$$\bar{\nu}_{cm^{-1}} = \frac{10^7}{\lambda_{nm}} \tag{1.6}$$

Wavenumbers (cm^{-1}) to wavelength (in nanometers):

$$\lambda_{nm} = \frac{10^7}{\bar{\nu}_{cm^{-1}}} \tag{1.7}$$

Wavelength (in nanometers) to frequency (in Hz):

$$\nu_{Hz} = \frac{2.9979 \times 10^{17} nm \cdot \sec^{-1}}{\lambda_{nm}} \tag{1.8}$$

Wavelength (in nanometers) convert to energy (in eV):

$$E_{p_{eV}} = \frac{1240}{\lambda_{nm}} \tag{1.9}$$

1.2 Applications of Raman Spectroscopy

Raman spectroscopy uses electromagnetic energy related to inelastic photon scattering to determine the molecular vibrational frequency of molecules. The energy band used for Raman is identical to that defined for mid infrared spectroscopy, which is designated for convention as 2.5 to 25.0 microns. The Raman spectral region extends from 2500 to 25,000 nanometers, most commonly referred to as 4000 to 400 cm^{-1}, and actually extends to 100 cm^{-1} in some cases.

Raman spectroscopy is useful for qualitative and quantitative analysis of pure materials and mixtures of various molecules. Advances in heuristic rules and the proliferation of multivariate software, both qualitative and quantitative chemometric tools, are quite common for modern Raman analysis and interpretation of spectra. The use of multivariate calibration algorithms and statistical methods (i.e., chemometrics) are used to model Raman spectra to chemical properties of a variety of inorganic and organic materials for analysis.[1–2]

The Raman method relies on the spectra-structure correlations existing between a measured Raman shift due to Stokes and anti-Stokes scattering and the fundamental molecular vibrations occurring at Raman shift frequencies. These harmonic molecular vibrations occur at wavenumber (cm^{-1}) shifts (or frequencies) depending upon the analyte molecule scatter based on the type of scattering molecules present within the sample, and the concentration of those molecules. This chapter introduces the use of Raman spectroscopy as a technique for qualitative identification, quantitative analysis, and structure-correlation assessment for Raman scatter properties and functional group frequencies.

1.2.1 *Early Work in Raman Spectroscopy*

The History of Raman begins with C.V. Raman and K.S. Krishnan discovering what is termed the "Raman scattering effect" in 1928.[3] Since lasers did not exist in 1928, these first experiments used the brightest sources available, those being sunlight, or a quartz mercury arc emission lamp. Both of these sources were filtered for specific wavelengths of light and to eliminate illumination beyond the green mid-visible region of 435.6 nanometers. The sunlight and mercury lamp were used to excite the newly discovered scattering effect then termed "a new radiation" as measured using common organic compounds (solvents).

During this time spectra in general were collected on glass photographic plates. The results of these experiments showed a series of emission or bright lines that had a wavelength greater (lower energy) than the emission sources, so the outgoing light was shifted over the incoming light. One of the first spectra demonstrated on the photographic plates was that of benzene.[4] The quantity of Raman, or the scattered light, as compared to the source is only approximately 1 part in 1 million to 1 part in 100 million ($1{:}10^6$ to $1{:}10^8$). For early experiments the exposure times to capture these scattered energy bands on a photographic plate were from several hours to almost 200 hours.

The original Raman work compiled a library of over 60 liquid and gas phase spectra demonstrating this "new radiation" phenomena. Figure 1.3 demonstrates the graphical appearance of a typical Raman photographic plate of this era (top) as compared to a Raman spectrum derived from a modern spectrometer (bottom). Raman spectrometers now use laser excitation, with notch filters to restrict the excitation light from entering the spectrometer detection system (i.e., monochromator, camera, or interferometer). The narrow line laser excitation provides a high density flux of photons in a short period, and when combined with quiet electronics and sensitive detectors allows real-time Raman spectra collection and imaging capabilities. The spectrometer provides real-time data of Raman scattering shift (in cm^{-1}) as compared to Raman intensity, as the abscissa and ordinates, respectively. Table 1.5 lists the various common Raman techniques utilizing modern design instrumentation.

Table 1.1 demonstrates the major differences between infrared and Raman spectroscopy. Although complementary these measurement techniques are significantly different in several key ways. Many of the absorption positions used for spectra-structure correlation are the same for infrared and Raman, however this table points to major distinctions in the techniques.

1.3 Basic Raman Spectroscopic Instruments (Optical Configurations)

Vibrational spectroscopy is possible because physical matter as gases, liquids, or solids responds in a unique way when exposed to specific frequencies of

Table 1.1 Major differences in measurement characteristics between Infrared and Raman measurement techniques.

Characteristic	Infrared	Raman
Theory of Action:	Absorption based (Beer's law)	Inelastic photon scattering (emission)
Greatest Sensitivity to:	Dipole vibrations: O-H, N-H, C=O	Polarizable vibrations: C=C, Aromatics
Sensitivity to aqueous samples:	Very sensitive to O-H	Very slight sensitivity
Molecular Structure:	End and attached groups	Molecular backbone
Sample Preparation:	Complex: thin deposition films, thin sheets, mulls, powders, gases, liquids	Minimal to none
Pathlength:	Approximately 0.1 mm	N/A
Fiber optic use:	Quite limited	Conducive to fiber use

electromagnetic radiation. The atoms and molecules that make up matter will absorb, and remit (scatter) certain amounts of the Raman laser energy as it collides with each molecule; this response is based on the vibrational frequency of the molecule in the case of the Raman effect. For Raman vibrational spectra a single frequency laser collides with molecules and causes three main responses: (1) elastic scattering or Rayleigh scattering, where an elastic scatter occurs at the same wavelength as the excitation laser; (2) negative inelastic scattering or Stokes frequency range demonstrating a negative Raman shift based on molecular vibrational frequencies; and (3) positive inelastic scattering or anti-Stokes frequency range demonstrating a positive Raman shift based on molecular vibrational frequencies of the excited material. (See Figures 1.4, 1.5 and 1.6.)

In the earliest spectroscopic experiments, scientists discovered that each atom or molecule will react in exactly the same way every time it is exposed to a particular type (i.e., energy or frequency) of radiation. This interaction results predictably in the absorption or remission of radiation at the different frequencies from the atoms or molecules. This phenomenon or behavior is unique to various types of matter and permits the science of spectroscopy to exist. For Raman spectroscopy it should be noted that no two different types of atoms or molecules react to laser radiation in exactly the same manner. However types of molecules that are very similar produce very similar Raman spectra and thus the spectral detail and signal-to-noise of a spectrophotometer measurement determine its ability to distinguish between very similar molecules from one another.

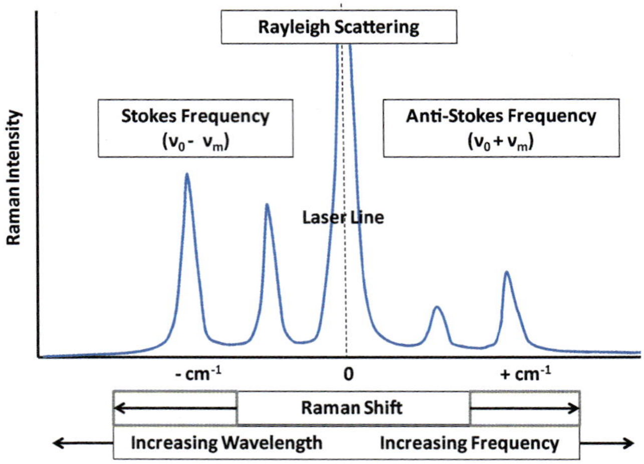

Fig. 1.4 Illustration of a Raman Spectrum showing Raman Shift. Note for this figure the wavelength increases to the left, and the frequency to the right; and the relative positions of the Stokes and Anti-Stokes shifts are shown. Note intensity of Stokes is greater than Anti-Stokes as a general rule.

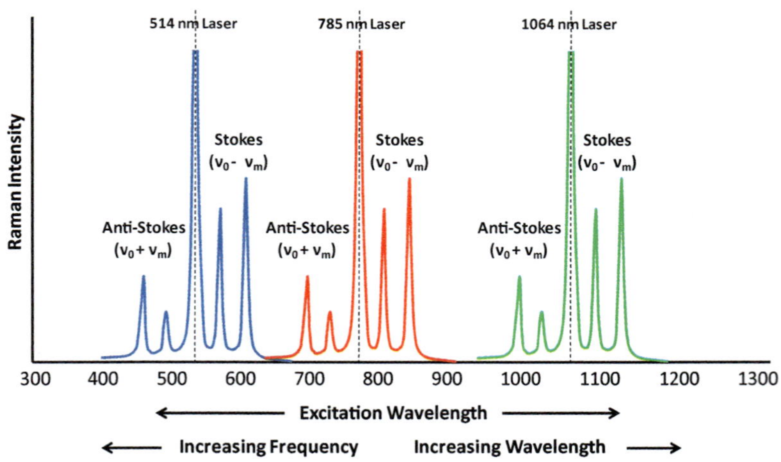

Fig. 1.5 Illustration of three sets of Raman spectra showing the Raman shift for three different laser sources along an increasing wavelength axis (in nanometers). Note for this figure the wavelength scale increases to the right, and the frequency to the left. Note the relative positions of the Stokes and Anti-Stokes Raman shifts are shown. Note the Raman intensity for all three lasers is normalized to be equivalent for this figure.

For Raman spectroscopy, a small fraction of the excitation energy is scattered by matter based on molecular content and the vibrational bands of the molecule(s) excited. When the vibration energy of a specific molecule type exhibits an inelastic collision with the excitation photons, a scattering phenomenon causes the energy of the exciting light to shift corresponding to the frequency of the molecular vibration.

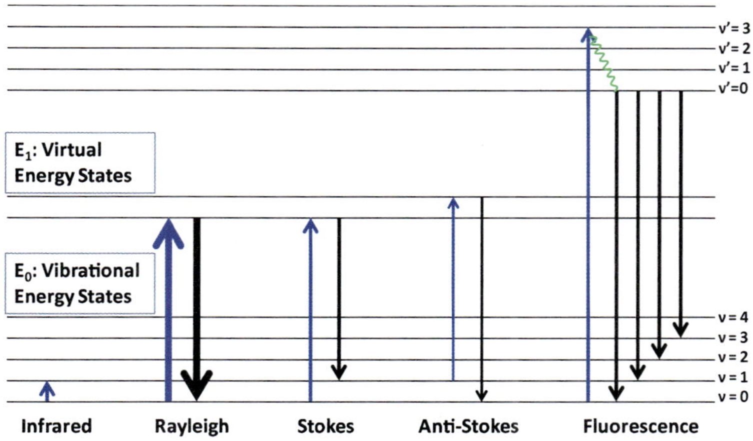

Fig. 1.6 Illustration of Energy States involved in Infrared, Raman, and Fluorescence phenomena; thickness of lines denotes relative intensity.

This shift is measured and recorded using a Raman spectrometer. The result is a spectrum produced by the measuring spectrometer instrument, corresponding to the energy shift of the scattered light in frequency or wavenumbers (abscissa, x-axis) versus the Raman intensity of the scattered light recorded as intensity (for the ordinate, y-axis).

Because the Raman scattered energy is predictable and reproducible, measured spectra are useful for both qualitative and quantitative work. Spectral analysis is used to reveal the types of molecules in a sample, and the amount of each type of molecule in a sample. Figures 1.7 through 1.9 show the basic optical configurations for various Raman spectrometer systems. For information on other Raman spectrometer designs and techniques please refer to the index and Chapter 3 of this volume covering topics related more specifically to instrumentation. Figure 1.7 illustrates a typical dispersive, monochromator-based laboratory Raman spectrometer, Figure 1.8 is a dispersive, monochromator-based laboratory Raman Microscope, and Figure 1.9 is a Laboratory Dispersive Michelson Interferometer-based Raman spectrometer microscope.

1.3.1 *Components of a Raman Spectrometer*

(a) General Components: Raman spectrometers share the same basic macro-components, which can be configured to achieve specific optical design configurations as shown in Figures 1.7 through 1.9. These components include: (1) an excitation source laser, these may be ultraviolet (UV), Visible (Vis), or Near

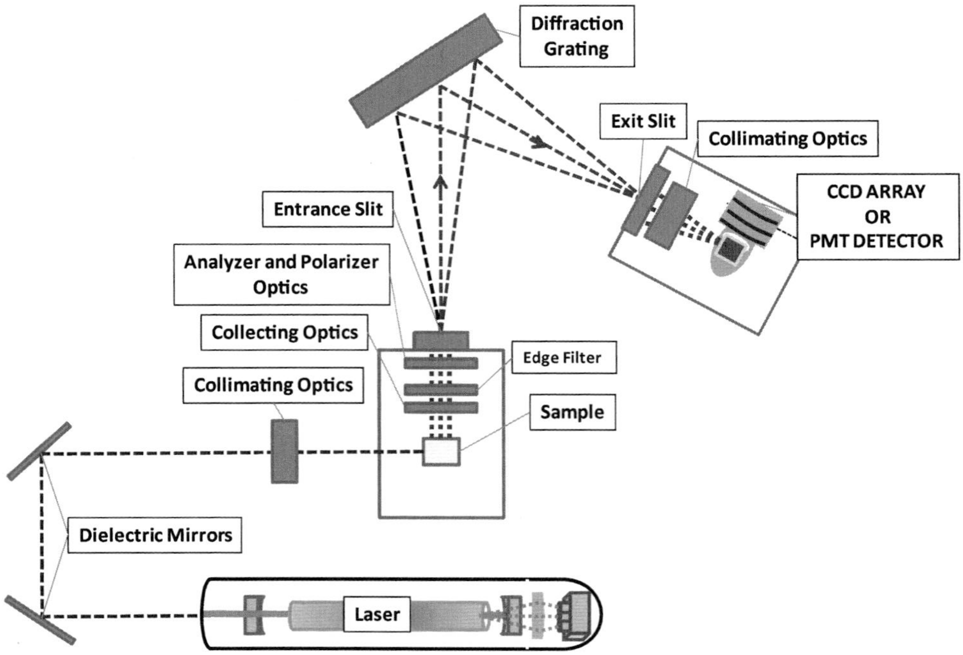

Fig. 1.7 Laboratory Dispersive Monochromator-based Raman spectrometer Design (Optics).

Infrared (NIR) lasers (see Table 1.2); (2) sample illumination (excitation) and scattered light collection optics; (3) a wavelength or frequency selection device (either a dispersive monochromator or an interferometer); and (4) a detector (either a photodiode array, CCD, or PMT (see Table 1.3)). The basic electronics include a power supply to distribute power to all motors, solenoids, shutters, and electronic components; a computer board for processing data, low-level instrument commands, and communications; a controller board for timed driving of motors, shutters, etc.; and a detector board for amplifying the Raman signal, and for converting an analog signal to a digital signal (A to D) for the basic instrument signal response (as a voltage). This output voltage with respect to wavelength or wavenumber position is used to generate the Raman shift intensity versus wavenumber spectral profiles.

(b) Laser Excitation Sources: The energy used in Raman spectrometry is typically produced by a laser using ultraviolet, visible, or near infrared energy. The higher the frequency the greater is the intensity of the Raman signal, but also greater is the fluorescence interference when measuring biological and natural samples. Note that higher energy excitation causes greater scattering, fluorescence, and signal intensity, as Figure 1.10 shows. The following Table 1.2 shows the variety of laser excitation

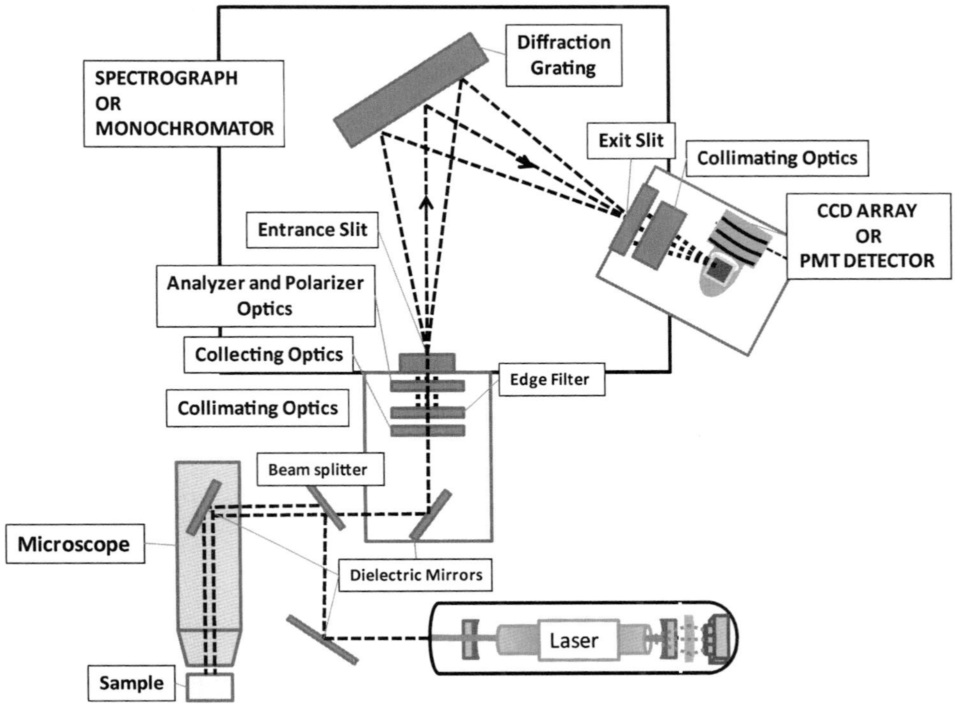

Fig. 1.8 Laboratory Dispersive Monochromator-based Raman spectrometer as a microscope Design (Optics).

sources for Raman spectrometers. More specific details of Raman excitation lasers may be found in Chapter 3 of this volume.

(c) Sample Holders: The most common sample holders in Raman analysis are small glass vials (liquids) or small glass cells or platforms for sample deposition. Chapter 4 of this volume describes the details of sample preparation related to Raman measurements.

(d) Detectors: Raman scattered energy from a sample and reference material is captured by focusing the scattered energy onto a detector, usually a visible range detector and computing the Raman shift as a means for displaying the Raman intensity (as counts) versus wavenumber shift (cm^{-1}). The most common detectors are photomultiplier tubes (PMT), charge couple devices (CCD), or photodiode array detectors (PDA) of various composition and design (Table 1.3).

The voltage signals generated by these detectors are amplified, converted from analog to digital signals (A-to-D), and used to quantify the amount of Raman intensity energy radiated from a sample. A more complete list of common detectors, common abbreviations, and their useful working wavelength ranges would include

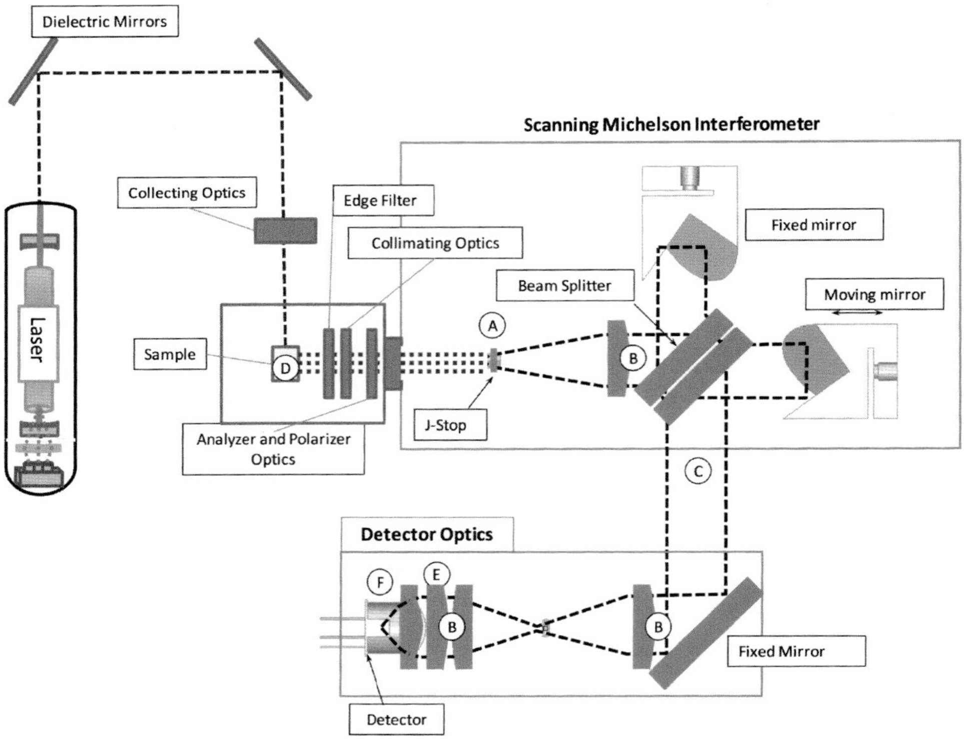

Fig. 1.9 Laboratory Michelson Interferometer-based Raman spectrometer microscope Design (Optics, Interferometer/FT-type). A = J-stop (aperture), B = collimating optics, C = Optical Train, D = Sample Cell, E = focusing optics, F = Raman Detector or CCD Detector.

Table 1.2 Most common commercial Raman laser excitation sources and excitation band properties.

Laser excitation type	Start (nm)	End (nm)	Difference (nm)
Visible (Laser diode), 455 nm	454.0	456.0	2 (0.1 stabilized)
Visible (Laser diode), 514 nm	513.0	515.0	2 (0.1 stabilized)
Visible (Laser diode), 532 nm	531.0	533.0	2 (0.1 stabilized)
Helium-neon laser (He:Ne), 632.8 nm	632.7	632.9	0.2 (0.1 stabilized)
Visible (Laser diode), 785 nm	784.0	786.0	2 (0.1 stabilized)
Neodymium yttrium aluminum garnet (Nd:YAG) laser, 1064 nm	1063.9	1064.1	0.2 (0.1 stabilized)

the following Table 1.3. The reader is referred to the index and Chapter 3 on Instrumentation for more details on the physics of detectors.

(e) Basic Measurement Modes: The measurement modes for Raman spectrometers are given in Table 1.4. A well designed and flexible research grade Raman spectrometer may be used for these multiple measurement techniques.

Table 1.3 Raman Spectrometer Detectors (Potential).

Detector sensitivity ranges	Start	End	Difference
Silicon (Si)	0.3	1.1	0.80
Indium Gallium Arsenide (InGaAs), typical	0.9	1.7	0.80
Indium Gallium Arsenide (InGaAs), extended	0.65	2.4	1.75
Indium Gallium Arsenide (InGaAs), IR extended	0.65	2.7	2.05
Germanium (Ge:X)	2	40	38.00
Germanium Gold (Ge:Au)	2	9	7.00
Germanium Cadmium (Ge:Cd)	2	24	22.00
Lead Selenide (PbSe)	1.7	5.5	3.80
Germanium Zinc (Ge:Zn)	2	40	38.00
Indium Antimonide (InSb)	1.8	6.8	5.00
Lead Telluride (PbTe)	1.5	4.5	3.00
Deuterated Triglycine Sulfate (DTGS/KBr)	0.83	25	24.17
Mercury Cadmium Telluride (MCT)	1	17	16.00
Pyroelectric Lithium Tantalate ($LiTaO_3$) (PLT)	1.5	30	28.50
Silicon Photodiode Detectors	0.19	1.1	0.91
Photomultipler Tube (PMT), S1 Type Photocathode	0.3	1.2	0.90
Photomultipler Tube, Cesium Galium Arsenide (CsGaAs) Photocathode	0.3	0.93	0.63
Photomultipler Tube, Cesium Antiminide CsSb Photocathode	0.25	0.75	0.50
Photomultipler Tube, Multialkali (Na-K-Sb-Cs) Photocathode	0.3	0.93	0.63
Photomultipler Tube, CsI Photocathode	0.2	0.9	0.70
Photomultipler Tube, S1 Type Photocathode	0.3	1.2	0.90
Photomultipler Tube, (CsInGaAs) Photocathode	0.3	1	0.70
Silicon (Si) CCD	0.2	1.1	0.90
Indium Gallium Arsenide (InGaAs) CCD	0.8	1.7	0.90
Ultraviolet range	0.19	0.36	0.17
Visible Range	0.36	0.78	0.42
Near Infrared Range	0.78	2.5	1.72
Infrared	2.5	25	22.50
Raman	2.5	25	22.50

(f) Raman Measurement Techniques: Raman spectroscopy provides multiple techniques for enhancing specific molecular signals and for obtaining images of samples where chemical composition is related to spatial structure of the sample. Table 1.5 lists the techniques used for Raman spectroscopy with brief descriptions. Other techniques are discussed within this volume and in greater detail, see for example the index and Chapters 2 and 3.

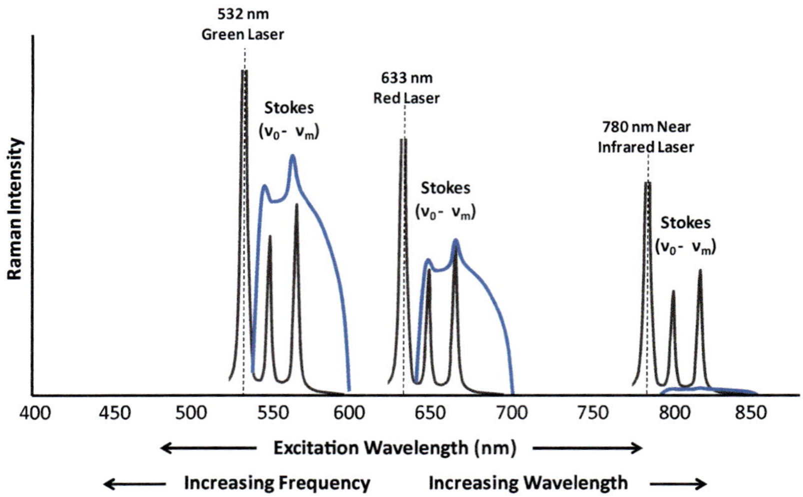

Fig. 1.10 Illustration of Fluorescence interference (as blue line) interfering with Stokes scattering as a function of laser excitation wavelength. The near infrared laser shows the lowest fluorescence interference for biological and other fluorescent materials, whereas the higher frequency green laser has a higher fluorescence response to biological samples.

1.3.2 *The Concepts Behind Raman Spectroscopy Measurements*

(a) Raman Spectroscopy for Quantitative Analysis: Raman spectroscopic measurements depend upon the principle that light energy interacting with a material will cause Raman scattering at a specific frequency, depending upon the molecules present and their vibrational states within the measured sample. The use of Beer's law, essential to quantitative analysis using absorption spectroscopy, does not apply to Raman spectroscopy. The Raman intensity (I_R) or amplitude of the signal at any particular frequency (or wavelength or wavenumber) is determined by the Raman scatter (R_S) of the molecule being measured and the number of molecules encountered by the beam path (i.e., laser and detector) of the measuring Raman instrument.

It is assumed that a change in spectral response is related to a concentration as described by the Raman relationship. Raman intensity (I_R) as Raman scatter (R_S) is described in the following section. The Raman intensity (i.e., amplitude, or signal intensity or strength) of an analyte being measured using a Raman instrument is described using Equations 1.10 through 1.14. This relationship is approximated to the product of the Raman scatter (R_S) of a specific type of molecular vibration at a given frequency; the concentration (c) of the molecules in the measurement beam; and the effective pathlength (L_E) of the sample holder within the measurement beam. Note that the effective pathlength is the actual optical path of the Raman

Table 1.4 Raman spectrometer measurement modes.

Instrument measurement mode	Description of measurement
Standard Scan	Raman scattering* vs. wavenumbers
Step-Scan Phase-Resolved	Amplitude Modulation, Phase Modulation, or Sample Modulation
Step-Scan Time-Resolved	Microsecond to nanosecond time resolution for timed events
Step-Scan Space-Resolved	Imaging spectra for spatial data acquisition for focal-plane imaging
Polarization Modulation	Modulation of light polarization for measuring polarization effects
Individual wavenumber step-scan	Individual Raman scattering* at selected wavenumbers(s)
Two-dimensional Correlation	Correlates multiple spectra for conditions of applied stress and relaxation to determine rheo-optical properties. Also for correlation of spectra from different regions such as NIR.
Quantitative methods	Concentration of analyte vs. absorbance
Qualitative methods	Correlation or spectral matching for Identification
Waveguide Raman	Sample measured along small diameter Teflon tube increases Raman detection sensitivity for aqueous samples
Fiber optic Raman	Raman laser sources may be applied through silica fiber optics for remote sensing
Kinetics	Kinetic rates of reaction (Repetitive fast scan)
Transmission Raman	Laser is projected through solids or turbid liquids for bulk analysis

*As Stokes scattering intensity (in counts).

source through the sample and is somewhat independent of the actual geometric pathlength of the sample holder. The relationship between measured Raman intensity (or spectral signal) and concentration of a molecule may be approximated for simplicity as:

$$I_R = R_S \cdot c \cdot L_E \tag{1.10}$$

where R_S is the molar Raman scattering coefficient in units of Liter $\cdot$ Mole^{-1} $\cdot$ cm^{-1}; c is the concentration of molecules in the Raman spectrometer beam in units of Mole $\cdot$ Liter^{-1} (Note: this is a scaled volume fraction unit); and effective optical pathlength or L_E (i.e., actual pathlength where Raman scattering is occurring). Note that (L_E) is the optical measured thickness of the sample in units of cm for the measured sample at a specific concentration. The Raman scattering coefficient (R_S) for any specific molecule type is calculated by careful measurements of the Raman Intensity (I_R) of a compound, generally diluted in a suitable organic solvent, and by

Table 1.5 Raman measurement techniques.

Instrument measurement mode	Description of measurement technique
Stimulated Raman	A dynamic Raman technique where two different wavelength (frequency) laser pulses are applied to a sample simultaneously with polarization (parallel or perpendicular). When the difference frequency of the excitation lasers matches a particular molecular vibrational frequency (also called the Raman shift), amplification of the Raman signal occurs. The polarization effectively measures molecular alignment and orientation properties.
CARS (Coherent Anti-Stokes Raman Spectroscopy)	Two lasers are used (i.e., near infrared pump beam and longer wavelength Stokes beam) to excite molecules to their virtual state and return to their vibrational state. Tuning of the excitation pump laser beam allows excitation of specific molecules at specific spatial locations with a 10^5 signal enhancement over conventional Raman. CARS imaging may be completed in real-time for specific molecular vibrations at known planar locations for image reconstruction without dyes.
RRS/RR (Resonance Raman Spectroscopy)	The laser excitation wavelength is matched to the specific electronic transition associated with the molecular vibration mode of interest. It provides a technique for "isolation" of specific molecular vibrational modes of interest for the study of specific bands of large molecules.
SERS (Surface-Enhanced Raman Spectroscopy)	Placing a sample within a gold or silver colloid increases the electric field for the molecules within the sample and thus the Raman scattering signal is enhanced by 10^{10} to 10^{11}.
SERRS (Surface-Enhanced Resonance Raman Spectroscopy)	This technique combines SERS and RRS/RR; the Raman excitation laser is matched to the maximum absorbance of the molecular vibration of interest. This technique enhances the signal of a specific vibration under study.
SRS (Spontaneous Raman Spectroscopy)	SRS is the modern name for the originally named phenomenon of Raman spectroscopy using continuous wave (CW) laser excitation. It is the original linear (or spontaneous) form of Raman spectroscopy where the intensity of the Stokes and Anti-Stokes signals is proportional to the pump excitation laser intensity, but much smaller by many orders of magnitude. It is classic Raman spectroscopy, where approximately 0.001% of incident laser energy is translated into Raman signal intensity.
SRS (Stimulated Raman Spectroscopy)	By applying a relatively high energy pulsed laser (in contrast to a traditional continuous wave (CW) laser), pulsed lasers have an enhanced effect on the Raman scattering signal to as much as 40–50% of the excitation laser energy. The overall Raman signal is enhanced by 4×10^4 to 5×10^4 over traditional spontaneous Raman spectroscopy, with the resultant significant increase in effective signal-to-noise ratio. This non-linear phenomena occurs in the Raman effect when the high energy pulsed laser excitation is used.

(*Continued*)

Table 1.5 (*Continued*).

Instrument measurement mode	Description of measurement technique
SORS (Spatially Offset Raman Spectroscopy)	Raman scattering is measured at a surface point and deeper into a sample. The two spectra are scaled and subtracted to yield a difference spectrum representing the depth profile Raman spectrum. Complex samples may be measured using multiple offset spectra combined with multivariate analysis of the spectral profile differences to determine the spectrum of the depth profile.
TERS (Tip-Enhanced Raman Spectroscopy)	TERS uses Raman spectroscopy and scanning probe microscopic techniques such as Atomic Force Microscopy (AFM). The AFM uses scanning probe tips as small as 10–20 nanometers and an enhanced Raman signal is collected along with the AFM profile of a surface or material. The enhanced Raman signal is combined with the AFM topography for a detailed nano-three-dimensional image of the spatially related chemistry of the measured specimen.
SPPERS (Surface Plasmon Polariton (SPP) Enhanced Raman Spectroscopy)	A small cylindrical lens is used for near field excitation of tiny samples or films deposited on the lens surface; measurement areas are as small as 100 microns in size with the thickness of a molecular monolayer are possible. The technique uses special excitation and collection optics to enable an enhanced small spatial signal. The technique is able to determine Raman scattering using an Argon ion laser source and standard CCD detector with the specialized sampling and collection optics.

applying the relationship:

$$R_S = \frac{I_R}{c \cdot L_E} \tag{1.11}$$

Note that the Raman scattering coefficient, as such, will be different for different Raman techniques, excitation lasers, and optical spectrometer systems. Modern Raman spectrometers utilize such assumptions for making spectroscopic measurement and generally display spectroscopic data as Raman scattering intensity (y-axis or ordinate axis) versus wavelength (nm, microns) or wavenumber (cm^{-1}) (as x-axis, or abscissa axis). As previously noted, Raman spectroscopy measures in mass per volume not in weight percent.[5] This principle holds since the Raman molecular signal is obtained by a scattering response from the number of molecules per unit volume of sample.

Note that Equations 1.10 and 1.11 are oversimplifications of the basic variables comprising the Raman intensity function for every measurement. For example, the Raman intensity (I_R) or amplitude of the Raman scattered signal at any particular frequency (or wavelength or wavenumber) is determined by other variables, such as the spectrometer throughput (τ), the self-absorption of the sample (α), the Rayleigh scattered radiation (ρ_S), the excitation laser intensity (I_0), the Raman scatter (R_S) of the molecule being measured at the frequency measured, the effective optical pathlength or L_E (i.e., actual pathlength where Raman scattering is occurring), and the number of molecules encountered by the beam path of the measuring Raman instrument (as c or concentration in Mole $\cdot$ Liter^{-1}). It is assumed that a change in spectral response is related to a concentration as described by the Raman relationship.

A more complete description of the resultant Raman intensity (I_R) as Raman scatter (R_S) with its dependency on the frequency of energy used (Greek nu as ν) is be described as:

$$I_R = \tau_\nu \cdot \alpha_\nu \cdot \rho^4_{S\nu} \cdot I_0 \cdot R_{S\nu} \cdot L_E \cdot c \quad (1.12)$$

Note for any particular sample measurement experiment with a specific instrument design, sample type, and excitation laser power and type; the instrument throughput, sample self-absorption, scattering, and Raman scatter could be considered fractions of signal loss (as a percent factor or fractions of less than 1.0 for each to reduce signal intensity). The effective pathlength depends upon laser power, excitation angle and angle of collection, as well as the sample matrix. Thus for each experiment where quantitative analysis is required, a set of calibration samples using varying analyte concentrations is recommended. A typical regression line is constructed for Raman intensity versus concentration, and the unknown test sample is estimated using regression. The regression technique for quantitative analysis is described elsewhere in this set of volumes and in greater detail in Chapter 6 of this volume.

A tracer (also known as a photometric standard, or internal standard) may be added at a fixed concentration to each calibration sample. This approach may be used to normalize the Raman scatter for samples of varying analyte concentration. In this approach, an internal standard (e.g., perchlorate in an aqueous solution) is added to each calibration sample at a fixed concentration (i.e., the same concentration for each calibration sample). Note the standard must not interfere with the analyte signal nor cause fluorescence interference. When a tracer or standard material is used for the calibration samples, the corrected Raman signal intensity (I_{RC}) for the analyte (I_{RA}) is normalized by dividing the Raman intensity for the analyte (I_{RA}) by the Raman intensity of the internal standard (I_{RT}). The corrected or ratioed Raman intensity $I_{RC} = (I_{RA}/I_{RT})$ is used for regressing the corrected intensity against the calibration sample concentration to construct a calibration curve for the unknown or test samples.

A more formalized equation for this ratioed relationship might be written as follows:

$$I_{RC} = \frac{\tau_\nu \cdot \alpha_\nu \cdot \rho_{S\nu}^4 \cdot R_{S\nu} \cdot L_E \cdot c}{\tau_\upsilon' \cdot \alpha_\nu' \cdot \rho_{S\nu}'^4 \cdot R_{S\nu}' \cdot L_E' \cdot c'} \tag{1.13}$$

A simplified equation useful for calculation of the estimated concentration for use with an internal standard is given as Equation 1.14. The internal standard for calibration may be combined with the internal standard using linear regression notation, and thus one may arrive at the following equation:

$$\hat{c} = k\left(\frac{I_{RA}}{I_{RT}}\right) + b + \varepsilon \tag{1.14}$$

where $\hat{c}$ is the estimated concentration of each unknown test sample; k is the regression coefficient; b is the y intercept; and ε is the unexplained error in the equation. Note k and b are computed from the original calibration regression equation using the corrected or ratioed Raman intensity (I_{RC}) for each calibration sample versus its concentration.

(b) Bandpass and Resolution: The terms *bandpass* and *resolution* are used to express the capability of a spectrometer to functionally separate spectral bands or lines that are separated by some finite spectral distance. If low resolution is used, certain spectral bands may overlap and remain unresolved. Low resolution measurements may yield the appearance of different band shapes than would be observed using fully resolved spectral bands. The bandpass of a spectrometer is used to describe the separation distance (in wavelength or wavenumbers) at which closely positioned Raman vibrational bands can be isolated (or resolved) by the spectrometer into distinct peaks. See Chapter 3 in this volume for more detailed information.

(c) Signal-to Noise: Simply stated, the *signal-to-noise ratio* (S/N) in a spectrometer is the numerical ratio of the amount of electrical current generated by the energy from the sample striking the detector at 100% signal ratioed to the standard deviation of replicate measurements at 100% signal. The measurements may be made under a variety of measurement conditions. Such varying Raman measurement conditions include differences in resolution, detector area, sampling device, concentration of analyte, measurement dwell time, power and wavelength of excitation laser, optical design, and spectrometer throughput.

A high signal-to-noise performance is optimum and decreases as throughput decreases and resolution increases for any given spectrometer. See Chapter 3 for details of signal-to-noise calculations for Raman spectrometers. Note that different

Raman techniques may increase signal to noise by increasing Raman signal as compared to the energy of the excitation laser. See Table 1.5 for a brief description of these techniques and the approximate signal enhancement factors associated with these various described Raman methods.

(d) Dynamic Range: The range of a specified analyte concentration over which a sensor response is directly proportional (or theoretically linear) to a change in concentration is termed the *dynamic range* of a Raman spectrometer. Dynamic range is noise limited for any spectrometer. Calculating the maximum theoretical dynamic range for an instrument, the relationship of detector non-linearity, and the maximum observable Raman intensity value requires knowledge of a few simple mathematical relationships. There is detailed information and computations shown in the various sections of these volumes, particularly the instrumentation chapters for each volume (e.g., Chapter 3).

1.3.3 *Practical Aspects for Raman Spectral Measurements*

A variety of sample presentation methods are available to the Raman spectroscopist; Chapter 4 covers many aspects of sampling for Raman applications. Raman is used to measure solids, liquids, gases, slurries, polymer films, film deposition samples, and *in vivo* biological samples. Basic sampling methods include samples in small glass (Pyrex) vials of 1 or 2 mm diameter, glass ampoules, waveguides such as small bore Teflon tubing, and direct application of laser excitation to solid samples.

Gases may be analyzed using multi-pass glass gas cells. Powders may be measured as is in powder holders or mixed with KBr to produce a pellet similar to those used for infrared measurements. Note that colored samples that absorb significant laser energy and heat rapidly require special consideration. Colored samples are discussed within Chapter 4, Sampling Considerations for Raman Spectroscopy.

1.3.3.1 *Basic Applications of Raman Spectroscopy*

Raman spectrometers are useful in analytical situations for identification of compounds in pure or mixture forms. Also oxidation and formation of double and triple bonds are readily observed using Raman spectra. The following descriptions of basic analytical Raman applications are helpful for clarification of uses and measurement techniques. Tables 1.4 and 1.5 briefly discuss the variety of Raman measurement modes and techniques useful for analysis.

(a) Quantitative Applications: The ability of Raman scattering to determine the molecular composition of materials and aqueous samples without sample preparation is an advantage of the technique.

- *Pharmaceutical companies* use Raman spectroscopy for remote and quality control identification of raw materials and analysis of finished dosage forms, even with layered composition or structures. It is used for reverse engineering of molecules; for molecular structure verification of new compounds; for production monitoring; for determination of contaminants; and for resolving complex production problems.
- *Chemical companies* use Raman spectroscopy to follow the progress of chemical reactions, identify raw materials, and monitor the kinetics and completion of chemical reactions.
- *Biotechnology companies* use Raman imaging to study biological structures and tissues and for bio-fermentation monitoring during batch processing. Raw materials and reaction monitoring may also be measured and controlled.
- *Universities,* technical schools, and other teaching environments use Raman spectroscopy to instruct students on laser instrumentation, vibrational spectroscopy, spectra-structure correlation, and qualitative and quantitative spectroscopy.
- *Food and Agricultural Producers* use Raman to measure raw materials for identification, and for natural and commercial food product production quality and nutrition.
- *Consumer Products companies* use Raman spectroscopy for identification of raw materials and quality control of finished materials. It is used for reverse engineering of products; for production monitoring; and for determination of yellowing contaminants, oxidation reactions, and double and triple bond formation due to unsaturation.

(b) Quantitative and Qualitative Applications: Raman spectroscopy is used to measure the fundamental molecular absorptions found in molecular stretching, bending, and wagging. The use of chemometric mathematical data processing can be used to calibrate for qualitative or quantitative analysis. See Chapter 6 in each Volume, and the index, for more detailed information on chemometric and regression methods.

(c) Selection of Measurement Technique and Pathlength: Samples are analyzed by laser excitation energy at various angles of excitation and collection for a variety of samples and sample holding devices, see Chapter 4 for details.

The physical nature and required information from the sample will determine which measurement technique is used. Samples can possess any of several different physical or optical properties that affect the type of analysis and sample presentation used. The following guidelines may be used for selecting a specific Raman sample presentation geometry based on the optical properties of a sample.

- *Gases (multiple pass glass cells):* Laser excitation through the sample
- *Clear solids (optical materials):* Direct laser excitation through the sample without sample preparation or through a glass vial containing the sample
- *Translucent or opaque solids:* Laser excitation and collection at 45°/45° angles, respectively to/from the sample surface
- *Reflecting optical surfaces (Specular sample):* Laser excitation and collection at 45°/45° angles, respectively to/from the sample surface
- *Clear liquids (glass vial of 1–5 mm thickness, elongated):* Laser excitation and collection at 0°/90° angles, respectively to/from the vial sample surface
- *Translucent or opaque liquids in glass vial:* Laser excitation and collection at 45°/45° angles, respectively to/from the sample surface
- *High optical density (highly absorbing solid sample):* Laser excitation and collection angles at 35° to 45°/0° to 45°, respectively to/from the sample surface
- *Mirrored surfaces*: Laser excitation and collection at 45°/45° angles, respectively to/from the sample surface
- *Colored Solid Materials*: Dilution, laser defocusing, cooled sample, sample rotation as liquid or solid phase; Laser excitation and collection at 45°/45° angles, respectively to/from the sample surface

Chapter 4 on sampling devices and accessories both illustrates and describes the various sampling methods used for Raman spectroscopy.

1.3.3.2 *Calibration and Basic Maintenance*

(a) General Concepts of Calibration: Materials are available for reference measurements to verify intensity and Raman frequency shift accuracy and alignment measurements for Raman spectrometers. Details of Raman testing methods are addressed in greater detail in other chapters of this volume, including Chapters 3 and 6, please see the index and these chapters for additional information. Note that emission lamps and chemical materials are both used for calibrating Raman instrument alignment. The following are practical methods used for basic alignment procedures. The basic methods for Raman spectrometer alignment and calibration include ASTM E1683-02(2014)e1, "Standard Practice for Testing the Performance of Scanning Raman Spectrometers";[6] ASTM E1840-96(2014), "Standard Guide for Raman Shift Standards for Spectrometer Calibration";[7] ASTM E2529-06(2014), "Standard Guide for Testing the Resolution of a Raman Spectrometer";[8] and ASTM E2911-13, "Standard Guide for Relative Intensity Correction of Raman Spectrometers".[9]

ASTM E1840-96(2014) is the established practice for using material peak frequencies for calibrating the Raman shift for laboratory Raman spectrometers. For this practice, eight spectrometers were used to collect locations of Raman peak frequencies. The agreement was determined to be ± 1 cm^{-1}. Tables and example spectra are given at the URL link to the McCreery Research Group.[10] Details of the following materials are given at their site.

1. Naphthalene (Mallinckodt #94848)
2. 1,4 Bis (2-methylstyryl) benzene (BMB) (Aldrich #25,740-0)
3. Sulfur (Aldrich #41,498-0)
4. 50/50 (v/v) Toluene/acetonitrile (Mallinckrodt Analytical Reagents)
5. 4-Acetamidophenol (active ingredients of Tylenol) (Aldrich A730-2)
6. Benzonitrile (Baker B883-07)
7. Cyclohexane (Mallinckrodt Analytical Reagents)
8. Polystyrene (Aldrich #18,243-5)

(b) Instrument Care: The proper procedure for using each Raman spectrometer is prescribed by the manufacturer.

(c) Sample Preparation and Handling: *Solvents and Operating Conditions.* Choosing the proper solvent is important for several reasons: (1) the solvent must interact with the analyte to dissolve it and as such will have some effect on the Raman spectrum, and (2) the solvent itself might have a unique fluorescence or Raman scattering response at the wavelength used for analyte determination.

Good solvents have their spectral properties delineated on the label: "Spectrophotometer grade", "Spectro-Grade", or "UV-Grade", and so forth. HPLC solvents may not be sufficient for sensitive spectroscopic work. Select a solvent that dissolves the sample but does not react with it. Use the purest solvents possible when performing spectroscopic work and make sure there is no residual materials left within or on the sample holding devices when making measurements. Note that pure water is an excellent solvent for Raman spectroscopy.

(d) Troubleshooting: Raman instruments may be quite complex. In general the user is able to perform only limited maintenance. When a problem does arise, though, its cause must be isolated to determine if the problem is caused by the instrument or the sample. Isolating and testing each variable individually allows diagnostics of a measurement problem.

Sample/Solvent Problem: If you suspect a problem with your sample or the solvent you are using, first try running the sample on a second instrument if available. It is always quicker to change instruments than to solve a complex instrument failure problem. If the problem persists on a different instrument, then the chemistry and sample must be checked. Re-measure one or two standards or samples in a different lot of solvent, using freshly cleaned sample cells or holders. Because most modern spectrometers are quite stable, the fault is often in sample handling or changes in sample chemistry.

Instrument Problems: If it has been determined that the samples are correct, then the instrument should be checked. Other than any special steps recommended by the manufacturer, the linearity, photometric accuracy, and wavenumber/wavelength accuracy checks should be completed. If the problem is not easily solved or performance seems to be not within published specifications, the best thing to do is call the instrument manufacturing company for assistance. See Chapters 3 and 6 for instrument performance details.

Avoiding Problems: Good laboratory techniques and reproducible sample work-up are good habits to develop. Each step of an analysis must be clearly stated and adhered to. Beyond that, routine maintenance is probably the best option for avoiding problems. Keeping a log record for each instrument allows the analyst to determine whether a spectrometer laser needs to be replaced or whether the detector may be aging. Cleaning is also quite important. Fumes, dust, and spills in and around the instrument will eventually destroy or cloud optical exposed surfaces.

1.4 Interpretive Raman Spectroscopy

Sections of this text are used by permission from J. Workman, *Handbook of Organic Compounds* (Academic Press - Elsevier, Boston 2000). Basic interpretive information is provided for infrared spectra in Volume 4 and this section is adapted for Raman spectroscopy.

1.4.1 *Group Frequencies*

A comparison of the intensity of Raman and MIR (Mid-Infrared) bands for C-H Stretching is presented in Table 1.6. Table 1.7 lists the various spectroscopic regions

Table 1.6 Relative Alkane Band Intensities: Raman vs. MIR for C-H stretching bands.

Band	Wavenumber region	Raman	Infrared
CH_3(Asymmetric)	2977–2857 cm^{-1}	vs	vs
CH_3(Symmetric)	2894–2874 cm^{-1}	vs	vs
CH_2(Asymmetric)	2930–2910 cm^{-1}	vs	vs
CH_2(Symmetric)	2865–2845 cm^{-1}	vs	vs

Note: The following abbreviations are used for Relative Intensities of Raman and infrared bands: w = weak, m = medium, m-s = medium-strong, s = strong, v = variable, vs is very strong.

Table 1.7 Spectroscopic regions of interest for chemical analysis.

Region	Wavenumbers (cm^{-1}) and nm	Characteristic measured
Ultraviolet	52,632–27,778 cm^{-1} (190–360 nm)	Electronic transitions: delocalized Pi electrons
Visible	27,778–12,821 cm^{-1} (360–780 nm)	Electronic transitions: color measurements
Near Infrared (NIR)	14,493–3,333 cm^{-1} (690–3000 nm) or 12,821–4,000 cm^{-1} (780–2500 nm)*	Overtone and combination bands of fundamental molecular vibrations, especially stretching and bending (some deformation as well)
Infrared (MIR)	4,000–400 cm^{-1} (2500–25,000 nm)	Fundamental molecular vibrations: stretching, bending, wagging, scissoring
Raman (R)	**4,000–400 cm^{-1} (2500–25,000 nm) Also 4000 to 100 cm^{-1}**	**Fundamental molecular vibrations: stretching, bending, wagging, scissoring**
Far-Infrared (FIR or Terahertz)	400–10 cm^{-1} (2.5 × 10 to 10^6 nm)	Molecular rotation

*Official ASTM International Definition.

of interest and the measured molecular (moiety) measured. The relative groups of atoms (termed functional groups) with common group names for Raman vibrational spectroscopy are shown below in Table 1.8. Note that a group of figures and tables illustrating band positions for Raman measured molecular vibrational bands are given in detail within Chapter 7 of this volume.

The broad band of Raman scattered energy directly indicates the energy of molecular vibration. Table 1.7 illustrates the region of the EMR (electromagnetic radiation) spectrum referred to as the Raman region. The table shows the molecular interactions associated with the energy frequencies (or corresponding wavelengths) of the various regions. As mentioned, specific molecular bonds most active in Raman spectroscopy are listed in Table 1.8.

Table 1.8 Raman active molecular bonds.

-OH, hydoxyl- [O-H]
R-OH, alcohols [O-H]
$-CH_2-OH$, primary alcohols
$R'R''CH-OH$, secondary alcohols
$R'R''R'''C-OH$, tertiary alcohols
Ar-OH, phenolics (or phenols) [O-H]: 947–980; 1400–1500; 2000-2010; 2770–2950
$-OCH_3$, methoxy (or ether group) 900–920
C-O-C, ethers
R-C=O -H, aldehydes 2195–2215
R-C=O–R′, ketones
C=C=O–R′, ketenes
R-C=O-OR’, esters
S-C=O -O-R′, thiol esters
R-C=O-C=O–R, anhydrides
-C-O-O-C-, peroxides
-O-C=O -C=O -O-, oxalates
-O-C=O –, carboxy-
-C=O -, carbonyl group
$-C{=}O\ -NH_2$, primary amides
-C=O -NHR, secondary amides
-C=O -NR’R″, tertiary amides
-C≡N, nitriles
C-C, paraffinics (or alkanes)
$CH_3-(CH_2)_N-CH_3$, normal paraffins
$R-CH_3$, methyl C-H
$R-CH_2R'$, methylene C-H
R-CHR’R″, methine C-H
C=C, olefinic group (or alkenes)
$-CH{=}CH_2$, vinyls
$R'R''C{=}CH_2$, vinylidenes
$>C{=}C{=}CH_2$, allenes
-C≡C-, acetylinics
R-C≡N, nitriles
C_NH_N, aromatics
C_NH_{2N}, naphthalenes (or cycloalkanes)
C-NO, nitroso- group
CONO, nitrites
R′R″NNO, nitrosamines
$R-NO_2$, nitro- group
$R-NH_2$, primary amines
R′R″NH, secondary amines
R-N=N-R′, azo- group
$-N{=}N^+{=}N^-$, $-N^--N^+{\equiv}N$; azides
NNO, azoxy group
R-O-N-N=O, organic nitrites
$-NO_3$, nitrates
ON=NO, nitroso- group
$H_3N^+-CH-COO^-$, amino acids
S-H, thio-
C-X, halogenated compounds

1.4.2 *Fingerprint Frequencies*

Fingerprint frequencies are due to molecular vibrations of the entire molecule rather than being specific to a particular functional group. Since the Raman or infrared spectrum is unique to an individual molecule it is termed the fingerprint of that molecule, much as the fingerprint of a person indicates the specific identity of that person. However, in identifying the unique "fingerprint vibrations" of a molecule, information is given as to the approximate molecular formula and composition of the molecule. (Incidentally, William James Hershel, grandson of F. William Hershel, was one of the first to identify the uniqueness of individual human fingerprints, aka fingermarks.)[11]

1.4.3 *Dividing the Raman Spectrum*

The group frequencies are given as those bands occurring below 6667 nm (above 1500 cm^{-1}); and the fingerprint frequencies are those above 6667 nm (below 1500 cm^{-1}). The group frequencies above 1500 cm^{-1} usually demonstrate bands of reasonable intensity, with reliable interpretation based upon frequency position; whereas the fingerprint frequencies contain bands for both group and fingerprint frequencies.

This mixed (or overlapping) band region below 1500 cm^{-1} may give inaccurate interpretation based only upon wavelength position. Therefore, in the fingerprint region the most useful interpretive information is based upon shape, e.g., broad, very broad, sharp, very sharp; or intensity, e.g., weak, medium, intense. The fingerprint region also is useful for interpretation relative to the absence of a particular band.

In summary, the group frequency region is used to derive a first assumption about the identity of the sample based on its Raman spectrum, and then the fingerprint region is used to verify the assumptions by the presence or absence of bands as well as their shapes and intensities. Group frequencies are used with general rules and comparative known (reference) spectra to make positive identification of a test sample spectrum. The ideal group frequencies have the characteristics of: (1) always being found above 1500 cm^{-1}, (2) have intense amplitude, (3) have identifiable and consistent position in frequency, (4) are reliable and always appear when a particular group is present, (5) are isolated without interference, (6) are narrow (in the range of $\pm$ 25 cm^{-1}), and (7) the group has more than one characteristic frequency (used for further verification of the group's presence).[12–14]

1.4.4 *Coupling of Vibrations*

Coupling indicates that the oscillators or molecular vibrations of two or more molecules are interactive, so that the original vibrational energy states (if the vibrations could occur independently of one another) result in split energy states due to the interaction of the vibrations. Coupling is divided into two basic orders, first and second (Fermi resonance).

First order coupling can be involved in several important Raman or infrared group frequencies. For example, CO_2 has the two separate (uncoupled oscillators) of C=O, each occurring at approx. 5405 nm ($\sim 1850\ cm^{-1}$). The interactive (coupled oscillators) energy states occur at 4257 nm (2349 cm^{-1}, asymmetric stretch) and 7463 nm (1340 cm^{-1}, symmetric stretch). First order coupling is involved for multiple group frequencies as listed.[15–16]

a. The stretches for all cumulated double bonds, X=Y=Z, e.g. , C=C=N.
b. The stretches in XY_2, including $-CH_2-$, and H_2O.
c. The stretches in XY_3 groups, including $-CH_3$.
d. The deformations of XY_3 groups, including $-CH_3$, CCl_3.
e. The N-H in-plane bend of secondary amides, e.g., R-CO-NH-R′.

1.4.5 *Fermi Resonance (or Second Order Coupling)*

Fermi resonance is the interaction or coupling of two vibrational energy states with resultant separation of the states where one of the states is an overtone or a sum tone.[17] An overtone vibration occurs as the integer multiple of a fundamental vibration in frequency space (intensity falls off rapidly with the higher multiples):

a. The first overtone of a fundamental vibration (ν_i) is equal to $2 \times \nu_i$.
b. The second overtone of a fundamental vibration is equal to $3\ \nu_i$.
c. The third overtone is equal to $4\ \nu_i$.

Note that overtones are a special case of sum tones where the frequencies are identical.

A sum tone is the general case of an overtone where the frequencies are not equal and where a variety of vibrational energy states can occur:

a. A binary sum tone is equal to the sum of two fundamentals, e.g., $\nu_i + \nu_k$.
b. A ternary sum tone is equal to the sum of three fundamental vibrations, e.g., $\nu_i + \nu_k + \nu_m$.
c. Other sum tones can occur such as the sum of an overtone and a fundamental vibration, e.g., $2\nu_i + \nu_k$.

Three requirements are stated for Fermi resonance.[18,20]

a. The zero-order frequencies must be close together (typically within 30 cm^{-1}).
b. The fundamental and the overtone or sum tone must have the same symmetry.
c. There must be a mechanism for the interaction of the vibrations.

(1) The vibrations cannot be separate (or localized) in distinctly different parts of the molecule.

(2) The vibrations must be mechanically interactive in order that the interaction of one vibration affects another.

The results of Fermi resonance are important for Infrared, Raman, and Near Infrared spectroscopy. Fermi resonance causes the following effects on spectral bands:

a. The resultant bands are moved in position from their expected frequencies.
b. Overtone bands are more intense than expected.
c. There may exist doublet bands where only singlets were expected.
d. Solvent changes can bring about slight shifts in frequency location of a band and intensities can be greatly changed.[19–21]

1.4.6 *Frequencies of Alkanes (Paraffins)*

1.4.6.1 *Normal hydrocarbons (n-alkanes)*

For Raman active molecular vibrations, the number of normal vibrations occurring within the spectral region for saturated hydrocarbons is 3N-6 , where N is the sum of carbon plus hydrogen atoms, e.g., n-hexane (C_6H_{14}) exhibits $3(20) - 6 = 20$ normal vibrational modes. Internal rotation and overlapping of bands complicate the interpretation of alkane (paraffinic) group spectra. C-H stretching vibrations can be expressed to a first approximation by the straight-forward use of Hooke's law describing the vibration between two masses connected by a vibrating spring.

The first consideration in the description of alkane group spectral interpretation would include the C-H stretching vibrations for methyl (-CH_3) and methylene (-CH_2-) groups. The *methyl C-H stretching vibrations* (Figure 1.11) are found in two basic modes:

a. The asymmetric (degenerate or out-of-phase)) methyl C-H stretch at (2967 ± 10 cm^{-1}).
b. The symmetric (in-phase) methyl C-H stretch at (2884 ± 10 cm^{-1}).
c. Three C-H stretching modes correspond to three coupled C-H stretching vibration oscillators.

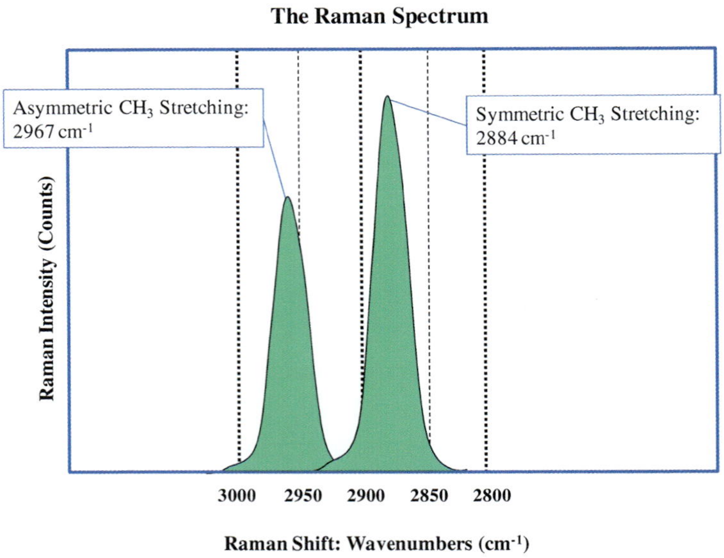

Fig. 1.11 Illustration of aliphatic methyl C-H stretching band locations as a Raman spectrum.

d. The molecular extinction coefficient for methyl C-H stretch remains constant irrespective of increasing alkane chain length.

The *methylene C-H stretching vibrations* (Figure 1.12) are also found in two basic modes:

a. The asymmetric (degenerate or out-of-phase) methylene C-H stretch at (2920 $\pm$ 10 cm^{-1}).
b. The symmetric (in-phase) methylene C-H stretch at (2855 $\pm$ 10 cm^{-1}).
c. Two methylene stretching modes result from two coupled C-H stretching vibration oscillators.
d. Increasing the chain length within alkane groups brings about a linear increase in the molecular extinction coefficient of methylene C-H stretching bands.

The carbon-hydrogen vibrational modes for alkanes also contain both methyl and methylene bending vibrations. The *methyl C-H bending vibrations* are found in two modes:

a. The asymmetric (degenerate or out-of-phase deformation) methyl C-H bending at 6803 nm to 6897 nm (1460 $\pm$ 10 cm^{-1}).
b. The symmetric (in-phase or "umbrella") methyl C-H bending at 7220 nm to 7326 nm (1375 $\pm$ 10 cm^{-1}). This functional group vibrational frequency is dependent upon the atom to which the methyl group is attached. Examples from

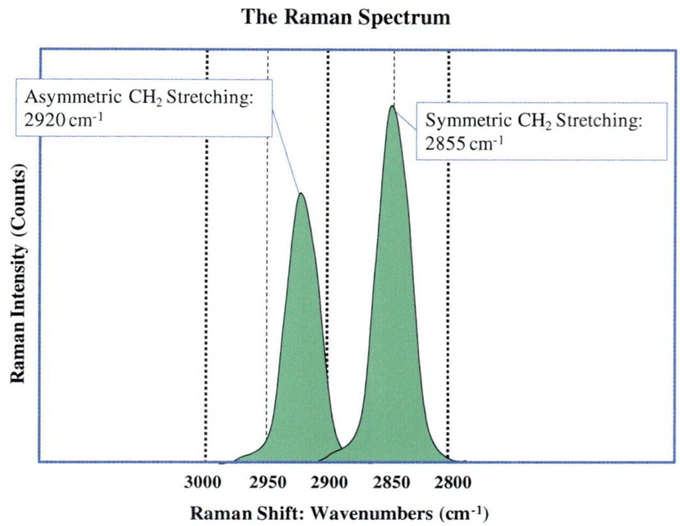

Fig. 1.12 Illustration of aliphatic methylene C-H stretching band locations as a Raman spectrum.

the literature illustrating band frequency location relative to the atom attached to methyl group (X-CH_3) include those shown in Table 1.9.[22]

c. There is no coupling of methyl C-H bending modes to the rest of the system.

The *methylene C-H bending vibrations* are found in four basic modes:

a. The symmetric (scissoring) methylene C-H bending at 6826 nm to 6920 nm (1455 ± 10 cm^{-1}) where no coupling is observed.
b. The in-phase twisting vibration at 7634 nm to 7752 nm (1300 ± 10 cm^{-1}) with observed coupling.
c. The coupled wagging vibrations at 7692 nm to 10,000 nm (1300 to 1000 cm^{-1}).
d. The in-phase rocking vibration for alkanes with four or more carbons is located at 13,699 nm to 14,085 nm (720 ± 10 cm^{-1}) where coupling is observed.

In addition, other bending modes of methyl and methylene can be observed.

The alkane carbon-to-carbon stretching, and carbon-to-carbon bending vibrations also occur as weak group frequencies.

1.4.6.2 *Branched Chain Hydrocarbons (iso-, gem-dimethyl, tertiary butyl, and Cyclic Groups)*

For Raman absorptions, the following spectra-structure correlations apply.

a. Methine groups (tertiary methyl, R_3C-H) are not important for the infrared region and do not exhibit unique infrared spectral characteristics.

Table 1.9 Raman and Infrared band frequency locations relative to the specific atom attached to a methyl group (X-CH_3).

Functional group	Fundamental frequency ($\pm$10 cm^{-1})	Fundamental wavelength (in nm)
Pb-CH_3	1165 cm^{-1}	8511–8658
Sn-CH_3	1190	8333–8475
Sb-CH_3	1200	8264–8403
Hg-CH_3	1200	8264–8403
Ge-CH_3	1235	8032–8163
As-CH_3	1250	7937–8065
I-CH_3	1252	7924–8052
Se-CH_3	1280	7752–7874
Br-CH_3	1305	7605–7722
Si-CH_3	1255 $\pm$ 5 cm^{-1}	7937–8000
P-CH_3	1300	7634–7752
B-CH_3	1320	7519–7634
S-CH_3	1325	7491–7605
Cl-CH_3	1355	7326–7435
Carboxyl C- CH_3	1355	7326–7435
C-CH_3	1378 $\pm$ 5 cm^{-1}	7231–7283
N-CH_3	1425	6969–7067
O-CH_3	1455	6826–6920
F-CH_3	1475	6734–6826

b. Isopropyl and gem-dimethyl groups exhibit a symmetric bending vibration doublet at 7246 nm to 7353 nm (1370 $\pm$ 10 cm^{-1}) with coupling absent.
c. Tertiary butyl substitution exhibits a symmetric bending vibration doublet from 7194 nm to 7326 nm (1390 to 1365 cm^{-1}) that does not exhibit mechanical coupling.
d. Cyclic hydrocarbons demonstrate methylene C-H stretching modes based upon ring size. In addition there are a number of spectral effects based upon lone electron pair interactions with carbon-hydrogen molecular orbitals. The reader is referred to a number of references describing the unique spectra of cyclic hydrocarbons with various substituted functional groups.

1.4.7 *Group Frequencies of Alkenes (Olefins)*

1.4.7.1 *C-H Stretch of Alkenes*

For Raman the olefins include bands for saturated hydrocarbons as well as the olefinic C=C stretching and olefinic C-H stretching and bending vibrations.

1.4.7.2 *Olefinic C-H Stretch*[23]

a. Olefinic C-H stretch occurs at 3226 nm to 3333 nm (3100 to 3000 cm^{-1}). C-H stretch for an unsaturated carbon occur below 3333 nm (above 3000 cm^{-1}), whereas the saturated carbon C-H bond stretching occurs above 3333 nm (below 3000 cm^{-1}). The sp-type C-H stretch for ≡C-H occurs at approximately 3030 nm (3300 cm^{-1}); the sp^2-type C-H stretch associated with aromatic and olefinic carbons occurs at 3226 nm to 3333 nm (3100 to 3000 cm^{-1}); and the sp^3-type stretch for saturated carbons is observed at the high frequency ranges between 3344 to 3571 nm (2990 and 2800 cm^{-1}).

b. Olefinic C-H stretch can occur as an end-group, e.g. R-R-C=C; or as an imbedded group associated with other carbons on either end, e.g., R-C=C-R. The absorption bands for these two olefinic types occur from 3268 nm to 3226 nm (3060 to 3100 cm^{-1}) for the end-group type, and from 3322 nm to 3279 nm (3010 to 3050 cm^{-1}) for the imbedded type. Note that compounds containing the vinyl group (-CH=CH_2) have weak absorption bands at approximately 3247 nm to 3300 nm (3080 cm^{-1} and 3030 cm^{-1}).

1.4.7.3 *Olefinic C=C Stretch*

a. Olefinic C=C stretch is both Raman and infrared active but the intensity is quite variable depending upon associated atoms to the C=C group. The table below summarizes the appearance of absorption bands associated with olefinic C=C stretch for vinyl-type molecules, as illustrated in Table 1.13.

Table 1.10 Showing locations of C=C stretch vibrations related olefinic containing molecules.

Olefinic molecule type (R-type substituents)*	C=C Stretch band location	Relative band intensity
trans, tri- and tetra substituted	1685 to 1665 cm^{-1}	weak
cis olefins	1665 to 1625 cm^{-1}	medium to strong
mono-substituted	1660 to 1630 cm^{-1}	medium to strong
trans, di- substituted	1680 to 1685 cm^{-1}	weak
cis, di-substituted	1669 to 1630 cm^{-1}	medium to strong
di-substitution on a single carbon	1669 to 1630 cm^{-1}	medium to strong
tri-substitution	1680 to 1685 cm^{-1}	weak
tetra-substitution	1680 to 1685 cm^{-1}	weak

*Substituents other than R-groups will change the frequencies of absorption.

b. Factors affecting the frequencies for the various olefinic groups include conjugation. Conjugation is the sharing of electrons between the C=C olefinic bond and atoms or molecules attached to either of the carbons in the C=C group. The associated groups lower the frequency by 75 nm to 115 nm (20 to 30 cm^{-1}). The conjugated groups that lower the frequency of the -C=C- stretch include nitrile (-C≡N), occurring at 6215 nm (1609 cm^{-1}); carbonyl (-CO-R), occurring at 6184 nm (1617 cm^{-1}); phenyl (-C_6H_5), at 6135 nm (1630 cm^{-1}); vinyl (-CH=CH_2), at 6169 nm (1621 cm^{-1}); and methyl (-CH_3), at 6057 nm (1651 cm^{-1}). These are important bands for Raman spectroscopy.

c. The attachment of =C to a variety of ring sizes affects the frequency of the C=C olefinic stretch. A six member phenyl group attached to a (=C) group yields a 6057 nm band (1651 cm^{-1}); a five member ring yields a 6035 nm band (1657 cm^{-1}); a four membered ring exhibits a band at 5959 nm (1678 cm^{-1}); a three-membered ring at 5615 nm (1781 cm^{-1}); and the allene two carbon C=C olefinic group ($H_2C=C=CH_2$) occurs at 5136 nm (1974 cm^{-1}).

d. Carbonyl groups attached to olefinic groups yields a variety of frequency shifts depending upon the conjugation between carbonyl, the olefinic group, and the other attached groups; the section on carbonyls discusses this in greater detail.

e. Cyclic olefins exhibit unique frequencies for -C=C- stretch depending upon the number of carbons within the cyclic olefin group, and the type of attached molecule or atom to the olefinic carbons. The following table illustrates the frequency for olefinic C=C stretch dependent upon the size of the cyclic olefin and the attached group types (Table 1.11).

An increase in the electronegativity of the attached group raises the frequency of the olefinic C=C stretch; resonance lowers the frequency of the C=C stretch

Table 1.11 Demonstrating C=C stretch frequency with respect to carbon size of cyclic olefin and type of attached group.

Attached atom or molecule to both sites of olefinic carbons as (-C=C-)	Number of carbons in cyclic olefin			
	6	5	4	3
-H	1653 cm^{-1} (6050 nm)	1611 cm^{-1} (6207 nm)	1565 cm^{-1} (6390 nm)	1648 cm^{-1} (6068 nm)
-CH_3	1683 cm^{-1} (5942 nm)	1686 cm^{-1} (5931 nm)	1695 cm^{-1} (5900 nm)	1887 cm^{-1} (5299 nm)

and there are interactions when the attached groups exhibit both electronegative and resonance features.

f. In the IR region the intensity of the fundamental olefinic carbon-carbon stretch, designated by $\tilde{\nu}$ (C=C) is highly variable. In some cases, the band is completely absent; in other cases the band may be very strong. If there is little change in the dipole moment of an olefinic containing molecule, the C=C stretching mode will be Raman inactive. Thus, if the C=C group is across the center of symmetry for a molecule the C=C stretch will be forbidden in Raman. A center of symmetry is where identical atoms or molecules are attached coaxially across the C=C olefinic group. As a general rule, the more symmetrically substituted the olefin, the weaker the Raman band is for the olefinic molecule. Bands present in the Raman spectrum for an olefin can indicate the presence of an olefinic group, but the absence of such bands does not give conclusive evidence for the absence of an olefin.

1.4.7.4 *Olefinic C-H Bend*

a. Two modes of bending occur in Olefinic hydrogen bonds, in-plane and out-of-plane. In-plane bend represents a change in the C-H bond angle along the same plane, occurring at 6920 nm to 8333 nm (1455 to 1200 cm^{-1}) as a weak and not very useful group frequency for the presence of olefinic hydrogen bonds. Out-of-plane bending occurs as the C-H bonds vibrate outside the flat olefinic group plane. Out-of-plane bending occurs at approximately 10,000 nm to 15,385 nm (1000 to 650 cm^{-1}) and represent excellent bands for identifying the presence of olefinic groups. The out-of-plane C-H bend related bands are group frequencies which are distinct, not being affected by conjugation or coupling with other vibrations in the olefinic molecule.

b. There is only one Raman active C-H bend of the four possible modes for olefinic bending vibrations. The four bending modes include *twisting or rotational motion*; opposite twisting at the individual olefinic carbons producing a *torsion* at the C=C bond; *wagging*, or a motion of the hydrogen bonds; and two forms of wag, one where the two groups of two each hydrogens move in opposite planes causing a rotational motion around a common center of axis; and the second wag where all four hydrogens move in or out of the plane together. The first type of wag is Raman only active and the second type of wag is the only infrared active type. Table 1.12 below summarizes the frequencies for C-H out-of plane bending given the number of R groups attached to the two olefinic carbons.

Table 1.12 Showing locations of C-H out-of plane bending vibrations for olefinic molecules.[24]

Olefinic molecule type (R-type substituents)*	C-H bend frequencies	Relative band Intensity/Shape
mono-substituted	995 to 985 cm^{-1} (10,050 to 10,152 nm)	very strong
	915 to 905 cm^{-1} (10,929 to 11, 050 nm)	very strong
trans, di-substituted cis olefins	970 to 960 cm^{-1} (10,309 to 10,417 nm)	very strong
cis, di-substituted	730 to 650 cm^{-1} (13,699 to 15,385 nm)	weak and broad
di-substitution on a single carbon	895 to 885 cm^{-1} (11,173 to 11,299 nm)	very strong
tri-substitution	840 to 790 cm^{-1} (11,905 to 12,658 nm)	strong
tetra-substitution	N/A	N/A

*Substituents other than R-groups will change the frequencies of absorption for the C-H out-of plane bend.

1.4.8 *Molecules with Triple Bonds (-C≡)*

1.4.8.1 *Acetylenes (-C≡ C-)*

a. Triple bonded carbon-to-carbon functional groups are termed acetylinic (or acetylenes). Due to symmetry about the C≡C bond the absorptions for C≡C are weak in the infrared, but strong in the Raman spectrum. For acetylene (H-C≡C-H) the C≡C stretch is forbidden in the infrared and thus not observed. For R-C≡C-H (mono-substituted) the C≡C stretch occurs at 4673 nm to 4762 nm (2140 to 2100 cm^{-1}) and is weak in intensity. For R-C≡C-R′ (di-substituted acetylenes) the C≡C stretch occurs from 4425 nm to 4566 nm (2260 to 2190 cm^{-1}) with very weak, sharp bands. The ≡C-H stretch band for mono-substituted acetylenes is medium and sharp in the infrared occurring at 3021 nm to 3040 nm (3310 to 3290 cm^{-1}).

b. The bending vibrations for ≡C-H occur as medium-strong bands at approximately 15, 385 nm (650 cm^{-1}).

1.4.8.2 *Nitriles (R-C≡ N)*

a. The frequency for unconjugated nitriles, where -C≡N is not associated with electron withdrawing groups, occurs at 4425 nm to 4464 nm (2260 to 2240 cm^{-1}). Where conjugation occurs such as Aromatic -C≡N, or >C=C-C≡N, the frequency is found from 4464 nm to 4525 nm (2240 to 2210 cm^{-1}).

b. Hydrogen cyanide produces the only differences to the above rules. In the case of HCN, frequencies occur at 4787 nm (2089 cm^{-1}) for gas phase and 4850 nm (2062 cm^{-1}) for liquid phase. In the case of acetonitrile, the frequency for C≡N stretch is found at 4444 nm (2250 cm^{-1}); for RCN the frequency of 4425 nm to 4464 nm (2260 to 2240 cm^{-1}) is observed.

Table 1.13 Frequencies for linear and non-linear A=B=C systems in the order of frequency.[24]

Frequency in cm^{-1} (Wavelength in nm)	Functional group	Vibrational mode	Bond angle (degrees)
2349 cm^{-1} (4257 nm)	→O=C=O→	asymmetric	180, linear
1340 cm^{-1} (7463 nm)	←O=C=O→	symmetric	
1362 cm^{-1} (7342 nm)	→O=S=O→	asymmetric	119, non-linear
1151 cm^{-1} (8688 nm)	←O=S=O→	symmetric	

Note bond angle is only one consideration for frequency difference for interacting oscillators, another important consideration is the mass of the central atom.

c. Related groups have different frequencies depending upon electron sharing of the associated atom or group with the C≡N bond. As examples, nitrile oxide -C≡NO occurs at 4255 nm (2350 cm^{-1}, approximately 100 cm^{-1} higher than C-C≡N); and thiocyanate RS-C≡N occurs at 4651 nm (2150 cm^{-1}, approximately 100 cm^{-1} lower than C-C≡N).

1.4.9 *Cumulated Double Bond Systems (A=B=C)*

As examples, Table 1.13 illustrates the frequencies for a linear and non-linear system in the order of frequency. Note that the linear system, carbon dioxide, has a much greater frequency differential between the asymmetric and symmetric modes due to the high interaction between two co-linear oscillators sharing a common central atom. This first order interaction between two (hypothetical) fundamental vibrations is termed *coupling*.

1.4.10 *Aromatic Systems*

a. Aromatic systems produce strong and sharp Raman bands due to the relatively rigid molecular structures. There are multiple vibrational modes present in aromatic systems including: (1) carbon-hydrogen stretching, (2) in-plane bending, and (3) out-of-plane bending. Carbon-carbon interactions include (4) ring stretching, and (5) ring bending modes. Internal vibrations caused by groups associated with the aromatic system also exhibit vibrational group frequencies. The total vibrational modes for aromatic systems is equal to 3N-6 vibrations, where N is the number of carbons in the ring structure. These vibrational modes are shown in several references.[25–28]

b. Carbon-hydrogen stretching vibrations occur from approximately 3226 nm to 3333 nm (3100 to 3000 cm^{-1}) as shown in Figure 1.13. Raman spectra of aromatics

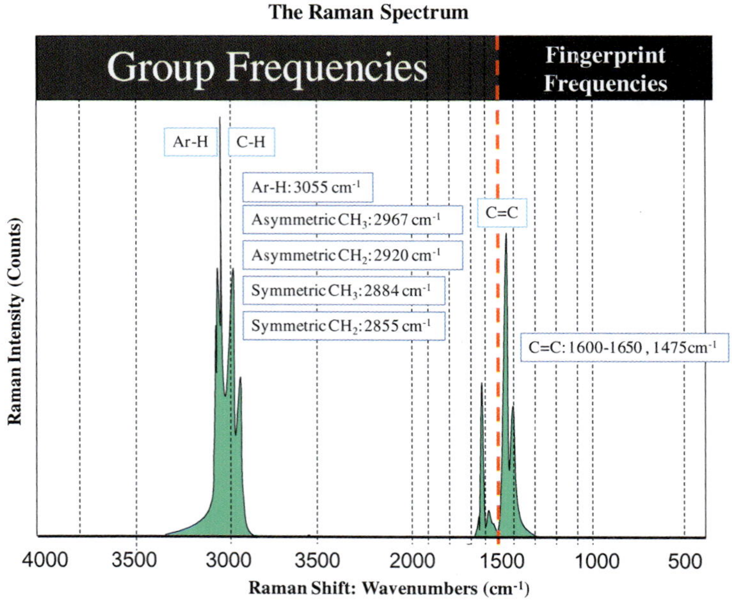

Fig. 1.13 Aromatic C-H stretching band as a Raman spectrum.

using high resolution instrumentation shows a pair of composite bands near 3257 nm (3070 cm^{-1}) and 3300 nm (3030 cm^{-1}). These composite bands arise as the combined interactions of several stretching modes discussed briefly above.

c. The carbon-carbon ring stretching occurs as a vibrational pair near 6250 nm (1600 cm^{-1}) as illustrated in Figure 1.13. This pair of frequencies is termed a degenerate pair and results from two stretching modes of the carbon-carbon ring stretching. However due to Fermi resonance one of the vibrations (near 6270 nm or 1595 cm^{-1}) is not observed in the infrared spectrum. When there is substitution on the aromatic ring, the degeneracy is lost and the pair of bands occur near 6250 nm (1600 cm^{-1}) and 6329 nm (1580 cm^{-1}). The 6329 nm (1580 cm^{-1}) band will be more intense if there is conjugation with the aromatic system by an electron withdrawing group, or if substitution by O or N has occurred. Para-disubstitution of a benzene ring by two identical groups yields to the center of symmetry rule, and thus the carbon-to-carbon ring stretching is infrared forbidden. It follows that if the para-disubstituted groups are not identical that some weak bands will occur near the 6250 nm (1600 cm^{-1}) region.

d. The carbon-hydrogen bending (wag) vibrations occur as in-plane bending and out-of plane bending. The in-plane bending vibrations are much less consistent than the out-of-plane bending. The in-plane bending vibrational frequency depends

Table 1.14 Out-of-plane bending vibrations for benzene.

Substitution on ring	Adjacent hydrogen atoms	Frequency (in cm^{-1})	Wavelength (in nm)
None	6	765–735	13072–13605
Mono-	5	770–735	12987–13605
Ortho (di-)	4	775–740	12903–13514
Tri-	3	810–750	12346–13333
Tetra-	2	860–800	11628–12500
Penta-	1	910–850	10989–11765

upon the number of C-H groups present and the substituent groups. Out-of-plane bending is much more consistent and intense in the infrared region. Out-of-plane bending is not particularly influenced by the substituents present, but the frequency for the bending vibrations depends upon the numbers of adjacent hydrogens. The typical presentation for the out-of-plane bending vibrations for benzene is given in Table 1.14.

The carbon-hydrogen bending vibrations give rise to a series of combination bands termed *sumtones* or *summation bands*. These sumtones, summation bands, or combination bands occur from 5000 nm to 5882 nm (2000 to 1700 cm^{-1}). The patterns are used in Raman spectroscopy to determine ring substituents. The pattern for the sumtones is more important than their exact position. These composite band patterns are given in most Raman or infrared texts.[29–30]

1.4.11 *Vibrational States in Benzene*

Hayward and Henry calculate the frequencies for all eight members of the C-H stretching overtone spectrum in benzene. The paper provides a limited generalization method for calculating the anharmonicity constants for molecules where nearly identical oscillators are involved, and when these oscillators have vibrational motion which is approximately independent of the rest of a polyatomic molecule. The authors report the overtone frequencies (in cm^{-1}) of $\nu = 1$–8 as 3053, 5940, 8737, 11411, 14003, 16458, 18807, and 21025, respectively. The equivalent wavelength positions (in nm) for these overtone bands are 3275, 1684, 1145, 876, 714, 608, 532, and 476.[26–28,31]

Heller and Mukamel describe a general theory for estimating the overtone absorption line shapes for polyatomic molecules. The authors expand on the molecular Hamiltonian and local mode theory, and provide details for overtone line shape estimation. Numerical estimation of benzene overtone line shapes are given along with the derivation and detailed explanation of both theory and methodology.[32]

A description of the aspects related to local mode combination bands for aromatic and aliphatic hydrocarbons is discussed in a paper by Burberry and Albrecht. The authors explain the origin of side bands in the local mode (LM) overtone vibrations as resulting from LM-LM combination bands of various C-H oscillators. The local mode overtones are described by theory and observation for methyl groups in tetramethylsilane (TMS) as well as benzene.[33]

Child discusses issues on the use of local modes for spectroscopic assignment and interpretation. Examples discussed include water, acetylene, methane, and silane. The author generally reviews the topic for pre-1985 fundamental literature, including 42 references.[34]

1.4.12 *Carbonyl Compounds (>C=O)*

a. Compounds containing carbonyl groups exhibit frequencies over a broad range of 5128 nm to 6452 nm (1950 to 1550 cm^{-1}). The position or frequency of the carbonyl C=O stretching vibration is affected by: (1) the isotope effects and mass change of substituted groups, (2) bond angles of substituted groups, (3) electronic (resonance and inductive) effects, and (4) interactions of these effects.

Substituents with higher mass decrease the C=O stretch frequency; increasing the mass or bond angles of substituents also decreases the frequency by up to 40 cm^{-1} at 1715 cm^{-1}; this is equivalent to an increase of 139 nm at 5831 nm. Similarly decreasing the substituent mass, or bond angles between the carbonyl carbon and its substituents increases the frequency by 25 cm^{-1} above the nominal 1715 cm^{-1} carbonyl C=O stretch frequency; equivalent to a decrease in wavelength of 84 nm at 5831 nm.

More electronegative (electron withdrawing) substituents will increase the carbonyl carbon-oxygen stretch frequency by up to 100 cm^{-1} above the 1715 cm^{-1} nominal frequency; a decrease of 321 nm at 5831 nm. Conjugation of the carbonyl group to aromatic or olefinic groups tends to lower the frequency for both C=O and C=C by 30 to 40 cm^{-1} dependent upon the ring size of the substituent. For example, carbonyl conjugation with aromatic ring sizes of 6, 5, 4, and 3 carbons produces C=O stretch frequencies at 5935 nm (1685 cm^{-1}), 5928 nm (1687 cm^{-1}), 5931 nm (1686 cm^{-1}), and 5963 nm (1677 cm^{-1}), respectively. Electronic effects such as resonance and inductive type produce a double-bond character and increases the frequency of the carbonyl stretch.

1.4.12.1 *Special cases of the carbonyl group*

The group called anhydrides (R-C=O -O- C=O -R), consists of two carbonyls connected by an oxygen. The frequency for the carbonyl stretch occurs from 5495 nm to 5714 nm (1820 to 1750 cm^{-1}); the in-phase (←C=O→ and ←C=O→) stretch for the two carbonyl groups in the anhydride occurs at 5495 nm (1820 cm^{-1}). In the out-of-phase carbonyl pair (←C=O→ and →C=O←) the frequency for the carbonyl is 5714 nm (1750 cm^{-1}). In cyclic anhydrides the in-phase frequency is 5348 nm to 5405 nm (1870 to 1850 cm^{-1}), and the out-of-phase occurs at 5556 nm to 5618 nm (1800 to 1780 cm^{-1}). Solvent effects like hydrogen bonding to the carbonyl oxygen lowers the carbonyl C=O frequency.

As in other carbonyl containing molecules smaller mass substituents or decreasing bond angle of carbonyl carbon substituent bonds increase the frequency for the carbonyl. Large inductive effects from the addition of largely electronegative substituents also increases the frequency for the carbonyl. When a halogen is attached to the alpha carbon attached to the carbonyl carbon the frequency is increased by 60 nm to 90 nm (20 to 30 cm^{-1}).

Acid dimers interact by strong molecular hydrogen bonding and lower the carbonyl frequency by 25 to 60 cm^{-1}. The interaction is between the carbonyl oxygen and the hydrogen associated with the acid molecules:

$$(R - C = O - OH \ldots O = C - R - OH).$$

1.4.13 *Amides (R-C=ONH-X)*

a. Amides are a special case of carbonyl compounds which are subject to frequency shifts due to hydrogen bonding. The frequencies for dilute solutions of amides can vary by as much as 250–275 nm (75 cm^{-1}) at ~5880 nm as compared to the same molecule with hydrogen bonding (such as a water or oil matrix/diluent preparation). Dilute solutions of amides exhibit carbonyl frequencies at 5764 nm (1735 cm^{-1}), (primary, R-C=O -NH_2); 5865 nm (1705 cm^{-1}), (secondary, R-C=O -NHR); and 6042 nm (1655 cm^{-1}), (tertiary, R-C= -NR_2). The respective hydrogen bonded vibrations occur at 5882 nm to 6061 nm (1700 to 1650 cm^{-1}) for primary; 5952 nm to 6116 nm (1680 to 1635 cm^{-1}) for secondary; and 6024 nm (1660 cm^{-1}) for tertiary amides.[35–36]

b. The N-H stretch for amides occurs as two bands between 2976 nm and 3150 nm (3360 and 3175 cm^{-1}). Geometrically, two forms of amides exist as either the trans or cis forms; the trans form demonstrating a frequency of 3008 to 3053 nm (3325 to 3275 cm^{-1}); and the cis form 3140 nm to 3190 nm (3185 to 3135 cm^{-1}). The geometries are as R-C=O -N-R′ -H where the R and R′ are either in the trans or cis configuration.[37]

The term "amide I" band refers to the C=O carbonyl stretch and from the section on carbonyls it was mentioned that the nominal C=O stretch frequency is 5831 nm (1715 cm^{-1}). Since the amide form of the carbonyl has the electronegative N atom, lowered by resonance with the nitrogen electrons and lowered by hydrogen bonding, the C=O nominal frequency for the amide I stretch is approximately 1650 cm^{-1} (6061 nm).

The term "amide II" refers to the NH_2 scissoring associated with primary amides (R-C=O -NH_2). The amide II frequency results from the combined interactions of the C-N stretch and the N-H in-plane bend; these vibrations interact to form bands at the 6494 nm (1540 cm^{-1}) region, and the 7634 nm to 8197 nm (1310 to 1220 cm^{-1}) region. A band at 6410 nm to 6515 nm (1560 to 1535 cm^{-1}) indicates a trans substituted amide; this band is lost in the cis-amide configuration.

The Carboxylate ion (O=C-O^-) has two identical C┈┈O, C┈┈O bonds; consisting of the in-phase ←C┈┈O →, ←C┈┈O →; and out-of-phase ←C┈┈O →, →C┈┈O← bands. The broad out-of-phase band occurs at 6192 nm to 6452 nm (1615 to 1550 cm^{-1}), whereas the broad in-phase band has a frequency of 7143 nm to 7692 nm (1400 to 1300 cm^{-1}). The in-plane band is not particularly useful in spectroscopy due to its variable intensity and position.

1.4.14 *Ethers (C-O-C)*

The ether group can be distinguished by the degree of saturation for the ether carbons. A saturated ether has the maximum number of hydrogens (6) associated with the ether carbons. Unsaturated ethers can either be single unsaturated or double unsaturated, depending upon the number of unsaturated methoxy carbons (=C-O-) within the ether group. Saturated ethers exhibit a very strong C-O stretch at a frequency of 1160 to 1050 cm^{-1} (8621 nm to 9524 nm); doubly unsaturated ethers exhibit a strong, slightly broadened C-O stretch frequency at 1275 to 1230 cm^{-1} (7843 nm to 8130 nm). Epoxides (C-O-C) demonstrate a symmetric stretch near 1255 cm^{-1} and an asymmetric stretch of 860 cm^{-1}(11,628 nm).

1.4.15 *Esters (R-CO-OR')*

The ester group exhibits the frequencies of a mixed ether and it does show a very strong band at 7937 nm to 8696 nm (1260 to 1150 cm^{-1}); and a strong band at 8333 nm to 10,000 nm (1200 to 1000 cm^{-1}). These two bands correspond to the C-O for RCO-OR' and the O-R′ of RCOO-R′.

1.4.16 *Anhydrides (R-C=O-C=O-R)*

Cyclic anhydrides show bands for doubly unsaturated ethers at 7937 nm (approximately 1260 cm^{-1}); open chain anhydrides show the saturated ether band at 9434 nm (1060 cm^{-1}).

1.4.17 *Peroxides (-C-O-O-C-)*

Raman active bands are extremely difficult to identify; there is an ether C-O stretch vibration near 9091 nm (1100 cm^{-1}) that can be useful for identification. Raman bands are stronger than infrared.

1.4.18 *X-H Functional Groups (O-H and N-H)*

X-H frequencies occur below 5000 nm (above 2000 cm^{-1}), except for Boron compounds which demonstrate bands from 4762 nm to 6173 nm (2100 to 1620 cm^{-1}). Interaction of X-H with other groups does not usually occur due to the high frequency of these bands and the low mass of hydrogen, therefore the frequency of X-H is specific and reliable for Raman spectroscopy. The X-H stretch follows the two-body harmonic oscillator approximation (discussed in Volume 3, Chapter 1).

The relationship of X-H to the harmonic oscillator allows the relationship to frequency of any specific X-H group to be related to the Periodic Table; the frequency of an X-H group increases as one goes upward in any column or to the right within any row of the Periodic Table. The first row of the Periodic Table gives H-H with a nominal fundamental stretch at 2404 nm (4160 cm^{-1}). The second row yields ν(Li-H) = 7353 nm (1360 cm^{-1}); ν(Be-H) = 5089 nm (1965 cm^{-1});ν(B-H) = 4000 nm (2500 cm^{-1}) and 4762 nm to 6173 nm (2100 to 1620 cm^{-1}); ν(C-H) = 3384 nm (2955 cm^{-1}); ν(N-H) = 2941 nm (3400 cm^{-1}); ν(O-H) = 2770 nm (3610 cm^{-1}); ν(F-H) = 2532 nm (3950 cm^{-1}). In the third row the more important ν(X-H) include ν(Si-H) = 4577 nm (2185 cm^{-1}); ν(P-H) = 4255 nm (2350 cm^{-1}); and ν(S-H) = 3876 nm (2580 cm^{-1}). Raman scattering intensity is weak for O-H and N-H as compared to the interaction of infrared absorption. The most important of these functional groups for the discussion of near infrared spectroscopy (see Volume 3) include O-H and C-H stretching.[37–46]

1.4.19 *Hydrogen Bonding (X-H.......Y)*

A hydrogen bonded molecular system has four possible modes: (1) O-H stretch found at 2941 nm to 4000 nm (3400 to 2500 cm^{-1}), (2) R-O-H in-plane bend at

7143 nm to 7692 nm (1400 to 1300 cm^{-1}), (3) the R-O-H out-of-plane bend at 15, 152 nm (660 cm^{-1}), and (4) the -O-H........O stretch at 57, 143 nm (175 cm^{-1}).

Specific frequencies for molecules containing O-H with hydrogen bonding include alcohols (R-O-H) with ν(O-H) = 2981 nm (3355 cm^{-1}), carboxylic acids (-COOH) with ν(O-H) = 3333 nm (3000 cm^{-1}), and amines (R-NH-X) with ν(N-H) = 3000 nm (3300 cm^{-1}). In general, hydrogen bonding has several main effects on frequency of X-H bands: (1) it lowers the frequency of the X-H stretch as compared to the gas phase and dilute liquid phase of a molecule; (2) it raises the X-H bend frequency; (3) it broadens all O-H stretch bands by up to 300 to 475 nm (300 to 500 cm^{-1}); and (4) it intensifies all bending and stretching X-H bands.[40–46]

1.4.20 *The O-H Stretch*

a. It is common knowledge among Raman or infrared spectroscopists that any fundamental band occurring above 2857 nm (3500 cm^{-1}) can be confidently assigned to the O-H stretch group. A sharp band exists for the stretching frequency of O-H for the infrared spectrum of O-H containing systems. Primary alcohols (-CH_2-OH) exhibit ν_s(O-H) = 2740 nm (3650 cm^{-1}); secondary alcohols (R-CHOH-R′) exhibit ν_s(O-H) = 2762 nm (3620 cm^{-1}); tertiary alcohols (R-CR'OH-R″) demonstrate ν_s(O-H) = 2770 nm (3610 cm^{-1}); phenols ν_s(O-H) = 2762 to 2793 nm (3620 to 3580 cm^{-1}); and carboxylic acids have ν_s(O-H) = 2817 to 2857 nm (3550 to 3500 cm^{-1}). To distinguish between primary, secondary, and tertiary alcohols using Raman spectra requires the use of these bands as well as the ν_s(C-O) = 8547 to 9615 nm (1170 to 1040 cm^{-1}). O-H is a poor Raman scatterer.

b. The O-H in-plane bend occurs as ν_B(O-H) = 7143 to 7692 nm (1400 to 1300 cm^{-1}). The C-O-H out-of-plane bend ν_B(O-H) = 13,158 to 15,385 nm (760 to 650 cm^{-1}).[40–46]

c. Carboxylic acids (-COOH) form dimers (hydrogen bonding between molecules) as (-C=O......H-O-C-) in condensed states that exhibit four distinct frequencies: (1) a very broad, very strong, O-H stretch at approximately 3333 nm (3000 cm^{-1}, at a lower frequency than alcohols); (2) Summation tones as one to three weak bands near 3846 nm (2600 cm^{-1}); (3) carbonyl (C=O) stretch vibration near 5797 nm (1725 cm^{-1}); and (4) a weak, broad band indicating a dimer condition near 10,471 nm (955 cm^{-1}).

1.4.21 *N-H Group Frequencies and Nitrogen Containing Compounds*[36–38]

a. The average frequencies for free N-H groups include primary amides (-C=O -NH_2, ν_s(N-H) = 2857 nm (3500 cm^{-1}) and 2933 nm (3410 cm^{-1}); and secondary amides (-C=O -NHC, ν_s(N-H) = 2899 nm (3450 cm^{-1}). Alkyl (aliphatic-NH-X) groups include alkyl-NH_2, ν_s(N-H) = 2946 nm (3395 cm^{-1}) and 3003 nm (3330 cm^{-1}); for alkyl-NH-X, ν_s(N-H) = 2981 nm (3355 cm^{-1}). In the case of aryl (aromatic-NH-X), the N-H stretch fundamentals occur as ν_s(N-H) = 2869 nm (3485 cm^{-1}) and 2946 nm (3395 cm^{-1}) for aryl-NH_2, and ν_s(N-H) = 2911 nm (3435 cm^{-1}) for aryl-NH-X. Note the N-H frequency is lower and less intense than the O-H frequency; the hydrogen bonding effects with N-H are weaker and of less effect than in O-H.

b. Primary amines (R- NH_2) demonstrate two N-H stretching bands due to coupling; the in-phase (or symmetric), and the out-of-phase (or asymmetric). Aliphatic amines, (C_nH_{2n+1})-NHX, demonstrate in-phase frequencies from 2999 to 3053 nm (3335 to 3275 cm^{-1}) with medium strength; out-of-phase frequency occurs from 2933 to 2985 nm (3410 to 3350 cm^{-1}), also with medium strength. Aromatic amines exhibit in-phase and out-of-phase frequencies of 2920 to 3003 nm (3425 to 3330 cm^{-1}), and 2849 to 2950 nm (3510 to 3390 cm^{-1}), both with medium strength. The scissoring for primary amines occurs from 6042 to 6289 nm (1655 to 1590 cm^{-1}), as a medium to strong, broad band; the NH_2 wagging is a strong, broad band near 12,5000 nm (800 cm^{-1}).

c. Secondary amines (R′R″-NH) exhibit a single, broad N-H stretch vibrational frequency near 3021 nm (3310 cm^{-1}); an N-H bend occurs as a weak-to-medium strength band near 6667 nm (1500 cm^{-1}).

d. To separate spectral features for the various amines, i.e., primary, secondary, and tertiary; it is helpful to point to obvious differences between the groups. For example, the number of N-H stretches for the three types are primary (2), secondary (1), and tertiary (0). For N-H bending vibrations, primary amines have a band near 6042 to 6289 nm (1655 to 1590 cm^{-1}); secondary amines, near 6667 nm (1500 cm^{-1}); and this feature is absent in tertiary amines. Lastly, the C-H stretch vibration for CH_2 is to be found for all three types of amines from 3503 to 3683 nm (2855 to 2715 cm^{-1}); for primary amines, the C-H stretch will be found at higher frequencies nearer to 3448 nm (2900 cm^{-1}).

e. Amine salts as R-$NH_3^+$and R′R″NH_2^+exhibit several weak bands near 3571 to 4545 nm (2800 to 2200 cm^{-1}). These salts exhibit a weak-to-medium, very broad band, centered near 4000 nm (2500 cm^{-1}), lower in frequency than the C-H stretch and up to 314 nm (500 cm^{-1}) in width at the baseline.

1.4.22 *Amines (R-NH-X)*

Amines are polar compounds, and both primary and secondary amines can form intermolecular hydrogen bonding. All classes of amines are water soluble (up to 6 carbon atoms) and will form hydrogen bonding in aqueous solutions.

References

1. H. Martens and T. Naes, *Multivariate Calibration* (John Wiley and Sons, Chichester, 1992).
2. H. Mark and J. Workman, *Chemometrics in Spectroscopy,* (Elsevier, Amsterdam, 2007).
3. C.V. Raman, A new radiation, *Indian J. Phys.* **2** (1928) 387–398.
4. R. Singh, C.V. Raman and the Discovery of the Raman Effect, *Phys. Perspect.* **4**(2002) 399–420.
5. H. Mark and J. Workman, Units of Measure in Spectroscopy, Part 1: Scaled Volume Fraction, *Spectroscopy* **29**(2) (2014) 1–10.
6. ASTM E1683-02(2014)e1, "Standard Practice for Testing the Performance of Scanning Raman Spectrometers", ASTM International, 100 Barr Harbor Drive, PO Box C700, West Conshohocken, PA, 19428–2959 USA (2009).
7. ASTM E1840-96(2014), "Standard Guide for Raman Shift Standards for Spectrometer Calibration", ASTM International, 100 Barr Harbor Drive, PO Box C700, West Conshohocken, PA, 19428–2959 USA (2009).
8. ASTM E2529-06(2014), "Standard Guide for Testing the Resolution of a Raman Spectrometer", ASTM International, 100 Barr Harbor Drive, PO Box C700, West Conshohocken, PA, 19428–2959 USA (2009).
9. ASTM E2911-13, "Standard Guide for Relative Intensity Correction of Raman Spectrometers", ASTM International, 100 Barr Harbor Drive, PO Box C700, West Conshohocken, PA, 19428–2959 USA (2009).
10. The Richard McCreery Research Group provides the greatest detail of information at the following link: http://www.chem.ualberta.ca/~mccreery/ramanmaterials.html
11. W.J. (William James) Hershel, Finger-Prints, Nature **23** (Nov. 25, 1880) 76.
12. W. de W. Abney and Lt. Col. Festing, On the Influence of the Atomic Grouping in the molecules of Organic Bodies on their Absorption in the Infra-red Region of the Spectrum, *Phil. Transact.* **172** (1882) 887–918.
13. L.J. Bellamy and D.W. Mayo, Infrared Frequency Effects of Lone Pair Interactions with Antibonding Orbitals on Adjacent Atoms, *J. Phys. Chem.* **80** (1976) 1217–1220.
14. L.J. Bellamy, Some Wider Uses of Group Frequencies, *Appl. Spectrosc.* **33** (1979) 439–443.
15. J. Grasselli, R. Hannah, D.W. Mayo, and F.A. Miller, *Infrared Spectroscopy*, (Bowdoin College, Brunswick, Maine, 1994, Volume 1 p. 33).
16. Bryan R. Henry, Use of Local Modes in the Description of Highly Vibrationally Excited Molecules, *Acc. Chem. Res.* **10** (1977) 207–213.

17. Von E. Fermi, Uber den Ramaneffekt des Kohlendioxyds, *Z. Physik* **71** (1931) 250–259.
18. J. Grasselli, R. Hannah, D.W. Mayo, and F.A. Miller, *Infrared Spectroscopy*, (Bowdoin College, Brunswick, Maine, 1994, Vol. 1 p. 34).
19. J.F. Bertrán and L. Ballester, Assignment of the forbidden level in the carbonyl resonance of cyclopentanone, *Spectrochimica Acta Part A: Molecular Spectroscopy* **39**(2) (1983) 123–125.
20. N.B. Colthup, L.H. Daly, and S.E. Wiberley, *Introduction to Infrared and Raman Spectroscopy*, 3rd edn. (Academic Press, Boston, 1990, pp. 31–33).
21. D. Lin-Vien, Daimay, N.B. Colthup, W.G. Fateley, and J.G. Grasselli, *The handbook of infrared and Raman characteristic frequencies of organic molecules,* 1st edn. (Academic Press Elsevier, Boston, 1991).
22. J. Grasselli, R. Hannah, D.W. Mayo, and F.A. Miller, *Infrared Spectroscopy* (Bowdoin College Press, Brunswick, Maine, **1** (1994) 44.
23. H.L. Fang and D. A. C. Compton, Overtone Spectroscopy of Nonequivalent C-H Oscillators in 1-Alkenes and Dienes, *J. Chem. Phys.* **92** (1988) 7185–7192.
24. G. Socrates, Infrared Characteristic Group Frequencies, 2nd edn. (John Wiley & Sons, Chichester, 1997, pp. 121–129).
25. N.B. Colthup, L.H. Daly, and S.E. Wiberley, *Introduction to Infrared and Raman Spectroscopy*, 3rd edn. (Academic Press, Boston, 1990, p. 262).
26. R.G. Bray and M.J. Berry, Intramolecular rate processes in highly vibrationally excited benzene, *J. Chem. Phys.* **71**(12) (1979) 4909–4922.
27. K.G. Spears and S.A. Rice, Study of the Individual Vibronic States of the Isolated Benzene Molecule, *J. Chem. Phys.* **55**(12) (1971) 5561–5581.
28. R.J. Hayward and B.R. Henry, Anharmonicity in Polyatomic Molecules, *Journ. of Mol. Spectrosc.* **46** (1973) 207–213.
29. J. Grasselli, R. Hannah, D.W. Mayo, and F.A. Miller, *Infrared Spectroscopy*, (Bowdoin College, Brunswick, Maine, 1994, Vol. 1, p. 82).
30. N.B. Colthup, L.H. Daly, and S.E. Wiberley, *Introduction to Infrared and Raman Spectroscopy*, 3rd edn. (Academic Press, Boston, 1990, p. 274).
31. B.R. Henry and W. Siebrand, Anharmonicity in Polyatomic Molecules. The CH-Stretching Overtone Spectrum of Benzene, *The Journal of Chemical Physics* **49**(12) (1968) 5369–5376.
32. D.F. Heller and S. Mukamel, Theory of Vibrational Overtone Line Shapes of Polyatomic Molecules, *J. Chem. Phys.* **70** (1979) 463–472.
33. M.S. Burberry and A.C. Albrecht, Local Mode Combination Bands and Local Mode Mixing, *J. Chem. Phys.* **71** (1979) 4631–4640.
34. M.S. Child, Local Mode Overtone Spectra, *Acc. Chem. Res.* **18** (1985) 45–50.
35. J. Grasselli, R. Hannah, D.W. Mayo, and F.A. Miller, *Infrared Spectroscopy* (Bowdoin College Press, Brunswick, Maine, **1** (1994) p. 117.
36. A. Leifer, and E.R. Lippincott, The infrared spectra of some amino acids, *Journal of the American Chemical Society* **79**(19) (1957) 5098–5101.
37. S.E. Krikorian and M. Mahpour, The identification and origin of NH overtone and combination bands in the near-infrared spectra of simple primary and secondary amides, *Spectrochimica Acta Part A: Molecular Spectroscopy* **29**(7) (1973) 1233–1246.
38. R.J. Hayward and B.R. Henry, Anharmonicity in Polyatomic Molecules: A Local-Mode Analysis of the XH-Stretching Overtone Spectra of Ammonia and Methane, *J. Mol. Spectrosc.* **50** (1974) 58–67.

39. B.T. Darling and D.M. Dennison, The Water Vapor Molecule, *Phys. Rev.* **57** (1940) 128–139.
40. E.R. Lippincott and R. Schroeder, One-Dimensional Model of the Hydrogen Bond, *J. Chem. Phys.* **23** (1955) 1099–1106.
41. J.N. Finch and E.R. Lippincott, Hydrogen bond systems–temperature dependence of OH frequency shifts and OH band intensities, *J. Chem. Phys.* **61** (1957) 894–902.
42. R. Schroeder and E. R. Lippincott, Potential Function Model of Hydrogen Bonds. II, *J. Chem. Phys.* **61** (1957) 921–928.
43. J.M. Stutman, A.S. Posner, and E.R. Lippincott, Hydrogen Bonding in the Calcium Phosphates, *Nature* **193** (1962), 368–369.
44. J.N. Finch and E.R. Lippincott, Hydrogen Bond Systems: Temperature Dependence of OH Frequency Shifts and OH Band Intensities, *J. Chem. Phys.* **24** (1956) 908–909.
45. R. Schroeder and E. R. Lippincott, Potential function model of hydrogen bonds. II, *The Journal of Physical Chemistry* **61**(7) (1957) 921–928.
46. A.S.N. Murthy and C.N.R. Rao, Spectroscopic Studies of the Hydrogen Bond, *Appl. Spectrosc. Rev.* **2** (1968) 69–191.

Chapter 2

Theory of Raman Spectroscopy, Light Interaction with Matter

Contents

2.1 Introduction 51
2.2 The Compton and Raman Effects 52
2.2.1 The Compton Effect 52
2.2.2 The Raman Effect 53
2.2.3 Raman Theory 53
2.3 Raman Measurement Techniques 56
2.3.1 Stimulated Raman 56
2.3.2 CARS (Coherent Anti-Stokes Raman Spectroscopy) 57
2.3.3 RRS/RR (Resonance Raman Spectroscopy) 57
2.3.4 SERS (Surface-Enhanced Raman Spectroscopy) 58
2.3.5 SERRS (Surface-Enhanced Resonance Raman Spectroscopy) . 58
2.3.6 SRS (Spontaneous Raman Spectroscopy) 58
2.3.7 SRS/SPSRS (Stimulated Raman Spectroscopy) 59
2.3.8 SORS (Spatially Offset Raman Spectroscopy) 60
2.3.9 TERS (Tip-Enhanced Raman Spectroscopy) 60
2.3.10 SPPERS (Surface Plasmon Polariton (SPP) Enhanced Raman Spectroscopy) 61
References 67

2.1 Introduction

Light interaction as spectrophotometry includes scattering, absorption, reflectance, and emission from atoms and molecules. A variety of spectrometer configurations are used to optimize the measurements of electromagnetic radiation as it interacts in these basic ways with matter. A more comprehensive list of light interaction with matter in the forms of solids, liquids, or gas phase samples would include absorption, elastic and inelastic scattering, specular reflection, diffuse reflection, transmission, diffuse transmission, interactance, dispersion, diffraction, refraction, opacity transmittance/scattering, color interaction, absorbance, absorptance, phosphorescence, and fluorescence.

2.2 The Compton and Raman Effects

The main topics relating to light interaction with matter for spectroscopy are discussed in significant detail throughout Chapter 2 for each volume of this set (i.e., Volumes 1–5). The light interactions for Ultraviolet, Visible, Near Infrared and Infrared spectroscopy are considered absorption interactions for the most part and adhere for practical purposes to Beer's law, explained in detail throughout these volumes (including a summary within the Front Material of each volume).

Raman scattering is unique among the molecular spectroscopy techniques and as such will require a different approach for explaining the principles of Raman, recalling that C.V. Raman and K.S. Krishnan referred to this discovery as "a new radiation".[1] Indeed the Raman effect was a surprise discovery, resulting from investigations into the Compton effect.

2.2.1 *The Compton Effect*

The Compton effect is a phenomenon where a photon interacts with a charged particle (e.g., electron), via inelastic scattering or collision, where there is a decrease in energy (i.e., increase in wavelength) for the photon. The interaction of X-rays or gamma radiation at an initial wavelength (λ_i) with an electron causes a surprising shift in the scattered energy wavelength (λ_S), as well as a change in the light scattering angle.[2] The Compton equation is given as:

$$\lambda_S - \lambda_i = \frac{h}{m_E \cdot c}(1 - \cos\theta) \tag{2.1}$$

And, the scattering wavelength is determined using:

$$\lambda_S = \left[\frac{h}{m_E \cdot c}(1 - \cos\theta)\right] + \lambda_i \tag{2.2}$$

And then the scattered energy angle of divergence may be computed using:

$$\theta = \cos^{-1}\left[1 - \left(\frac{\lambda_S - \lambda_i\,(M_E \cdot c)}{h}\right)\right] \tag{2.3}$$

where θ is angle for scattered photons following collision; λ_i is the initial wavelength of the incident photons; λ_S is the scattered photon wavelength; h is Planck's constant ($6.62606957 \times 10^{-34}$ J s); M_E is the mass of the electron at rest (8.19×10^{-14} J); and c is the speed of light (2.99792458×10^{8} m s^{-1}). Note that $h/M_E \cdot c = 2.43 \times 10^{-12}$ m.

2.2.2 *The Raman Effect*

The Raman effect or Raman scattering is demonstrated when a high flux of photons strikes an atom or molecule and an inelastic collision with the photons and the atom(s) causes both an increase (anti-Stokes shift) and decrease (Stokes shift) in photon radiation energy. This shift in energy is proportional to the vibrational energy (i.e., frequency) of the vibrations in the molecule(s) collided with.[3–5] The Raman effect may be described using a classical interpretation of vibrational spectroscopic phenomena. This classical interpretation is most often illustrated using a simple mass on a spring model to describe the physics of molecular vibrational modes. Such a description of the behavior of molecular vibrations and their interactions with photons involves several concepts and terms. These classic terms include: Normal Mode Theory (The Ideal Harmonic Oscillator) or Hooke's law, The Anharmonic Oscillator, the Harmonic Oscillator or Normal Mode, The Selection Rule, and Local Mode Theory (as the Morse Oscillator). These topics, useful for description of photon interaction with molecules for near infrared, infrared and Raman spectroscopic methods are described in detail (with illustrations) in Volume 3, Sections 1.2.4 through Section 1.2.5.

2.2.3 *Raman Theory*

Note that the Raman effect is characterized as inelastic scattering of a photon from a molecular vibration occurring at a chemical bond. The incident photons from an intense excitation laser beam excite a small percentage of the photons into a virtual energy state. Note that for spontaneous Raman scattering approximately 0.001% of incident laser energy is translated into Raman Stokes and anti-Stokes signal intensity.

When the laser light strikes the vibrating bond there are several potential events that may occur. If the molecule relaxes to the ground state and emits a photon of equal energy, then elastic Rayleigh scattering occurs; if the molecule relaxes to a higher energy state and emits a photon with less energy than the incident photon then Stokes shifted Raman scattering occurs; if the molecule is in an excited state and is further excited by the laser photon to a higher energy state it may relax to the ground state and emit a photon of higher energy than the incident photon, resulting in Anti-Stokes scattering. Since most molecules at standard temperature and pressure are in the ground state then most of the inelastic scattering will occur as Stokes shifted photons and so that is what is measured for typical Raman spectroscopy.

Other events may occur, such as absorption and resonance Raman or resonance fluorescence. These resonance phenomena occur when the laser or excitation photon energy is close to the transition energy or vibrational states of the molecular bond. Resonance fluorescence is an inelastic scattering that is more intense than the inelastic Raman scattering. Typical or relaxed fluorescence is different in that it is an absorption and re-emission process involving photon emission from the lowest vibrational level of an excited electronic (not phonon) state, see Volume 1 for greater details regarding fluorescence phenomena. Note that Raman transitions occur in approximately 10^{-12} s (i.e., picoseconds), and fluorescence transitions are much slower at 10^{-9} s (i.e., nanoseconds). Figure 2.1 illustrates the quantum energy transitions for infrared, Rayleigh and Raman Scattering, and Fluorescence as a Jablonski type diagram. Figure 2.2 shows main phenomena of interest for Raman spectroscopy measurements, including the signals for Rayleigh and Stokes Raman scattering, and fluorescence.

To continue the discussion of Raman theory, note that for Rayleigh elastic scattering from the laser excitation source refer to Equation 2.4 illustrating the Rayleigh scattering phenomenon for this case:

$$\rho_S \propto \frac{I_0}{\lambda_e^4} \tag{2.4}$$

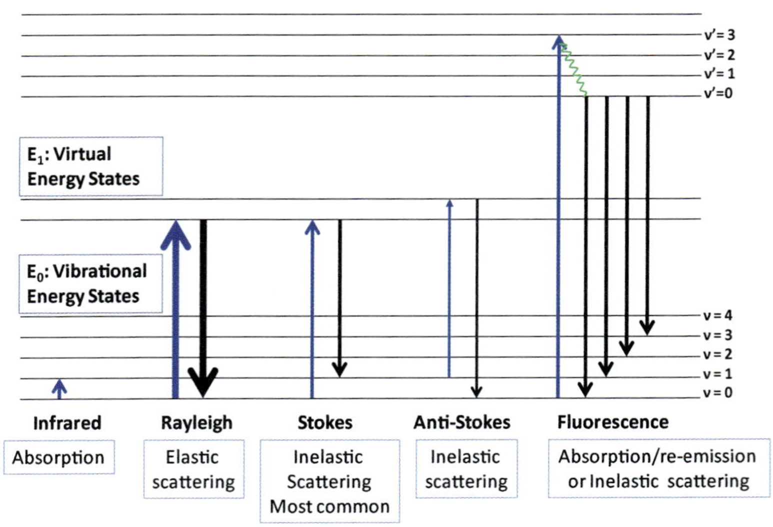

Fig. 2.1 Illustrations of quantum energy transitions for Infrared, Rayleigh and Raman Scattering, and Fluorescence (Jablonski Type Diagram). Thickness of lines indicates intensity. The blue arrows indicate increasing quantum energy state and the black arrows a decrease in quantum energy state.

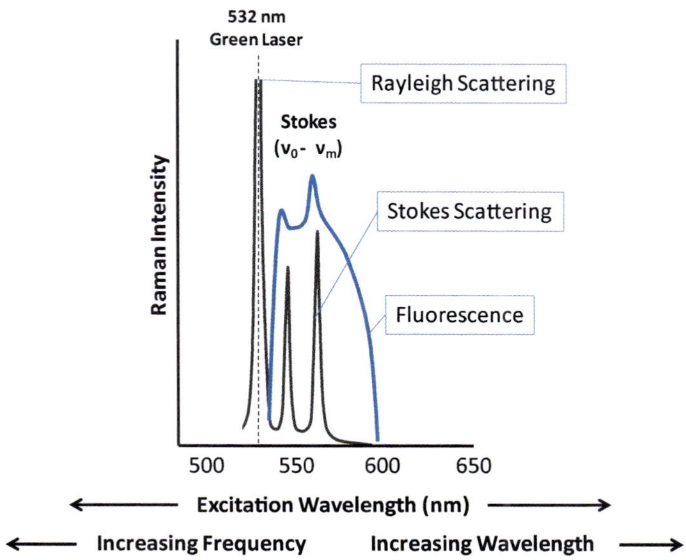

Fig. 2.2 Illustration of the major signal phenomena of interest for Raman spectroscopy measurements, including the signals for Rayleigh and Stokes Raman scattering, and fluorescence. Example is for a 532 nm excitation laser.

where ρ_S is the Rayleigh scattered intensity; I_0 is the excitation laser intensity; and λ_e is the wavelength of the excitation laser energy. So note that sensitivity for Raman scattering measurements (I_R) is proportional to the inverse of the fourth power of the wavelength of the excitation energy (λ_e) as:

$$I_R \propto \frac{1}{\lambda_e^4} \tag{2.5}$$

And fluorescence (F), as an interferent to Raman scattering, is proportional to the inverse of the wavelength of the excitation energy (λ_e) as:

$$F \propto \frac{1}{\lambda_e} \tag{2.6}$$

Since only about 0.001% of the available excitation energy is converted to Stokes signal for typical spontaneous Raman spectroscopy, then a rule of thumb calculation for the Raman signal intensity would be as:

$$I_R \cong I_0 \cdot 10^{-5} \tag{2.7}$$

So for a typical Raman instrument with diode laser power of 300 mW the Stokes portion of the Raman scattering might be in the 300 mW $= 10^{-3}$ W $\times 10^{-5} = 10^{-8}$ W Stokes signal.

These equations are simplistic, since the Raman intensity (I_R) or amplitude of the Raman scattered signal at any particular frequency (or wavelength or wavenumber) is determined by multiple variables, such as the spectrometer throughput (τ), the self-absorption of the sample (α), the Rayleigh scattered radiation (ρ_S), the excitation laser intensity (I_0), the Raman scatter (R_S) of the molecule being measured at the frequency measured, the effective optical pathlength or L_E (i.e., actual optical pathlength where Raman scattering is occurring), and the number of molecules encountered by the beam path of the measuring Raman instrument (as c or concentration in Mole $\cdot$ Liter^{-1}). It is assumed that a change in spectral response is related to a concentration as described by the Raman relationship.

A more complete description of the resultant Raman intensity (I_R) as Raman scatter (R_S) with its dependency on the frequency of energy used (Greek nu as ν) is described as Equation 2.8.

$$I_R = \tau_\nu \cdot \alpha_\nu \cdot \rho_{S\nu}^4 \cdot I_0 \cdot R_{S\nu} \cdot L_E \cdot c \quad (2.8)$$

Please see Chapters 1 and 3 of this volume for more detailed discussions of the variables associated with the Raman scattering signal.

Table 2.1 describes the potential measurement modes using a Raman spectrometer transposed from Chapter 1. Raman instruments involve a wide variety of technique even more diverse than other vibrational spectroscopy techniques.

2.3 Raman Measurement Techniques

Raman measurement methods and technique are extremely diverse. The Raman phenomenon continues to be enhanced and studied such that new techniques are discovered on a routine basis. Improvements include enhancement of signal, isolation of signal, micro-sampling, and imaging over ever smaller spatial areas. The following text, Figures 2.3 through 2.15, and Table 2.2 illustrate the variety of applied Raman measurement techniques used for industrial laboratory, and academic research work.

2.3.1 *Stimulated Raman*

This is a dynamic Raman technique where two different wavelength (frequency) laser pulses are applied to a sample simultaneously with polarization (parallel or perpendicular). When the difference frequency of the excitation lasers matches a particular molecular vibrational frequency (also called the Raman shift), amplification of the Raman signal occurs. The polarization effectively measures molecular

Table 2.1 Raman spectrometer measurement modes.

Instrument measurement mode	Description of measurement
Standard Scan	Raman scattering* vs. wavenumbers
Step-Scan Phase-Resolved	Amplitude Modulation, Phase Modulation, or Sample Modulation
Step-Scan Time-Resolved	Microsecond to nanosecond time resolution for timed events
Step-Scan Space-Resolved	Imaging spectra for spatial data acquisition for focal-plane imaging
Polarization Modulation	Modulation of light polarization for measuring polarization effects
Individual wavenumber step-scan	Individual Raman scattering* at selected wavenumbers(s)
Two-dimensional Correlation	Correlates multiple spectra for conditions of applied stress and relaxation to determine rheo-optical properties. Also for correlation of spectra from different regions such as NIR
Quantitative methods	Concentration of analyte vs. absorbance
Qualitative methods	Correlation or spectral matching for Identification
Waveguide Raman	Sample measured along small diameter Teflon tube increases Raman detection sensitivity for aqueous samples
Fiber optic Raman	Raman laser sources may be applied through silica fiber optics for remote sensing
Kinetics	Kinetic rates of reaction (Repetitive fast scan)
Transmission Raman	Laser is projected through solids or turbid liquids for bulk analysis

*As Stokes scattering intensity.

alignment and orientation properties. An enhanced Stokes signal of 10^7 is possible using the Stimulated Raman technique (Figure 2.3).[6–8]

2.3.2 *CARS (Coherent Anti-Stokes Raman Spectroscopy)*

Two lasers are used (i.e., near infrared pump beam and longer wavelength Stokes beam) to excite molecules to their virtual state followed by their return to the vibrational state. Tuning of the excitation pump laser beam allows excitation of specific molecules at specific spatial locations with a 10^5 signal enhancement over conventional Raman. CARS imaging may be completed in real-time for specific molecular vibrations at known planar (spatial) locations for image reconstruction without the use of dyes (Figure 2.4).[9–11]

2.3.3 *RRS/RR (Resonance Raman Spectroscopy)*

The laser excitation wavelength is matched to the specific electronic transition associated with the molecular vibration mode of interest. It provides a technique

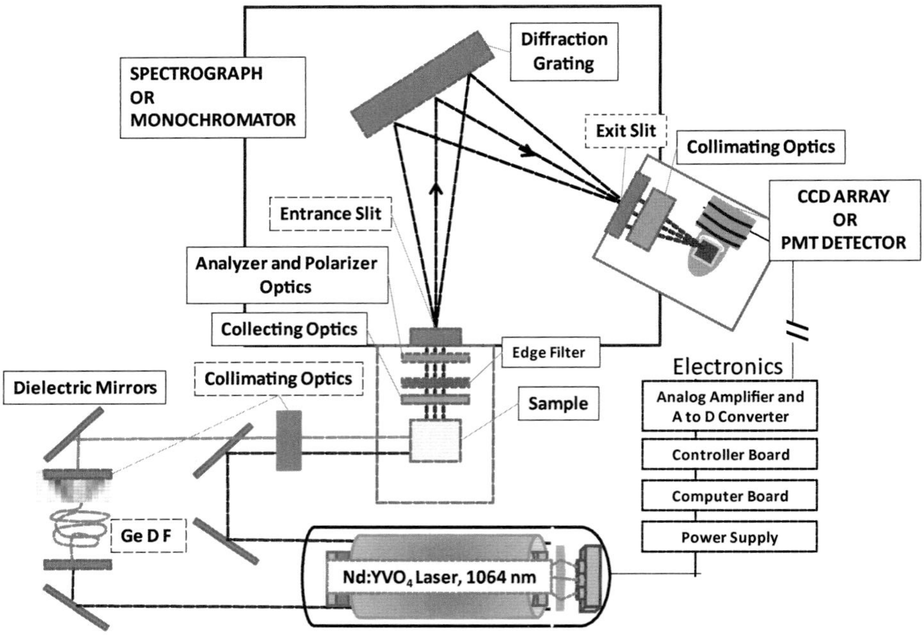

Fig. 2.3 Example of a measurement configuration used for stimulated Raman measurements (SRS). Ge D F refers to a Germanium Doped Fiber for producing a variable wavelength beam.

for "isolation" of specific molecular vibrational modes of interest for the study of specific bands of large molecules, such as synthetic polymers, and natural biopolymers (Figure 2.5).[12–14]

2.3.4 *SERS (Surface-Enhanced Raman Spectroscopy)*

Placing a sample within (or on) a gold or silver colloid increases the electric field for the molecules within the sample and thus the Raman scattering signal is enhanced by a factor of 10^{10} to 10^{11} (Figures 2.6 and 2.7).[15–17]

2.3.5 *SERRS (Surface-Enhanced Resonance Raman Spectroscopy)*

This technique combines SERS and RRS/RR; the Raman excitation laser is matched to the maximum absorbance of the molecular vibration of interest. This technique enhances the signal of a specific vibration under study (Figures 2.8 and 2.9).[18–20]

2.3.6 *SRS (Spontaneous Raman Spectroscopy)*

SRS is the modern name for the originally named phenomenon of Raman spectroscopy using continuous wave (CW) laser excitation. It is the original linear

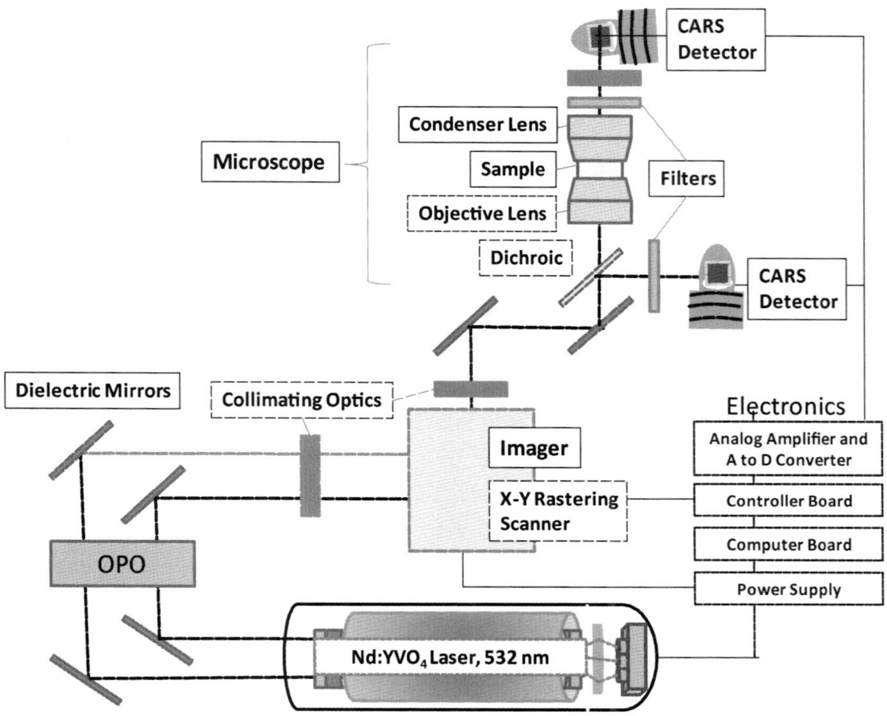

Fig. 2.4 Example of a measurement configuration used for CARS (Coherent Anti-Stokes Raman Spectroscopy) measurements. OPO refers to a dual-wavelength optical parametric oscillator. The combined laser beams (i.e., Stokes beam and pump beam from the OPO) are raster-scanned over the measurement sample surface area.

(or spontaneous) form of Raman spectroscopy where the intensity of the Stokes and Anti-Stokes signals is proportional to the pump excitation laser intensity, but much smaller by many orders of magnitude. It is classic Raman spectroscopy, where approximately 0.001% of incident laser energy is translated into Stokes Raman signal intensity (Figure 2.10).[21–23]

2.3.7 *SRS/SPSRS (Stimulated Raman Spectroscopy)*

By applying a relatively high energy pulsed laser (in contrast to a traditional continuous wave (CW) laser), pulsed lasers have an enhanced effect on the Raman scattering signal to as much as 40–50% of the excitation laser energy. The overall Raman signal is enhanced by 4×10^4 to 5×10^4 over traditional spontaneous Raman spectroscopy, with the resultant significant increase in effective signal-to-noise ratio. This non-linear phenomena occurs in the Raman effect when the high energy pulsed laser excitation is used (Figure 2.11).[6]

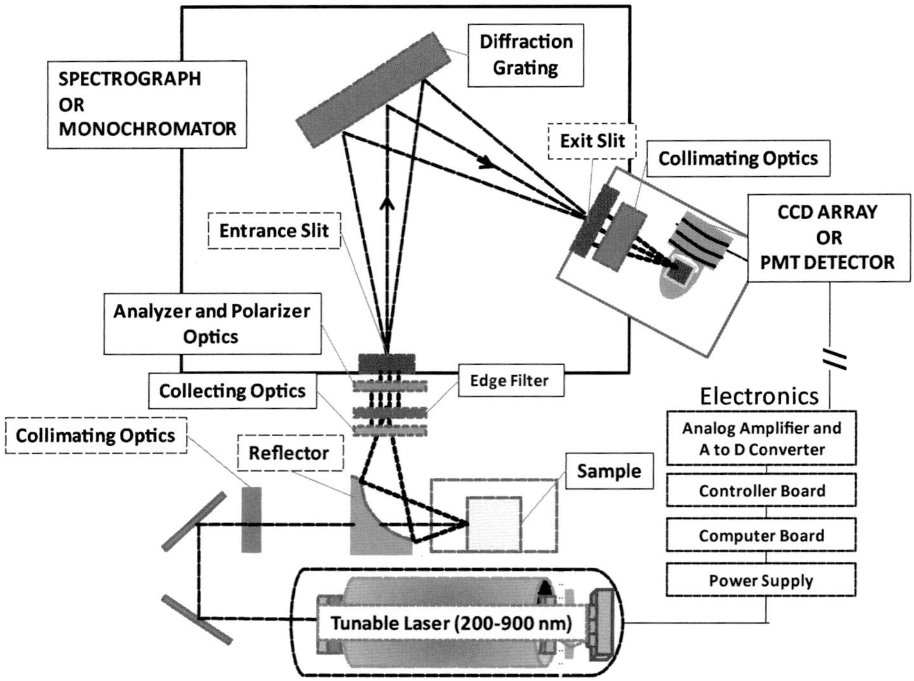

Fig. 2.5 Example of a measurement configuration used for RRS/RR (Resonance Raman Spectroscopy) measurements. The reflector is generally ellipsoid in shape and has a hole allowing the laser beam to pass through the reflector to the sample.

2.3.8 *SORS (Spatially Offset Raman Spectroscopy)*

Raman scattering is measured at a surface point and then at a deeper level into a sample. The two spectra are scaled and subtracted to yield a difference spectrum representing the depth profiled Raman spectrum. Complex samples may be measured using multiple offset spectra combined with multivariate analysis of the spectral profile differences to determine the spectrum of the depth profile (Figures 2.12 and 2.13).[24–26]

2.3.9 *TERS (Tip-Enhanced Raman Spectroscopy)*

TERS uses Raman spectroscopy and scanning probe microscopic techniques such as Atomic Force Microscopy (AFM). The AFM uses scanning probe tips as small as 10 to 20 nanometers and an enhanced Raman signal is collected along with the AFM profile of a material surface. The enhanced Raman signal is combined with the AFM topography for a detailed nano-three-dimensional image of the spatially related chemistry for the measured specimen (Figure 2.14).[27–29]

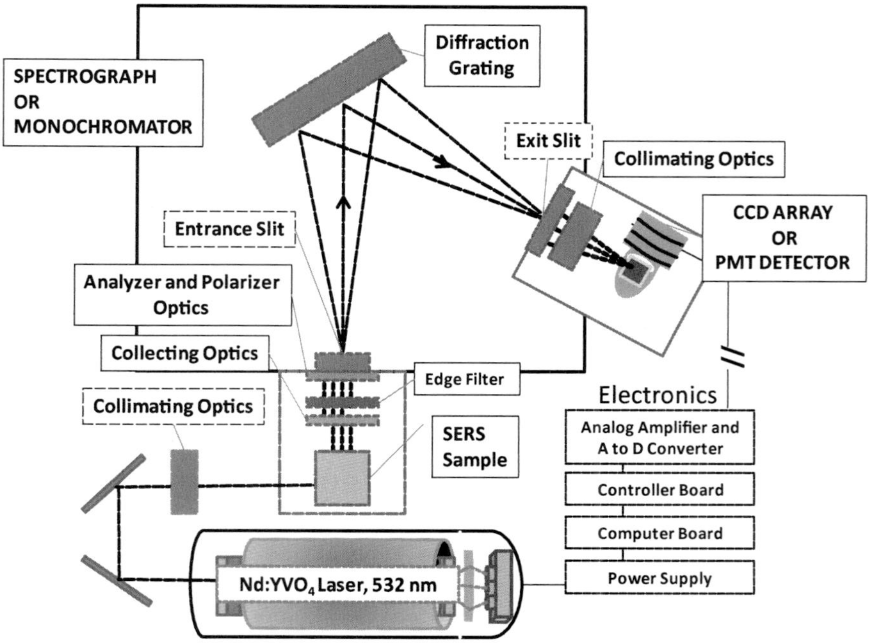

Fig. 2.6 Example of a measurement configuration used for SERS (Surface-Enhanced Raman Spectroscopy) measurements. The sample resides on a gold or silver colloid for signal enhancement.

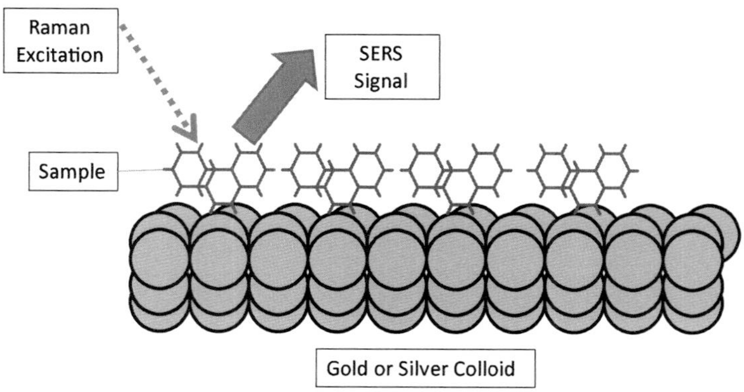

Fig. 2.7 Example of sample interaction for SERS (Surface-Enhanced Raman Spectroscopy) measurements; the purpose is for signal enhancement by several orders of magnitude.

2.3.10 *SPPERS (Surface Plasmon Polariton (SPP) Enhanced Raman Spectroscopy)*

A small cylindrical lens is used for near field excitation of tiny samples or films deposited on the lens surface; measurement areas are as small as 100 microns in

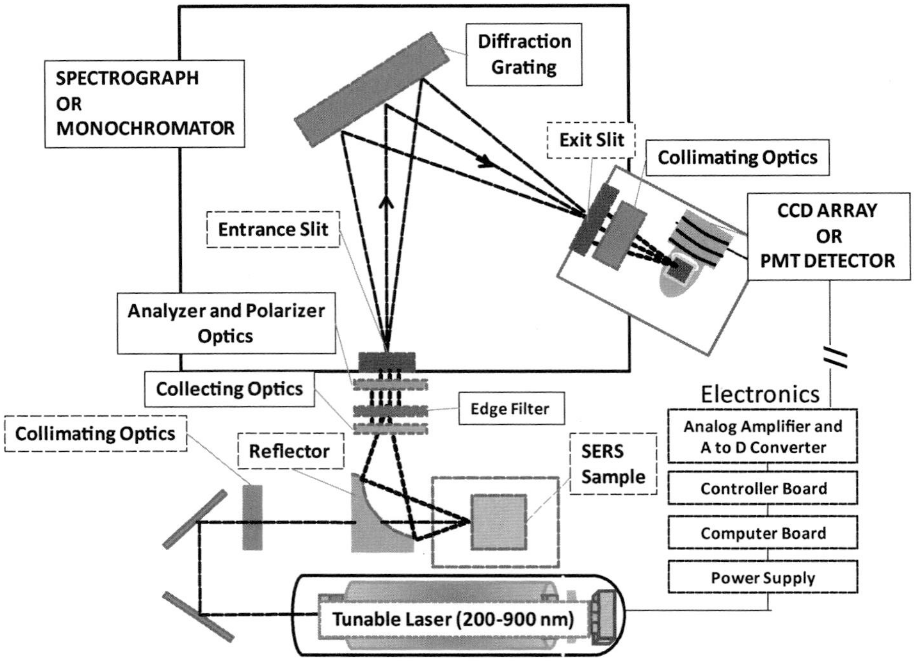

Fig. 2.8 Example of a measurement configuration used for SERRS (Surface-Enhanced Resonance Raman Spectroscopy) measurements.

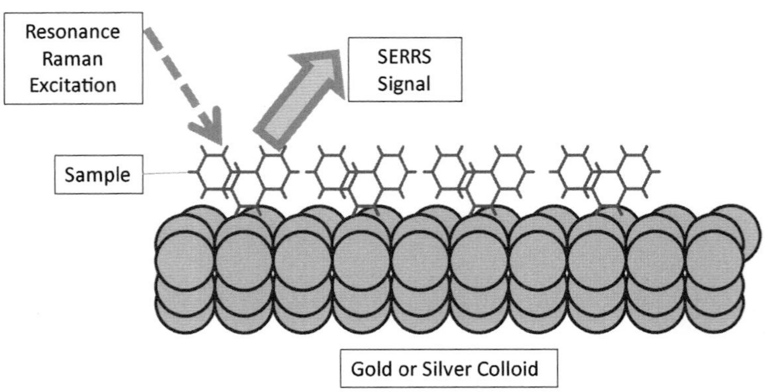

Fig. 2.9 Example of sample interaction used for SERRS (Surface-Enhanced Resonance Raman Spectroscopy) measurements. Like SERS, the sample resides on a gold or silver colloid for signal enhancement. The tunable laser is used to match the excitation wavelength to the maximum absorbance of the molecular vibration of interest; this causes a significantly enhanced signal.

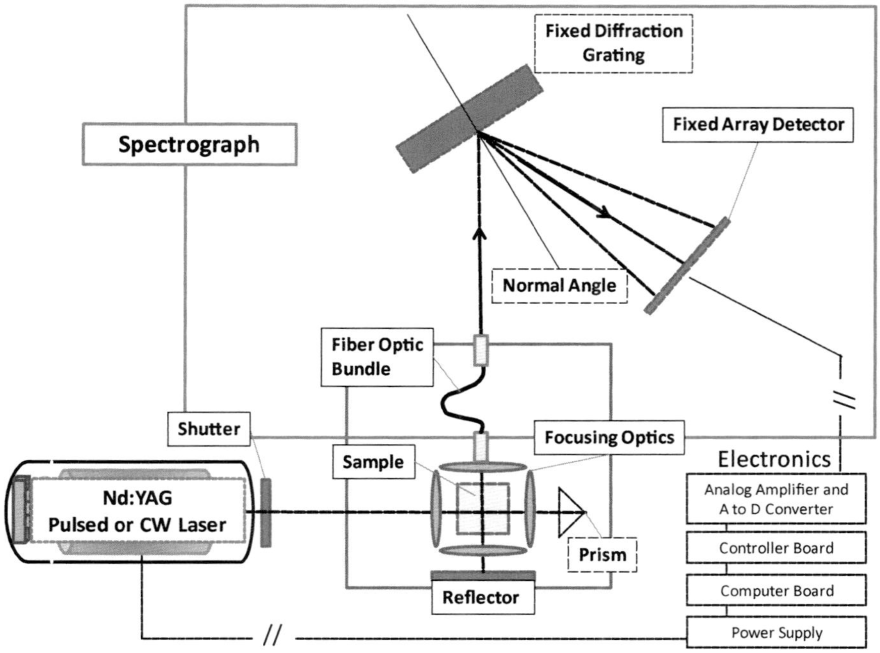

Fig. 2.10 Example of a measurement configuration used for SRS (Spontaneous Raman Spectroscopy) measurements (Enhanced signal sampling optics are shown).

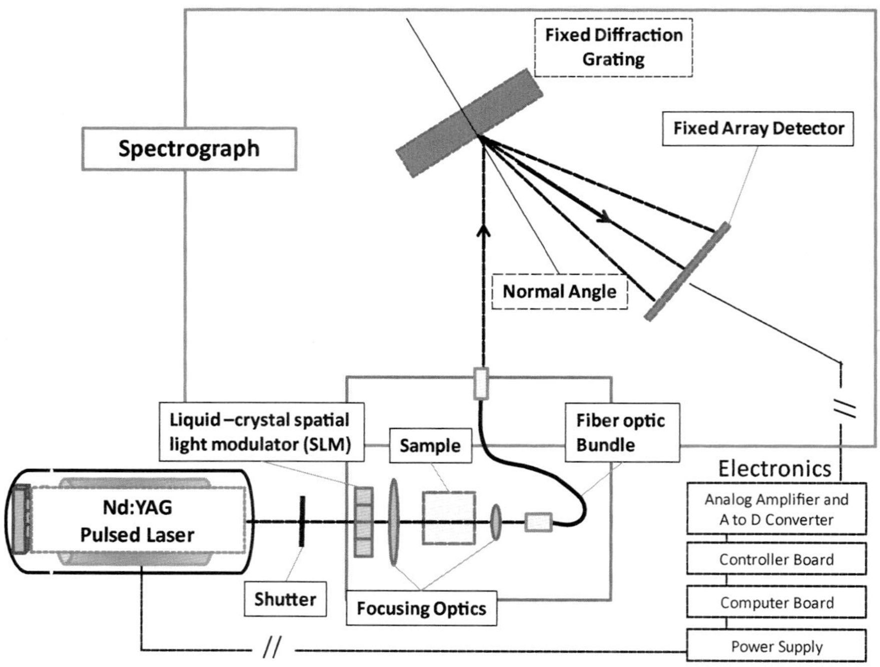

Fig. 2.11 Example of a measurement configuration used for SRS/SPSRS (Stimulated Raman Spectroscopy) measurements, spectrograph configuration shown.

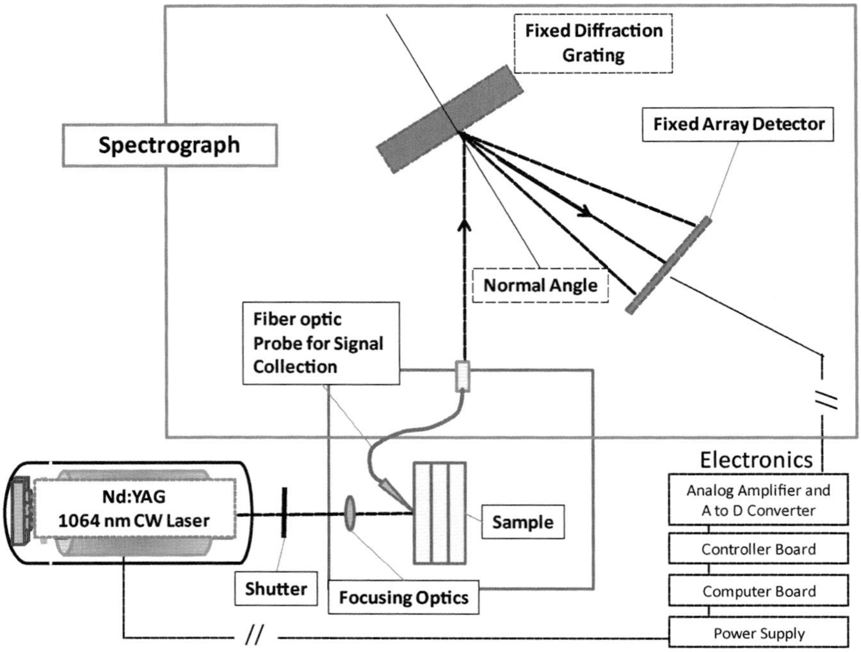

Fig. 2.12 Example of a *zero position* measurement configuration used for SORS (Spatially Offset Raman Spectroscopy) measurements.

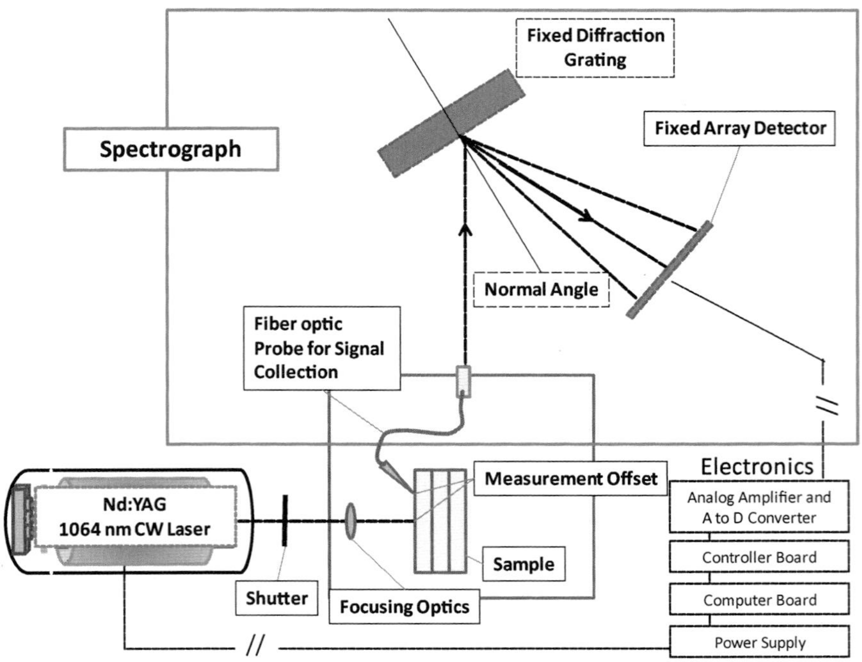

Fig. 2.13 Example of an *offset position* measurement configuration used for SORS (Spatially Offset Raman Spectroscopy) measurements.

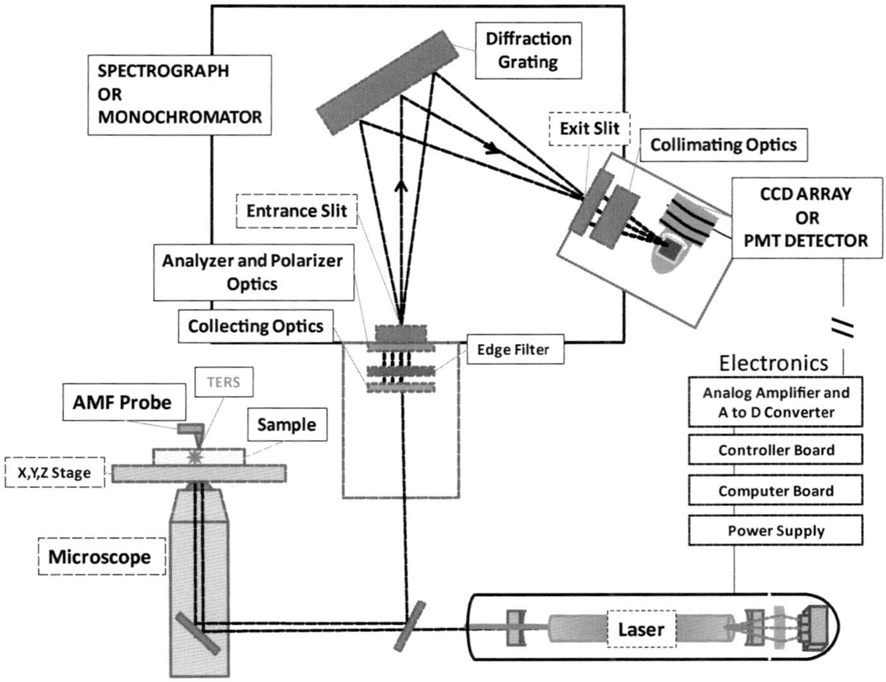

Fig. 2.14 Example of a bottom illumination measurement configuration used for TERS (Tip-Enhanced Raman Spectroscopy) measurements.

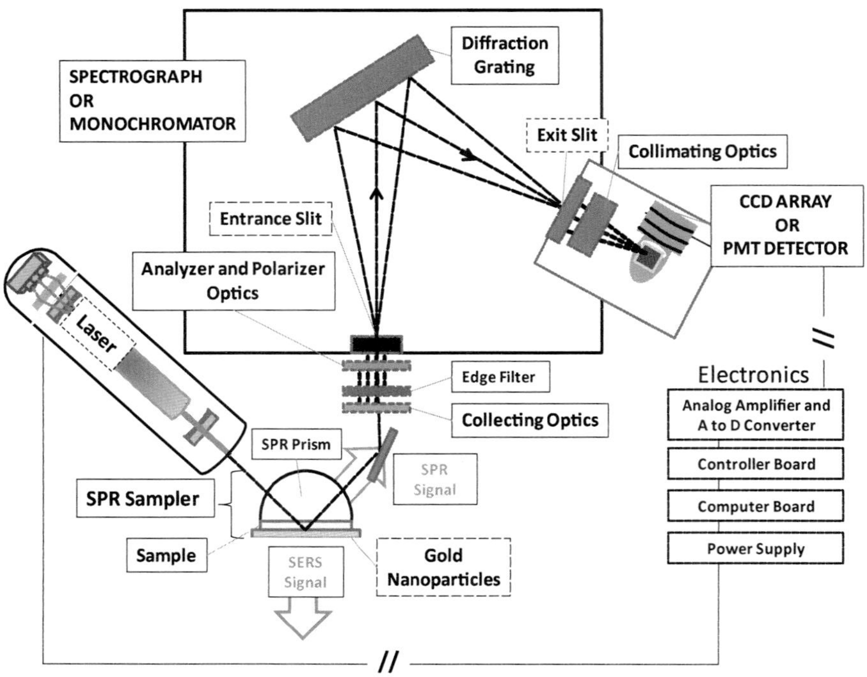

Fig. 2.15 Example of a measurement configuration used for SPPERS (Surface Plasmon Polariton (SPP) Enhanced Raman Spectroscopy) measurements.

Table 2.2 Purpose for specific Raman measurement techniques.

Instrument measurement mode	Purpose of Raman technique
Stimulated Raman	Used for amplification of the Raman signal. When combined with polarization (s and p) the technique effectively measures molecular alignment and orientation properties of molecules.
CARS (Coherent Anti-Stokes Raman Spectroscopy)	Used to enhance signal at specific spatial locations with a 10^5 signal enhancement over conventional Raman. CARS imaging is possible in real-time for mapping molecular vibrations at known locations for image reconstruction. The use of dyes is not required.
RRS/RR (Resonance Raman Spectroscopy)	This technique provides "isolation" of specific molecular vibrational modes of interest for the study of specific Raman active bands from large and complex molecules.
SERS (Surface-Enhanced Raman Spectroscopy)	This method enhances Raman signal by 10^{10} to 10^{11} for measurement of low levels of analyte.
SERRS (Surface-Enhanced Resonance Raman Spectroscopy)	This Raman technique enhances the signal of a specific molecular vibration under study.
SRS (Spontaneous Raman Spectroscopy)	This technique is classic Raman spectroscopy, where approximately 0.001% of incident excitation laser energy is translated into Raman signal intensity.
SRS/SPSRS (Stimulated Raman Spectroscopy)	The Raman scattering signal is enhanced to 40–50% of the excitation laser energy. The overall Raman signal is enhanced by 4×10^4 to 5×10^4 over traditional spontaneous Raman spectroscopy.
SORS (Spatially Offset Raman Spectroscopy)	This technique produces a Raman depth profile spectrum.
TERS (Tip-Enhanced Raman Spectroscopy)	An enhanced Raman signal is combined with AFM topography for a nano-scaled, three-dimensional image of the spatially related chemistry for the sample.
SPPERS (Surface Plasmon Polariton (SPP) Enhanced Raman Spectroscopy)	For measurement of sample areas as small as 100 microns in diameter, having the thickness of a molecular monolayer of sample material.

size with the thickness of a molecular monolayer are possible. The technique uses special excitation and collection optics to enable an enhanced small spatial signal. The technique is able to determine Raman scattering using an Argon ion laser source and standard CCD detector with the specialized sampling and collection optics (Figure 2.15).[30–32]

Table 2.2 is a summary of the comparative methods used for Raman spectroscopy. The table identifies the specific purpose of each Raman technique as the various methods are used to enhance signal, isolate specific molecular structures, and to profile spatial and chemical information.

References

1. C.V. Raman, A new radiation, *Indian J. Phys.* **2** (1928) 387–398.
2. A.H. Compton (May 1923), A Quantum Theory of the Scattering of X-Rays by Light Elements, *Phys. Rev.* **21**(5) (1923) 483–502.
3. D.A. Long, *Raman spectroscopy*, 1 edn., Vol. 206 (McGraw-Hill, New York, 1977).
4. E.B. Wilson, *Molecular vibrations: the theory of infrared and Raman vibrational spectra*. 1 edn. (Courier Dover Publications, 1955).
5. J.R. Ferraro and K.N. Nakamoto, *Introductory Raman Spectroscopy*, 1 edn. (Academic Press, Boston, 1994).
6. H. Frostig, O. Katz, A. Natan, and Y. Silberberg, Single-pulse stimulated Raman scattering spectroscopy, *Optics letters* **36**(7) (2011) 1248–1250.
7. Y.-R. Shen, *The principles of nonlinear optics*, 1st edn. (Wiley-Interscience, New York, 1984).
8. A.M. Weiner, D.E. Leaird, G.P. Wiederrecht, and K.A. Nelson, Femtosecond multiple-pulse impulsive stimulated Raman scattering spectroscopy, *JOSA B* **8**(6) (1991) 1264–1275.
9. C.L. Evans and X.S. Xie, Coherent anti-Stokes Raman scattering microscopy: chemical imaging for biology and medicine, *Annu. Rev. Anal. Chem.* **1** (2008) 883–909.
10. A. Zumbusch, G.R. Holtom, and X.S. Xie, Three-dimensional vibrational imaging by coherent anti-Stokes Raman scattering, *Physical Review Letters* **82**(20) (1999) 4142–4145.
11. W.M. Tolles, J.W. Nibler, J.R. McDonald, and A.B. Harvey, A review of the theory and application of coherent anti-Stokes Raman spectroscopy (CARS), *Applied Spectroscopy* **31**(4) (1977) 253–271.
12. H. Kim, K.M. Kosuda, R.P. Van Duyne, and Peter C. Stair, Resonance Raman and surface-and tip-enhanced Raman spectroscopy methods to study solid catalysts and heterogeneous catalytic reactions, *Chemical Society Reviews* **39**(12) (2010) 4820–4844.
13. T. Suzuki, K. Tanimura, and N. Itoh, Resonance-Raman-scattering spectroscopy of the self-trapped excitons in alkali halides, *Physical Review B***49** (1994) 7233–7241.
14. A. Quatela, A. Agresti, S. Mastroianni, S. Pescetelli, T.M. Brown, A. Reale, and A. Di Carlo, Fabrication and reliability of dye solar cells: A resonance Raman scattering study, *Microelectronics Reliability* **52**(9) (2012) 2487–2489.
15. K. Kneipp, Y. Wang, H. Kneipp, L.T. Perelman, I. Itzkan, R.R. Dasari, and M.S. Feld, Single Molecule Detection Using Surface-Enhanced Raman Scattering (SERS), *Physical Review Letters* **78** (1997) 1667–1670.
16. S. Nie and S.R. Emory, Probing single molecules and single nanoparticles by surface-enhanced Raman scattering, *Science* **275**(5303) (1997) 1102–1106.
17. M. Moskovits, Surface-enhanced Raman spectroscopy: a brief retrospective, *Journal of Raman Spectroscopy* **36**(6–7) (2005) 485–496.
18. P. Hildebrandt and M. Stockburger, Surface-enhanced resonance Raman spectroscopy of Rhodamine 6G adsorbed on colloidal silver, *The Journal of Physical Chemistry* **88**(24) (1984) 5935–5944.
19. A.M. Michaels, M. Nirmal, and L.E. Brus, Surface enhanced Raman spectroscopy of individual rhodamine 6G molecules on large Ag nanocrystals, *Journal of the American Chemical Society* **121**(43) (1999) 9932–9939.
20. K. Kneipp, Y. Wang, R.R. Dasari, and M.S. Feld, Approach to single molecule detection using surface-enhanced resonance Raman scattering (SERRS): A study using Rhodamine 6G on colloidal silver, *Applied Spectroscopy* **49**(6) (1995) 780–784.

21. T.T. Basiev, A.A. Sobol, Y.K. Voronko, and P.G. Zverev, Spontaneous Raman spectroscopy of tungstate and molybdate crystals for Raman lasers, *Optical Materials* **15**(3) (2000) 205–216.
22. M.G. Raymer and J. Mostowski, Stimulated Raman scattering: Unified treatment of spontaneous initiation and spatial propagation, *Physical Review A* **24** (1981) 1980–1993.
23. T.T. Basiev, A.A. Sobol, P.G. Zverev, V.V. Osiko, and R.C. Powell, Comparative spontaneous Raman spectroscopy of crystals for Raman lasers, *Applied Optics* **38**(3) (1999) 594–598.
24. P. Matousek, I.P. Clark, E.R.C. Draper, M.D. Morris, A.E. Goodship, N. Everall, M. Towrie, W.F. Finney, and A.W. Parker, Subsurface probing in diffusely scattering media using spatially offset Raman spectroscopy, *Applied Spectroscopy* **59**(4) (2005) 393–400.
25. P. Matousek, M.D. Morris, N. Everall, I.P. Clark, M. Towrie, E. Draper, A. Goodship, and A.W. Parker, Numerical simulations of subsurface probing in diffusely scattering media using spatially offset Raman spectroscopy, *Applied Spectroscopy* **59**(12) (2005) 1485–1492.
26. N. Stone, R. Baker, K. Rogers, A.W. Parker, and P. Matousek, Subsurface probing of calcifications with spatially offset Raman spectroscopy (SORS): future possibilities for the diagnosis of breast cancer, *Analyst* **132**(9) (2007) 899–905.
27. B. Pettinger, B. Ren, G. Picardi, R. Schuster, and G. Ertl, Nanoscale probing of adsorbed species by tip-enhanced Raman spectroscopy, *Physical Review Letters* **92**(9) (2004) 096101–096101.
28. B. Pettinger, G. Picardi, R. Schuster, and G. Ertl, Surface-enhanced and STM-tip-enhanced Raman Spectroscopy at Metal Surfaces, *Single Molecules* **3**(5–6) (2002) 285–294.
29. B. Ren, G. Picardi, and B. Pettinger, Preparation of gold tips suitable for tip-enhanced Raman spectroscopy and light emission by electrochemical etching, *Review of Scientific Instruments* **75**(4) (2004) 837–841.
30. I. Abdulhalim, M. Zourob, and A. Lakhtakia, Surface plasmon resonance for biosensing: a mini-review, *Electromagnetics* **28**(3) (2008) 214–242.
31. Q-Han Park, Optical antennas and plasmonics, *Contemporary Physics* **50**(2) (2009) 407–442.
32. S. Raza, N. Stenger, S. Kadkhodazadeh, S.V. Fischer, N. Kostesha, A.-P. Jauho, A. Burrows, M. Wubs and N.A. Mortensen, Blueshift of the surface plasmon resonance in silver nanoparticles studied with EELS, *Nanophotonics* **2**(2) (2013) 131–138.

Chapter 3

Raman Instrumentation

Contents

3.1 Instrumentation for Measuring Raman Spectra 70
3.1.1 Raman Measurement Theory 70
3.1.2 Basic Optical and Electronic Designs for Raman Spectrometers 72
3.1.2.1 Monochromator Instruments 72
3.1.2.2 Interferometer Instruments 73
3.1.3 Dispersive Raman Spectrometer with Array Spectrograph . . . 75
3.1.4 Raman Spectrometer Components 76
3.1.4.1 Laser Excitation Sources 76
3.1.4.2 Detector: Performance and Sensitivity 83
3.1.4.3 Theoretical Limit of D* for a Detector System 84
3.1.4.4 Signal-to-noise Ratio (SNR) in a Raman Spectrometer[19] 85
3.1.5 Electronic Components used in Spectrometry 89
3.1.6 Properties of Spectrophotometers 90
3.1.6.1 Entrance and Exit Pupils (Apertures) 90
3.1.6.2 Band Pass (Bandpass) and Resolution 90
3.1.6.3 Peak Positions 92
3.1.6.4 Attenuation Losses over Time 93
3.1.7 Raman Instrument Comparison and Evaluation Methods 94
3.1.8 Advanced Raman Measurement Techniques 95
References 96

3.1 Instrumentation for Measuring Raman Spectra

Raman spectrometers used for research, process control, and industrial laboratory work are available in a variety of optical and mechanical design configurations. Each of these configurations uses the basic tenets of the Raman measurement technique, similar optical and electronic components, and the same basic theory for Raman spectroscopy.

3.1.1 *Raman Measurement Theory*

Raman Spectroscopy for Quantitative Analysis: Raman spectroscopic measurements depend upon the principle that light energy interacting with a material will cause Raman scattering at a specific frequency depending upon the molecules present in the measured sample. The use of Beer's law, essential to quantitative analysis using absorption spectroscopy, does not apply to Raman spectroscopy.

As has been discussed previously (Front Material and Chapter 2 of this volume), the Raman intensity (I_R) or amplitude of the Raman scattered signal for any sample is determined by several recognized variables, such as the spectrometer throughput (τ), the self-absorption of the sample (α), the Rayleigh scattered radiation (ρ_S), the excitation laser intensity (I_0), the Raman scatter (R_S) of the molecule being measured at the frequency measured, the effective optical pathlength or L_E (i.e., actual pathlength or number designated as an excited volume of material), and the number of molecules encountered by the beam path of the measuring Raman instrument (as c or concentration in $\text{Mole} \cdot \text{Liter}^{-1}$, or mass per unit volume).[1] It is assumed that a change in spectral response is related to a concentration change as explained using the Raman relationship. Raman intensity (I_R) as Raman scatter (R_S) dependent upon the excitation frequency (as ν) is described as:

$$I_R = \tau_\nu \cdot \alpha_\nu \cdot \rho_{S\nu}^4 \cdot I_0 \cdot R_{S\nu} \cdot L_E \cdot c \tag{3.1}$$

Note for any particular sample measurement experiment with a specific instrument design, sample type, and excitation laser power and type; the instrument throughput, sample self-absorption, scattering, and Raman scatter could be considered fractions of signal loss (as a percent factor or fractions of less than 1.0 for each variable to reduce signal intensity). The effective pathlength depends upon laser power, excitation angle and collection, and the sample matrix. Thus for each experiment where quantitative analysis is required, a set of calibration samples using varying analyte concentrations in the same matrix is recommended.

A typical regression line is constructed for Raman intensity versus concentration, and the unknown test sample is estimated using regression. The regression technique for quantitative analysis is described elsewhere in this set of volumes and in greater detail in Chapter 6 of this volume. Chapter 1, and Section 1.3.2, entitled, "The Concepts Behind Raman Spectroscopy Measurements," gives a more detailed discussion of the basics of Raman measurements.

A typical sample measurement as (I_{RC}) or Raman Corrected Intensity is made by taking the ratio of the I_R (Sample) — I_R (Dark Signal) to the I_R (Blank Sample) — I_R (Dark Signal). This procedure compensates for the extraneous noise from the instrument operation, both mechanical and electronic. It also has an effect to mitigate sampling errors when a pure solvent is measured as the blank signal using the identical sampling procedure. Raman spectra may be measured using no correction or by adding such a correction depending upon the experiment.

Such measurement schema are illustrated in Figures 3.6 and 3.7. The corrected Raman spectrum equation would then be as:

$$I_{RC} = \frac{I_R(Sample) - I_R(Dark\_Signal)}{I_R(Blank\_Sample) - I_R(Dark\_Signal)} \tag{3.2}$$

A laboratory Raman spectrometer is designed to maximize the efficiency of the instrument such that the maximum percentage of laser excitation energy is returned to the detector from the sample as useful Raman signal. To maximize the energy from the sample through the optical components of the spectrometer and onto the detector requires well-aligned optics, properly focused optics, and maximum throughput design criteria. The optical throughput for a Raman spectrometer is dependent upon multiple factors, such as the laser power and area of excitation, the apertures present within the light path, the optical elements lens transmittance, mirror reflectance losses, the entrance and exit apertures, and the detector area and efficiency.

Several terms are useful in any discussion of spectrometry, these include *selectivity* or the specific sensor response to the component (or analyte) of interest; *sensitivity* or the *quantifiable level of response* from a sensor with respect to the concentration of a specified component (analyte) of interest; and *detection limit* or the smallest analyte concentration difference that can be detected above the background noise level of the Raman spectrometer. In the absence of fluorescence, the minimum signal that can be reliably detected is three times the background noise level. Note that Chapter 4 of this volume includes the details of sampling systems and sampling configurations for Raman measurements.

3.1.2 *Basic Optical and Electronic Designs for Raman Spectrometers*

3.1.2.1 *Monochromator Instruments*

The single monochromator spectrometer is configured with one reflection grating/one monochromator for optimization of throughput, but with limited resolution. It is a design for laboratory or process Raman systems. The optical configuration for such a design is shown in Figure 3.1a. Figure 3.1b shows an analogous design using a transmission grating system. Raman instruments may be constructed for a variety of applications using an ever increasing number of monochromators. The addition of each grating increases resolution, but decreases energy throughput. This decrease in throughput may be enhanced by using more powerful laser excitation and improved quality and higher throughput optics. Such improvements increase performance and add significantly to the cost of construction. Figures 3.2 and 3.3 illustrate the double and triple monochromator Raman spectrometer optical designs, respectively. Figure 3.4 illustrates the optical configuration for a dispersive Raman microscope system.

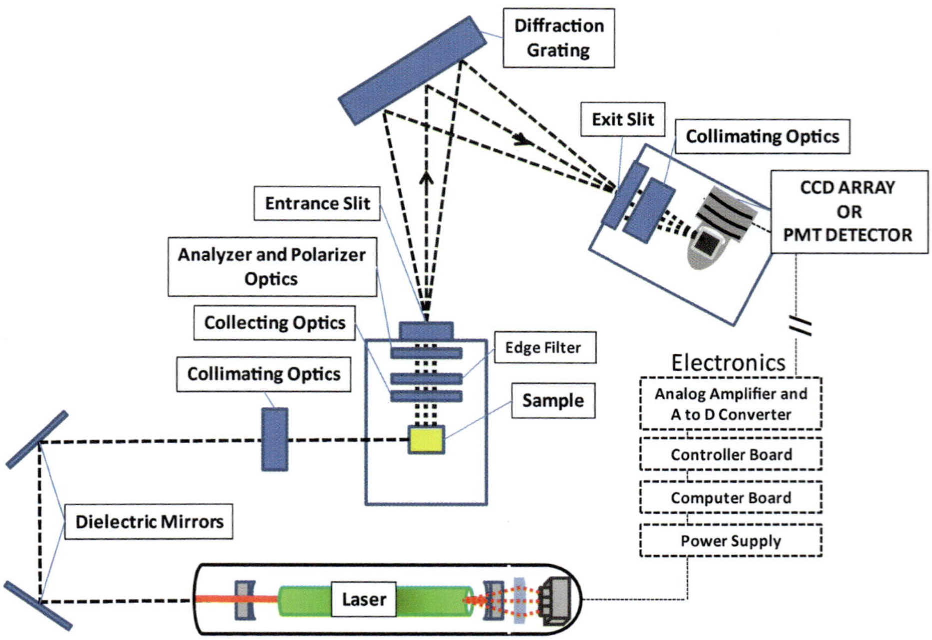

Fig. 3.1a Single reflection monochromator Raman spectrometer configuration.

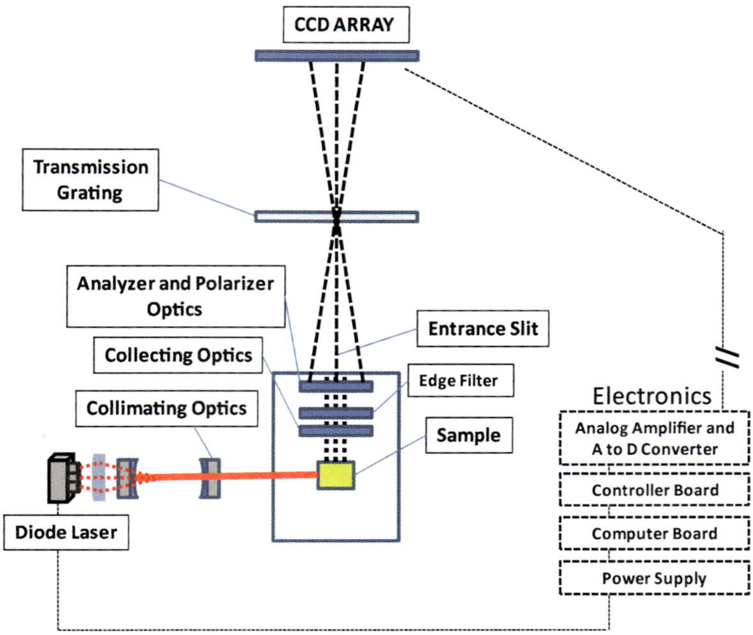

Fig. 3.1b Single transmission monochromator Raman spectrometer configuration.

3.1.2.2 *Interferometer Instruments*

Interferometry is used to measure the frequency of light passing through the optical interferometer. The frequency of measurements made using typical interferometric designed spectrometers are generally more precise (i.e., repeatable) than those made using standard dispersive type instruments. Dispersive instruments may be as precise but require significant design enhancements to match an interferometer, noting that interferometers include a laser to control the moving mirror position and as such have a built-in primary reference position for each scan. However, the typical transmission or reflectance data for an interferometer may be less accurate than a monochromator. The interferometer is designed to have constant line width and resolution in the frequency space, whereas the monochromator is designed to have relatively constant resolution in the wavelength space.

The most common type of interferometer is the Michelson type shown in Figure 3.5.[2] A parallel beam of light from the laser excitation source is directed into a bean splitter (BS) at an angle of 45 degrees from normal. The beam splitter consists of a 50% reflective surface. The transmitted portion of light from the beam splitter is directed to a highly reflective moveable mirror (MM), whereas the light

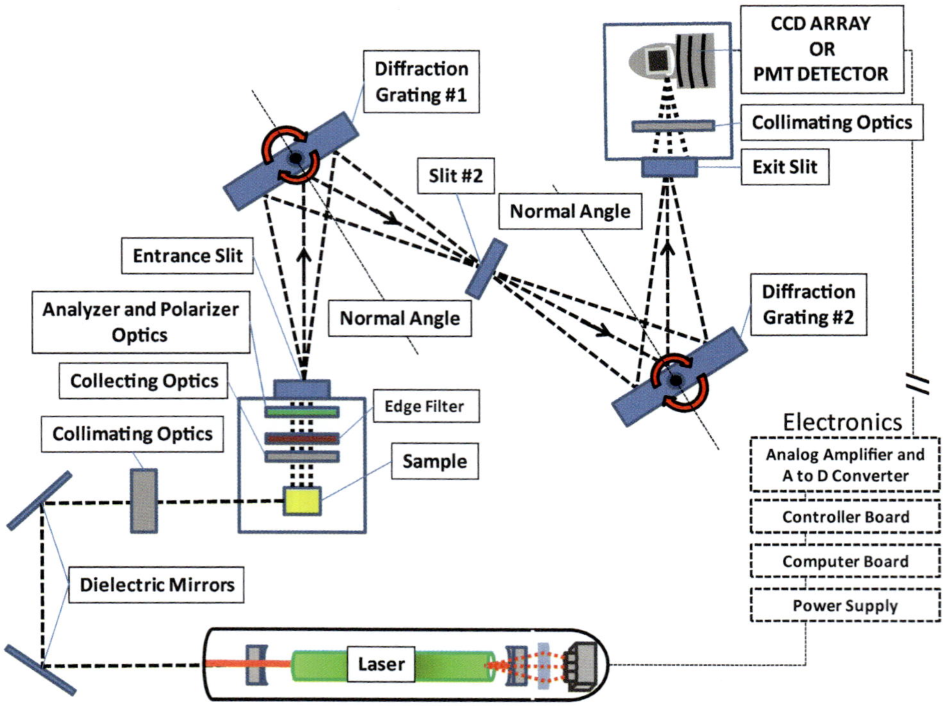

Fig. 3.2 Double monochromator Raman spectrometer configuration.

reflected from the beam splitter strikes a fixed mirror (FM). From the perspective of the detector (D) the following phenomena are "observed".

If the pathlengths to FM and MM are the same, an observer at the D position will see a bright fringe. If the pathlengths are not the same the phase difference will determine whether the fringe observed at the detector is light or dark. The fixed mirror is only adjusted for tilt to keep it normal to the incident beam. As the MM is moved, the path becomes different and the phase changes. A movement of MM (d) is related to the number of fringes in the interference pattern ($\#f$) observed at D by the relationship:

$$\frac{d}{\#f} = \frac{\lambda}{4} \tag{3.3}$$

when $\lambda =$ the wavelength of parallel incident light entering the interferometer, and d is expressed in units identical to wavelength units. The optical design configuration for an interferometer based instrument is shown in Figure 3.5.[2–6] A detailed explanation of the operation and theory of the interferometer is given

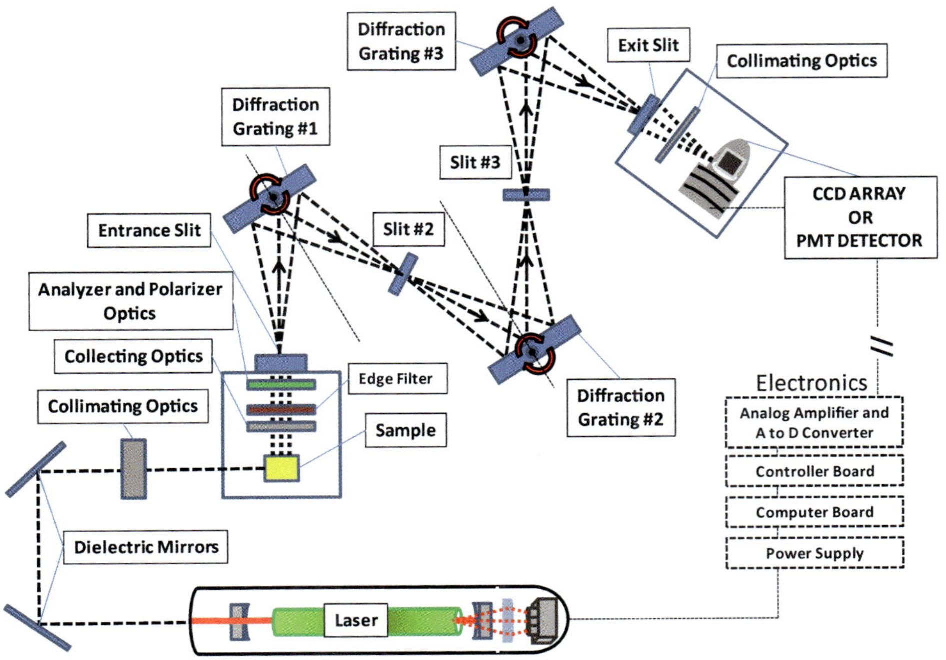

Fig. 3.3 Triple monochromator Raman spectrometer configuration.

in Volume 4, Chapter 3, and the reader is referred to the index and references for additional source material.

A different computational process is used to convert the measured interferogram to a Raman spectrum as compared to an absorption spectrum, such as UV-Vis, NIR, or MIR spectra. Note that Beer's law does not apply to Raman spectra so the interferogram is converted directly to the Raman spectrum. The interferogram is converted to the Raman signal spectrum as Raman intensity versus frequency, and then applicable correction is applied, such as baseline correction and normalization to derive the completed Raman spectrum. This process without and with background correction is illustrated in Figures 3.6 and 3.7, respectively.

3.1.3 *Dispersive Raman Spectrometer with Array Spectrograph*

Raman spectrometers are also constructed using a diode array configuration for solid-state operation where no moving parts may be required. Military, medical, or space applications would generally infer such a design configuration. In this design type the dispersive grating and array detectors are fixed and the dispersed Raman signal is detected over the array and spectra recorded in real time across all frequencies. Such a design is illustrated in Figure 3.8.

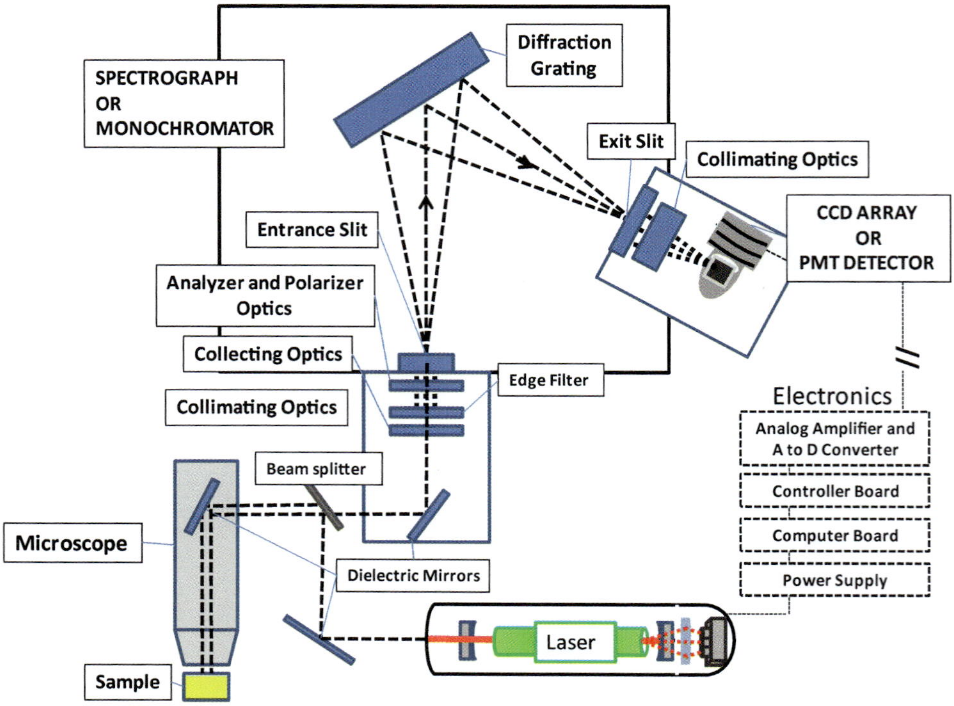

Fig. 3.4 Dispersive Raman microscope optical configuration.

3.1.4 *Raman Spectrometer Components*

3.1.4.1 *Laser Excitation Sources*

Note that Laser safety in the laboratory should comply with the American National Standard ANSI Z136.1-2007 or newer standard entitled, "American National Standard for Safe Use of Lasers". This standard on safe use of lasers is available from the Laser Institute of America, 13501 Ingenuity Drive, Suite 128, Orlando, FL 32826, ISBN-13: 978-0-912035-65-9 & ISBN-10: 0-912035-65-X.

Laser excitation is used to produce the Raman effect as inelastic scattering collisions with the various moieties in molecular structures. The laser sources may be comprised of gas lasers or diode lasers of relatively low power and constant frequency. Table 3.1 lists common laser excitation sources along with reference data. References for laser technology useful for Raman spectroscopy include, but are not limited to reference 7. There are many high quality manufacturers of Raman components.

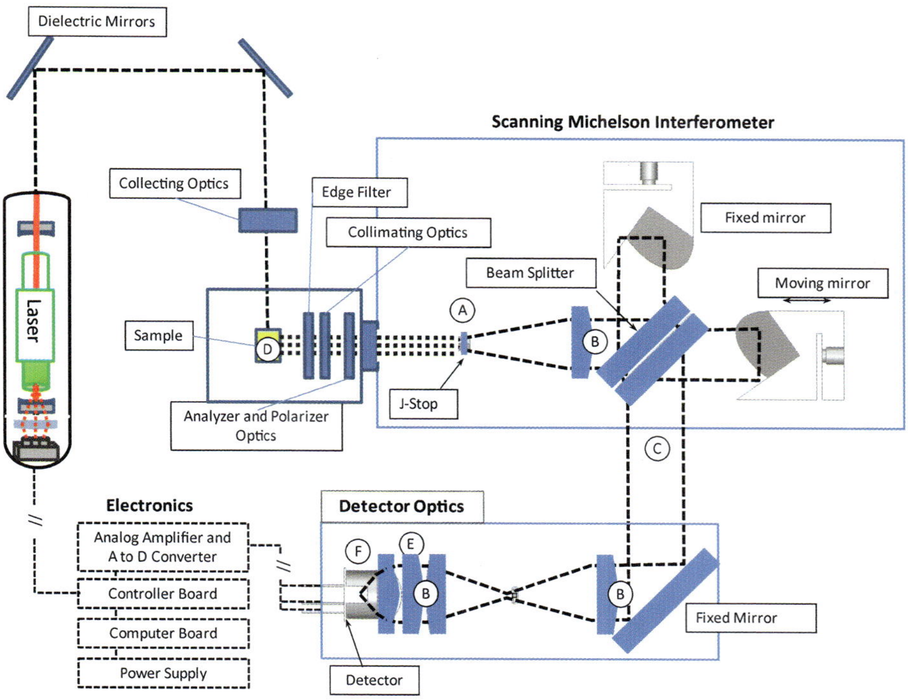

Fig. 3.5 Fourier transform Raman spectrometer (interferometer) configuration. Interferometer based system (Interferometer/FT-type). Laser = Excitation source, A = J-stop (aperture), B = Collimating optics, C = Optical train, D = Sample cell, E = Focusing optics, F = Raman detector.

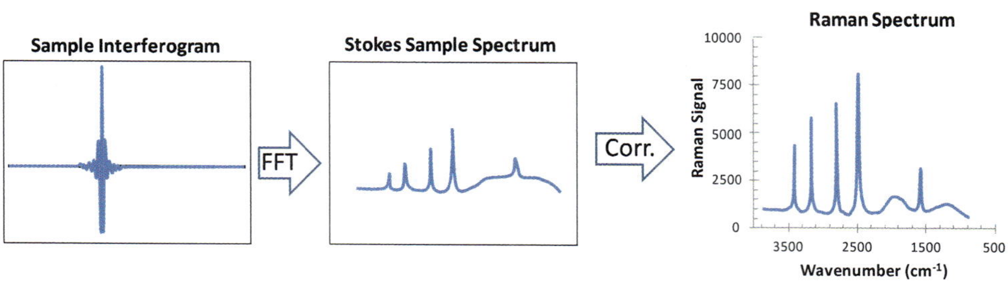

Fig. 3.6 Illustration of the conversion of an interferogram to a Raman spectrum without dark correction. This process does not involve the ratioing of sample and dark or reference spectra. Background corrected Raman spectra are shown in Figure 3.7.

Laser excitation sources for Raman spectrometers generally consists in diode lasers, gas lasers, solid-state flashlamp pump lasers, or solid-state VCSEL (vertical-cavity surface-emitting laser) pump lasers. These laser types are illustrated in Figures 3.9 through 3.12, respectively. Raman lasers have the following two

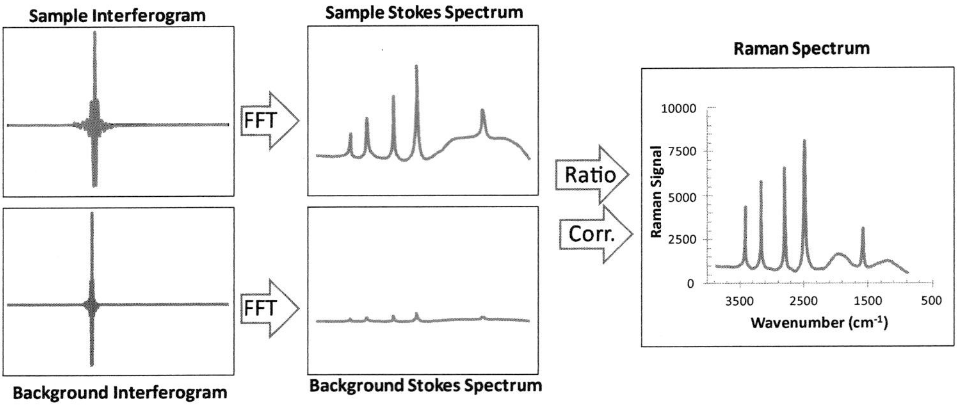

Fig. 3.7 Illustration of the conversion of an interferogram to a Raman spectrum with background reference correction. Ratioing the sample spectrum to the reference spectrum ratios the measurement signal noise out from the spectral signal. The reference consists of a blank solvent spectrum used to compensate for electronic noise or scattering signals due to sampling and instrument operation. A dark spectrum is subtracted from both the sample and reference spectra before the ratio calculation is applied as sample divided by background/reference. For the dark signal, a shutter blocks the excitation laser during the collection of the dark spectrum. Note in the figure (Corr.) indicates baseline correction or normalization correction in addition to ratioing step (see also Equation 3.2).

properties: 1) they must emit high energy (as high photon flux) at a narrow and reproducible band of energy, and 2) the energy output must also be consistent for power output, since the Raman signal is proportional to the input energy or power from the excitation laser source.

New laser technology is in continuous development. Laser technology for use in Raman spectroscopy continues to advance, for example, a variety of all silicon lasers have been developed.[8,9] A visible solid-state barium nitrate laser has also been developed that yields high efficiency in a compact footprint.[10]

Lasers most commonly used are shown in Figures 3.9 through 3.12. Laser power is described in terms of watts, or watts per square area; and also for fiber guided laser sources in terms of loss of power per meter. The unit conversions for these units are as follows: the unit of measure for laser power is the watt (W), which is defined as one joule (J) per second; the watt in SI units (Système International d'Unités) is as 1 Kg $\cdot$ $m^2 \cdot s^{-3}$, also expressed as 1 x 10^7 erg $\cdot$ s^{-1} in CGS units (i.e., the centimeter–gram–second) system. The watt is defined as the relationships in Equation 3.4.

$$watt(W) = \frac{J}{s} = \frac{Kg \cdot m^2}{s^3} = \frac{N \cdot m}{s} = \frac{Pa \cdot m^3}{s} = \frac{C \cdot V}{s} = \frac{10^7 ergs}{s} \tag{3.4}$$

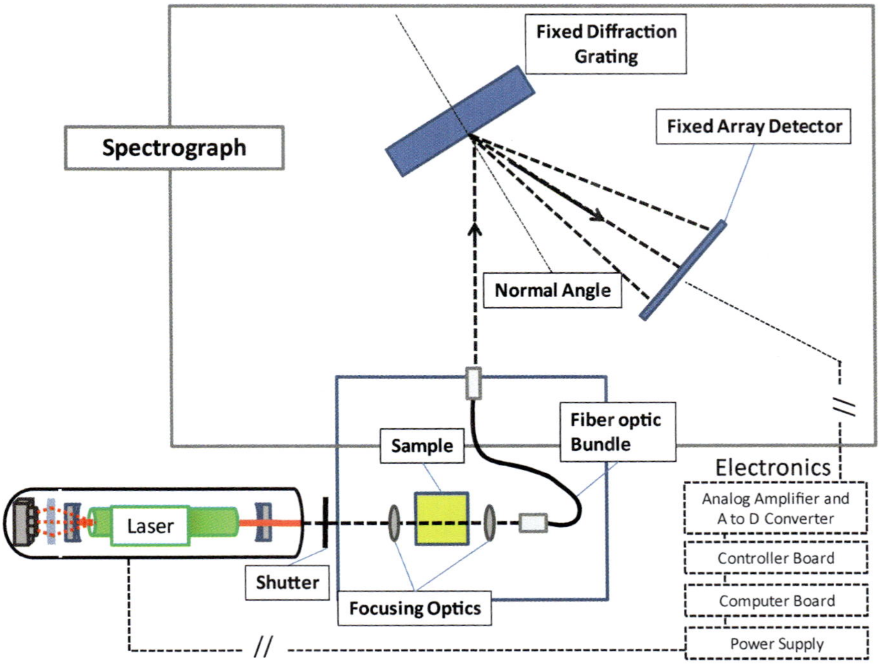

Fig. 3.8 Array based Raman spectrometer showing solid-state design. Both the grating and detector elements are fixed.

Table 3.1 Typical Raman laser excitation sources.

Laser emission ranges (μm)	Start	End
Ultraviolet (Laser Diode), 244 nm	0.2435	0.2445
Ultraviolet (Laser Diode), 257 nm	0.2565	0.2575
Ultraviolet (Laser Diode), 325 nm	0.3245	0.3255
Ultraviolet (Laser Diode), 364 nm	0.3635	0.3645
Visible (Laser diode), 455 nm	0.4545	0.4555
Visible (Laser diode), 473 nm	0.4725	0.4735
Visible (Laser diode), 488 nm	0.4875	0.4885
Visible (Laser diode), 514 nm	0.5135	0.5145
Visible (Laser diode), 532 nm	0.5315	0.5325
Helium-neon laser (He:Ne), 632.8 nm	0.6327	0.6329
Visible (Laser diode), 660 nm	0.6595	0.6605
Visible (Laser diode), 785 nm	0.7845	0.7855
Near infrared (VCSEL), 850 nm	0.8295	0.8305
Near infrared (Laser diode), 980 nm	0.9795	0.9805
NIR: Neodymium yttrium aluminum garnet (Nd:YAG) laser, 1064 nm	1.0639	1.0641

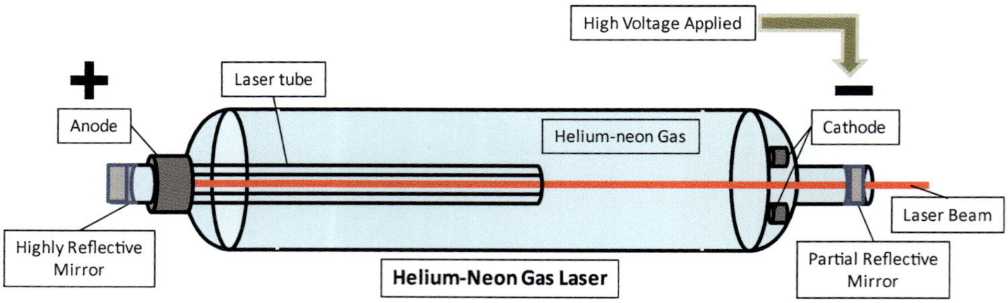

Fig. 3.9 Helium:Neon gas laser illustration. High voltage excites the He:Ne gas electrons and light is emitted as the electrons relax to the ground state.

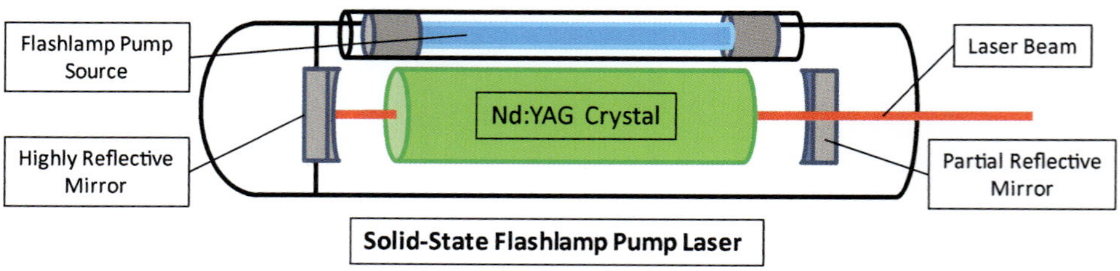

Fig. 3.10 Solid-state flashlamp pump laser illustration. The flashlamp pump source excites the electrons in the Nd:YAG crystal and light is emitted as the electrons relax to the ground state.

Where J = joule (unit of energy); Kg = kilogram (unit of mass); m = meter (unit of length); s = second (unit of time); N = newton (unit of force); Pa = pascal (unit of pressure); C = coulomb (unit of electrical charge); V = volts (unit of electrical potential); Ω = ohm (unit of electrical resistance), A = amperes (unit of electrical current); and ergs = *ergon* (unit of mechanical work).

Note there are continuous wave (CW) and pulse lasers. CW lasers have a constant or continuous power output, whereas pulse lasers have a very short pulse of power with an intermittent sequence. Pulsed lasers may emit pulses of the order: 10−3 s (ms, millisecond); 10−6 s (μs, microsecond); 10−9 s (ns, nanosecond); 10−12 s (ps, picosecond); or 10−15 s (fs, femtosecond).

For laser attenuation, when passing through an optical fiber, the unit of decibels per unit length is used by convention. A decibel (dB) is a logarithmic unit comparing two power levels before and after attenuation of the laser power as:

$$A_{db} = 10 \cdot \log_{10}\left(\frac{P_I}{P_{I_0}}\right) \Rightarrow P_I = 10^{A_{db}/10} \cdot P_{I_0} \tag{3.5}$$

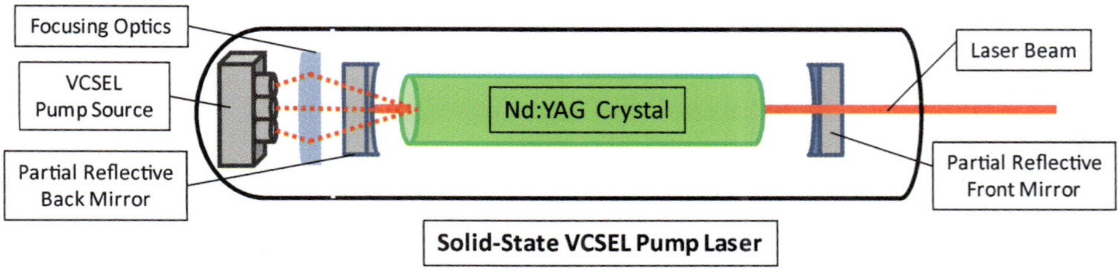

Fig. 3.11 Solid-state VCSEL (Vertical-Cavity Surface-Emitting Laser) pump laser illustration. The VCSEL pump source excites the electrons in the Nd:YAG crystal and light is emitted as the electrons relax to the ground state.

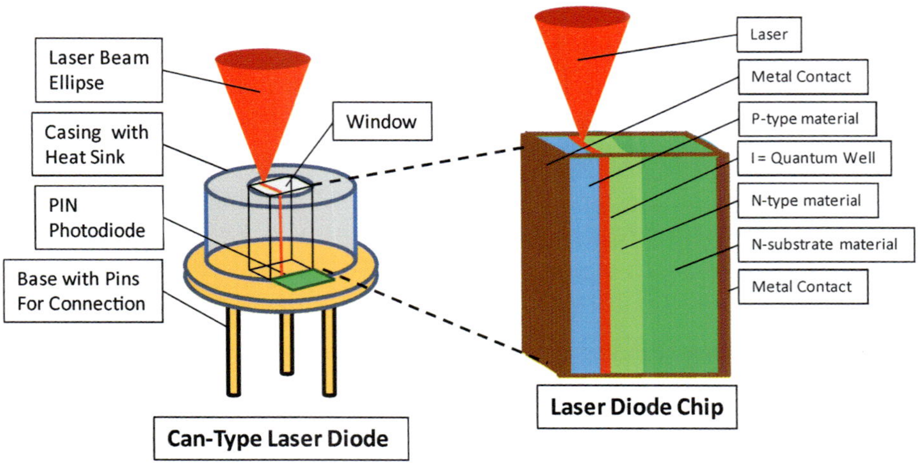

Fig. 3.12 Solid-state P-I-N (p-i-n) junction semiconductor diode laser illustration. Electrical stimulation excites the electrons in the quantum well (I) layer and light is emitted as the electrons relax to the ground state. This is a common laser device used in modern Raman spectrometers.

where A_{db} is the attenuated power in watts; P_I is the power output at the end of the fiber in watts; and P_{Io} is the initial laser power at the entrance of the fiber.

A diode laser is common in modern Raman systems and consists of a solid-state device capable of lasing, which is dependent upon its specific materials and construction. A laser emits coherent energy at a single wavelength by the process of electrons being excited to a higher energy state and then rapidly relaxing to their ground state. During the relaxation process energy is emitted from the electrons as laser energy and by using highly reflective mirrors and lenses the energy is directed along a single coherent path. If a pulse of high energy is directed into a dense electronic cloud we will have stimulated emission of radiation and lasing potential. Thus, light amplification by stimulated emission of radiation (laser) energy is generated. Gas lasers and pulse lasers use light stimulation of the electrons, diode lasers use a different stimulation process. The various forms of gas lasers are illustrated in Figures 3.9 through 3.11.

For laser diodes rather than light being used on a dense electron cloud, an electrically pumped semiconductor is used where the lasing material is the p-i-n junction of a semiconductor diode. The structure is optimized such that the electrical signals are combined in the I region of the junction, and produce light. Fabrication of laser diodes uses a bandgap semiconductor structure. The semiconductors are produced using crystal growth techniques. First an N doped substrate is grown, then the I doped laser active layer, and then the P doped cladding is added, and finally a contact layer is formed (Figure 3.12).

Raman fluorescence depends upon the excitation wavelength of the laser used. The intense fluorescence as a background signal increases with the frequency (decreased wavelength) of the laser excitation. In Figure 3.13, one may observe the same sample using different excitation wavelengths. The upper figure demonstrates the most fluorescence using the highest frequency laser, while the bottom figure demonstrates almost no fluorescence when using a near infrared excitation laser.

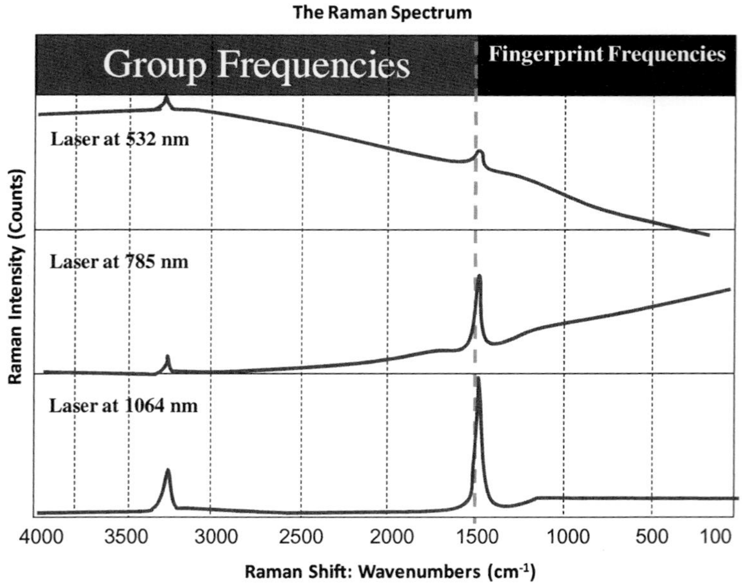

Fig. 3.13 A simulated spectrum showing the different amounts of fluorescence background depending upon laser excitation source used. The upper figure exhibits the greatest fluorescence, while the bottom figure demonstrates the least.

3.1.4.2 *Detector: Performance and Sensitivity*

Molecular spectroscopy involves two basic detector types: thermal and photon counting. Thermal or infrared detectors include bolometers, pyroelectric detectors, and thermoelectric detectors. Thermal detectors convert infrared photons to heat energy and then to electrical signal. The type of detection used for Raman spectrometers include the photon counting type. These include intrinsic, extrinsic, free carrier, and quantum well types. The basic operation of a photon counting detector is to convert photons to electrons and then to electrical signals. Such detectors have high sensitivity and fast response time relative to thermal types. For photon counting detectors and their properties refer to Tables 3.2 and 3.3 as well as the index for additional details. References for Raman detector technology include 11 through 14.

The basic physics of detector device performance or sensitivity involves two basic concepts. First the term specific detectivity or D-star (D*) is essential. This D* value is defined as the detectivity of a radiation detector as a function of the square root of the product of the active detector element area (A in cm^2) and the bandwidth (ω) in units of cycles per second (Hertz, Hz); divided by the noise equivalent power (NEP) in Watts of the detector element. This is summarized by the following formula:

$$D^* = \frac{\sqrt{A \cdot \omega}}{NEP} = \mathrm{cm} \cdot \mathrm{Hz}^{1/2} W^{-1} \tag{3.6}$$

The D* is reported in units of $\mathrm{cm} \cdot \mathrm{Hz}^{1/2}\mathrm{W}^{-1}$ units and is used to compare photodetector responsivity as a function of wavelength. These are referred to as *Jones* units, for the namesake of R. Clark Jones.[15–18]

For photo-detectors, detector responsivity (R) or input-output gain, is also used to express detector sensitivity in units of Amperes per Watt (A/W) as a function of the frequency or wavelength of light incident to the detector. This is a measure of the electrical output (or detector efficiency or sensitivity) as a function of optical energy input. This relationship is expressed as follows:

$$R = Q\left(\frac{\lambda(\mu m) \cdot A}{1.23985(\mu m \cdot W)}\right) = \text{Amps/Watt} \tag{3.7}$$

This computational form is derived from the responsivity expression, equivalent to the quantum efficiency times the electron charge divided by Planck's constant times the frequency of the optical signal; the responsivity changes as a function of the wavelength of the incident energy.

Note that responsivity (R) in Amps/Watt may be converted to Quantum Efficiency (Q) on a scale of 0 to 1 and is expressed as (Watts · Wavelength (in nm))/Amps.

This conversion of R to Q may be expressed as:

$$Q = \frac{1.23985R \cdot W}{A} \tag{3.8}$$

3.1.4.3 *Theoretical Limit of D* for a Detector System*

Noise from a detector often results from multiple factors, such as associated electronic and amplifier noise. However there is a theoretical noise associated with the background fluctuation attributable only to the detector itself, irrespective of whether it is a photovoltaic or photoconductive detector. The limits of D* are somewhat different for photovoltaic or photoconductive detectors and may be expressed in terms of the wavelength of light (λ), Quantum efficiency (Q), and the flux of background radiation (F). The constants for Planck and the speed of light are also included in the expressions.
For the photoconductive detector:

$$D^* = \frac{\lambda\sqrt{Q}}{hc\sqrt{2F}} \text{ in units of cm} \cdot \text{Hz}^{1/2}\text{W}^{-1} \tag{3.9}$$

For the photovoltaic detector:

$$D^* = \frac{\lambda\sqrt{Q}}{2hc\sqrt{F}} \text{ in units of cm} \cdot \text{Hz}^{1/2}\text{W}^{-1} \tag{3.10}$$

This D* value is defined as the detectivity of a radiation detector as a function of the wavelength, the square root of the quantum efficiency (Q), and the background radiation flux (F). The background radiation (F) is computed in Watts per area-squared based on the detector temperature, typically in degrees Kelvin. In computing the *Signal to noise for a Detector System*, a higher shunt impedance indicates a lower overall dark noise for the detector such that a detector with the same peak response, and responsivity profile, with higher shunt resistance will yield a higher signal to noise for measurement.

Raman signal is captured by using a variety of detectors, such as photomultiplier tube (PMT) detectors, Charged Coupled Devices (CCD) detectors, or Indium Gallium Arsenide (InGaAs) detectors (see Tables 3.2 and 3.3). The voltage signals generated by these detectors are amplified, converted from analog to digital signals, and used to quantify the amount of energy emerging from a sample. A more complete list of common detectors, common abbreviations, and their useful working wavelength ranges would include the following table. The reader is referred to the index for more details on the physics of photon detectors.

Table 3.2 Raman detectors.

Detector sensitivity ranges (μm)	Start	End
Silicon (Si)	0.3	1.1
Indium Gallium Arsenide (InGaAs), typical	0.9	1.7
Indium Gallium Arsenide (InGaAs), extended	0.65	2.4
Indium Gallium Arsenide (InGaAs), IR extended	0.65	2.7
Indium Antimonide (InSb)	1.8	6.8
Lead Telluride (PbTe)	1.5	4.5
Deuterated Triglycine Sulfate (DTGS/KBr)	0.83	25
Mercury Cadmium Telluride (MCT, HgCdTe)	1	17
Silicon Photodiode Detectors	0.19	1.1
Photomultipler Tube (PMT), S1 Type Photocathode	0.3	1.2
Photomultipler Tube, Cesium Galium Arsenide (CsGaAs) Photocathode	0.3	0.93
Photomultipler Tube, Cesium Indium Galium Arsenide (CsInGaAs) Photocathode	0.3	1
Photomultipler Tube, Cesium Antiminide CsSb Photocathode	0.25	0.75
Photomultipler Tube, Multialkali (Na-K-Sb-Cs) Photocathode	0.3	0.93
Photomultipler Tube, CsI Photocathode	0.2	0.9
Photomultipler Tube, S1 Type Photocathode	0.3	1.2
Photomultipler Tube, Cesium Galium Arsenide (CsGaAs) Photocathode	0.3	0.93
Photomultipler Tube, Cesium Indium Galium Arsenide (CsInGaAs) Photocathode	0.3	1
Silicon (Si) CCD	0.2	1.1
Indium Gallium Arsenide (InGaAs) CCD	0.8	1.7
Raman Range	2.5	25

Table 3.3 Raman detectors (as Photon Counting Types).

Detector type	Detector materials used
Intrinsic	PbS, PbSe, PbSnTe
	HgCdTe (MCT)
	InGaAs, InAs, InSb, InAsSb
Extrinsic	Ga, SiAs, GeCu, GeHg
Free Carrier	PtSi, Pt_2Si, IrSi
Quantum Well	GaAs/AlGaAs, InGaAs/AlGaAs

3.1.4.4 *Signal-to-noise ratio (SNR) in a Raman Spectrometer*[19]

In the following discussion, the SNR is defined as 100/N where N is the RMS noise. RMS noise[20] is usually measured in an interference and fluorescence free region, as these features add increased noise to the measurement that is not associated with instrument performance or design. At the lowest and highest frequency regions of the spectrum, SNR is considerably higher. This is due to the fact that the greatest sensitivity of performance for most detectors is near the center of the region of measurement and the sensitivity decreases on both ends of the detector range.

The quantitative relationships between SNR, resolution, and measurement time in Raman spectrometry are referred to as "trading rules". The factors that affect the SNR and dictate the trading rules are expressed in Equation 3.11, which gives the theoretical SNR of a spectrum measured with a rapid-scanning Michelson interferometer. The derivation of Equation 3.11 is given by Griffiths and de Haseth, more specifically related to FT-IR, but also applicable to interferometer based Raman (FT-Raman):[21]

$$SNR = \frac{U_V(T)\Theta(\Delta\nu)t^{1/2}\zeta}{NEP} \tag{3.11}$$

where $U_\nu(\mathrm{T})$ = the spectral energy density at wavenumber ν from a blackbody source at a temperature T; θ = the optical throughput of the spectrometric system; $\Delta\nu$ = is the resolution of the interferometer; t = is the measurement time in seconds; ζ = the efficiency of the interferometer; and NEP = Noise equivalent power.

Note NEP represents the quantity of light incident onto the detector that is equal to the noise level of the detector; therefore the NEP represents the light incident when the signal-to-noise (S/N) is equal to 1.0. The equation to determine NEP for an infrared detector is as:

$$NEP = \frac{ID}{S/N \cdot \sqrt{nb}} \tag{3.12}$$

where I is the incident energy to the detector in Watts/cm^2; D is the detector active area in cm^2; S is the signal in volts (V); N is the RMS noise in volts (V), note that S/N is set to 1.0 for this calculation; and nb is the noise bandwidth in Hz.

Although cumbersome, Equation 3.11 may be used to evaluate SNR under actual measurement conditions, for example see (reference 20). For this chapter, however, generalization will be inferred from this equation.

As shown in Equation 3.11, the SNR of a spectrum is proportional to the square root of the measurement time ($t^{1/2}$). For measurements made with a rapid scanning interferometer operating at a constant mirror velocity at a given resolution, the SNR increases with the square root of the number of scans being averaged.

Most Raman spectrometer operational software allows a wide range of spectral resolution options (low cost instruments may offer only a single or highly limited set of resolution settings). However, several factors must be considered when determining the optimum resolution for measuring a Raman spectrum. These factors include: 1) the ability to distinguish between the spectral features of analytes and those of interfering species or background fluorescence; 2) the trade-offs between resolution, Raman intensity, and SNR; 3) practical considerations, such as measurement time, and computational time required to process the interferogram (i.e., higher resolution includes more data points); and 4) the data requirements of the interferogram file for data processing and storage. The proper selection of instrument resolution minimizes spectral overlap while maintaining a balance between the SNR, analysis time, and data processing and storage requirements.

Fluorescence may cause problems in many measurements as inferred in item 1 in the preceding paragraph. The higher the frequency (e.g., ultraviolet or visible) of the exciting laser, the more likely that fluorescence interference will occur. The selection of a near infrared laser is helpful to minimize the fluorescence problem associated with measuring biological samples that tend to have highly fluorescing compounds inherent to their properties. For normal laboratory use item 3 (data processing and storage) is not a problem. However accumulation of many thousands of spectra for quality or process measurements become a significant database management challenge.

The following table is a useful guide to the noise-reduction factor versus number of scans. For example, 100 scans should provide a 10 times improvement in signal to noise as compared to 1 scan, and 400 scans should produce a 20 time improvement. Table 3.4 shows theoretical improvement, which may not be obtained in practice due to interference, or to non-random optical, mechanical, and electronic noise disturbances.

Spectroscopists should be aware that every instrument has its upper-limit in reduction of noise as function of the number of replicate scans. This limit can be determined experimentally by measuring signal-to-noise (SN or SNR) versus the number of scans measured and comparing the computed values to the theoretical values shown above. Additional details for measuring signal-to-noise instrument performance are included within this chapter and the index.

Table 3.4 Theoretical improvement factor for signal-to-noise of spectra with repeat measurements.

Number of scans	Expected S/N ratio improvement
1	1
4	2
16	4
100	10
400	20
1600	40
2500	50
3600	60

a) The Influence of Resolution on SNR

Optical throughput and the interferometer efficiency of the spectrometer are assumed to be constant for measurements made at both high and low resolution. Under these conditions, changing the nominal resolution from 2 to $4\,cm^{-1}$ will improve the SNR and decrease the number of data points, and scan time by one-half. The SNR is proportional to the square root of the measurement time (and most of the time proportional to the number of scans). As a consequence, the measurement time or number of scans required to maintain the original baseline noise must be increased by a factor of 4 each time the increment between data points is doubled, i.e. decreasing the resolution by one-half.

Resolution may also affect the peak intensity of the Raman bands being measured. For example, the special case, where a weak spectral feature has a 1/2-width (FWHH) much less than the instrumental resolution (that is, a narrow band), the peak intensity will approximately decrease by half when increasing the resolution by a factor of 2; this is due to a decrease in energy throughput for the Raman spectrometer at higher resolution settings. Therefore, the SNR ratio would appear to be the same for measurements taken at the higher and lower resolution settings, provided the measurement times are equal. For broad bands, the peak intensity will not be affected very much by changes in resolution, and the lower resolution measurement would be preferred because it would yield improved signal-to-noise.

Instruments capable of medium and high-resolution measurements are equipped with adjustable or interchangeable apertures (in some software these are referred to as the Jacquinot (J) stop). Note that energy throughput is halved going from a setting of 0.5 to $0.25\,cm^{-1}$ on most instruments. This results in an additional degradation in the SNR by one-half, which requires increasing the number of co-averaged scans by another factor of 4 to obtain the original SNR ratio. Thus, to maintain the original SNR, the total measurement time increases by a factor of 16.

For intense Raman bands, whether they are broad or narrow, calibration curves of intensity measured at different resolutions and plotted versus concentration are useful to ascertain the optimum resolution for quantitative analysis.

Many chemometric analytical schemes tolerate lower resolution successfully. In general, the minimum limit of detection should be found for measurements made at the lowest possible resolution that adequately resolves the spectral features of the analyte from those of interfering species. Increasing the resolution beyond this point degrades the SNR.

3.1.5 *Electronic Components used in Spectrometry*

For references describing the electronic components of optical spectrophotometers, see modern textbooks on electronic instrumentation including references 22–24. Other volumes in this set discuss more details regarding electronic components for instrumentation, please see index for specific topics. Two special categories of electronic devices deserve mention when discussing spectrometers. The first category includes detectors and detector electronics that produce a current or voltage signal proportional to the amount of energy striking the detector. Detector stability is provided by appropriate electronic circuitry allowing the detector signal to be selectively amplified with the minimum amplification of noise or amplifier structured signal, thus electronic circuitry is used to enhance the signal-to-noise ratio of the detector signal.

Digital microcomputers comprise the second essential electronic element for modern Raman spectrometers. With the addition of appropriate software, sophisticated instrument control and data processing the usefulness and user friendliness of Raman instruments can be made faster, cheaper, and more accurate. A basic overview of electronic components is shown in Figure 3.14, and their associations with instrumentation as shown in Figures 3.1 through 3.5, and 3.8 within this chapter.

The controller board circuitry will contain the electronics to drive any servo motors for shutters or other moving parts, including monochromator grating motors and encoders, and for interferometer motions. The controller board also generally

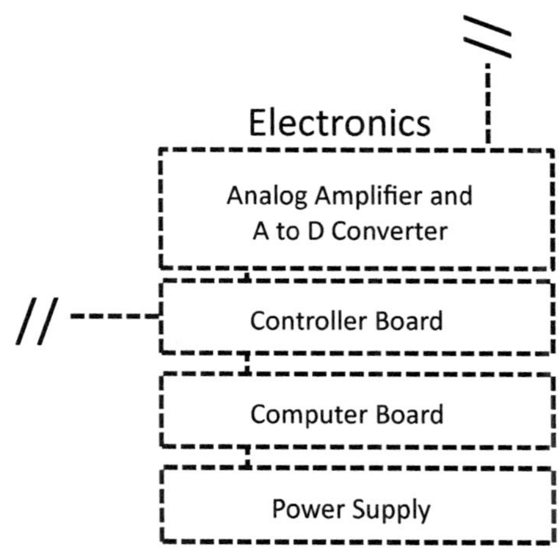

Fig. 3.14 Basic electronic systems for spectrometer systems.

includes power regulation for lasers, detectors, and sample motion accessories. The controller circuitry may also include diagnostic and temperature telemetry and storage of data or application programming interface (API) control algorithms. These are commands that may be used by the graphical user interface (GUI) software to drive and control the instrument functions. The power supply provides adequate and stable electrical power for the computer board and controller board electronics.

3.1.6 *Properties of Spectrophotometers*

3.1.6.1 *Entrance and Exit Pupils (Apertures)*

The entrance pupil refers to the size and location of the entrance aperture between the light/energy source and the remainder of an optical system. The exit pupil refers to the size and location of the exit aperture within an optical system or train just prior to the detector. Both the entrance and exit pupils affect the energy throughput and spectral resolution of a spectrometer based upon the clear area of each aperture.

3.1.6.2 *Band Pass (Bandpass) and Resolution*

Resolution using Raman spectrometers is often tested using ASTM E2529-06(2014), "Standard Guide for Testing the Resolution of a Raman Spectrometer, Book of Standards Volume: 03.06".[25] This practice assesses the spectral resolution of Raman spectrometers using either low-pressure arc lamp emission lines or a calibrated Raman band of calcite.

Note that the terms bandpass and resolution are used to express the capability of a Raman spectrometer to separate spectral bands or lines which are separated by some finite distance on the spectral energy scale. For an instrument that disperses energy over a prespecified spectral region of the electromagnetic spectrum the bandpass of a spectrometer is used to describe which portion of the spectrum can actually be isolated by the spectrometer in a "pure" wavelength form. The spectrometer bandpass is dependent upon the dispersion of the grating and the entrance and exit slit widths; or the construction of the interferometer, including the moving mirror velocity and distance traversed and the J-stop aperture (for a discussion on this aspect specific to interferometers, please refer to Volume 4, Chapter 3).

If the ideal Raman spectrometer were used to measure a bright line spectrum at a single wavelength (λ_1), the spectrum would appear as a single line (Figure 3.15). What really occurs when such a spectrum is measured using a conventional spectrometer is a broad band spectrum as shown in Figure 3.16. The spectrum assumes a somewhat Lorentzian shaped curve. This characteristic broadening of

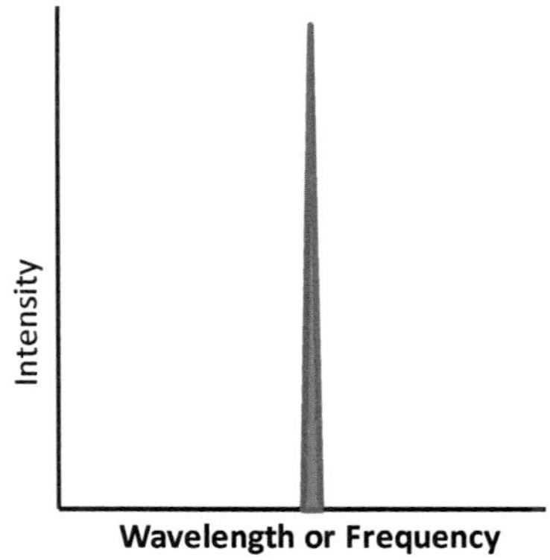

Fig. 3.15 The bright line emission spectrum at a single wavelength as it would appear in an ideal spectrometer.

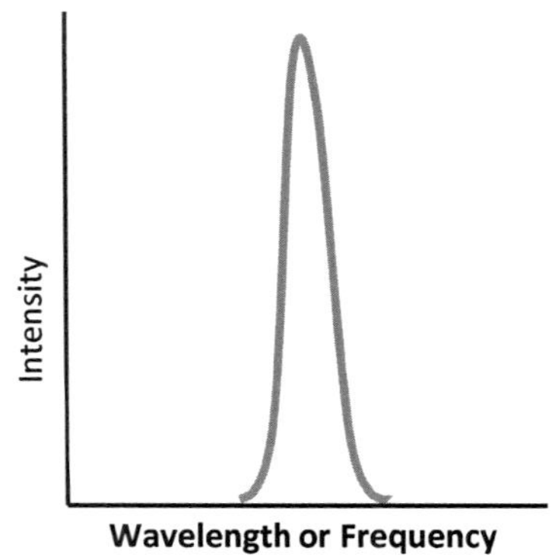

Fig. 3.16 Spectrum of a bright line emission source (e.g. deuterium lamp). The characteristic broadening is an illustration of the bandpass of a spectrometer.

a line spectrum through the spectrometer is an illustration of the spectrometer bandpass. The actual bandpass for any instruments assigned a value by determining the Full Width at Half Maximum (FWHM) height of the bell-shaped spectrum. Thus for the band in Figure 3.16 the FWHM could be empirically determined by finding the wavelength where maximum intensity occurs, and measuring the peak height at this position. This height measurement is divided in half and the bandwidth measured at this height on the band as illustrated in Figure 3.17. The actual shape of a band is the result of several instrumental characteristics including the overall quality of the optics and detector systems, the mirror motions, and the aperture of the J-stop (interferometer) or the entrance and exit slits (monochromator).

The resolution of a spectrometer can be defined as the minimum distance between two peaks that can be detected by the spectrometer under designated

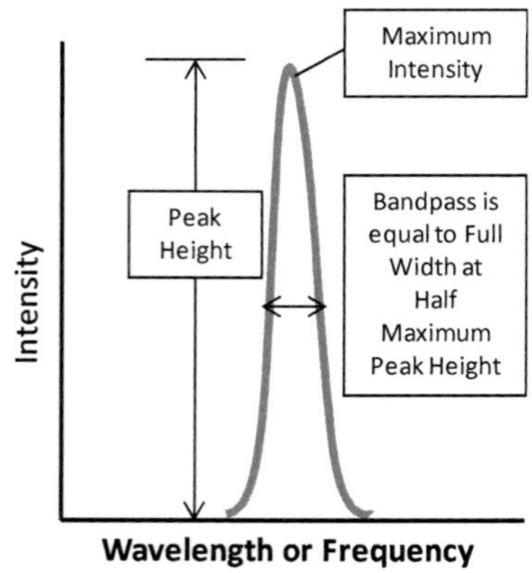

Fig. 3.17 Illustration of the determination of bandpass using the bell-shaped peak obtained by using a bright line source projected through a monochromator optical system.

operational performance settings. Resolution is calculated by multiplying the slit width (generally expressed in mm) by the dispersion of the monochromator (in nm per mm). Due to practical issues and non ideal optics, the actual resolution of a spectrometer must be slightly greater (poorer) than the theoretical value.

To summarize, bandpass and resolution are identical in practice. Only the resolution specification of a spectrometer is the expression of bandpass under the specified measuring conditions of an instrument dependent upon the slit width settings.

3.1.6.3 *Peak Positions*

There are several sections on selecting peak positions using spectroscopy covered in these volumes. The reader is referred to the topic of "peak picking" or "peak selection" in the index for greater treatment of the subject. Note that often the peak positions for specific functional groups or compound spectra do not agree precisely within the literature. There are many causes for this and a brief list of the important aspects for peak selection are given here. This entire subject could use a fundamental and extensive review article for each spectral region with careful referencing of source material. However one must remember that identifying peak positions accurately is a complex matter and subject to at least the following issues:

- Reference compound purity
- Matrix/solvent surrounding the analyte
- Temperature of sample
- Optical focus of the instrument at the sample (may change bandshapes)
- Hydrogen bonding (i.e., degree of hydrogen bonding)

- Lineshape and halfwidth of instrument (fixed or variable resolution over scan range)
- Lineshape and halfwidth of spectral band (spectral bands may be narrow or broad and resolve differently)
- Type of instrument used (FT or dispersive). FT instruments have constant resolution in frequency or wavenumber (cm^{-1}) space, but not in wavelength (nm) space. Alternatively, dispersive instruments have relatively constant resolution in wavelength (nm) space, but not in frequency or wavenumber (cm^{-1}) space. This creates a distortion and artifacts in bands and band positions when comparing spectra measured using these instrument types.
- Data preprocessing type (degree of smoothing and smoothing type changes peak positions.)
- The specific peak picking algorithm used (there are several common methods — all producing slightly different values)
- The data resolution (If there are not equal numbers of data points for a scan range, then smoothing or interpolation algorithms will give different comparative peak positions.)
- The accuracy of the data recording and if any shifts occur while translating or moving data formats affect peak positioning
- The instrument calibration method used prior to scanning samples (Different calibration methods and reference standards used to calibrate instruments produce different peak positions. One should note the accuracy and precision of wavelength tests prior to scanning samples for research or peak identification purposes.)
- Some peak positions in the literature are actually measured and some are calculated from measurements of the fundamental or other harmonics; this should be noted (the method for calculation should also be noted).

So all being said and done, much of the literature does not include this detailed information, and many instruments are not well calibrated prior to making measurements and collecting spectral libraries for peak identification.

3.1.6.4 *Attenuation Losses over Time*

Attentuation is a term often used to describe the light transmittance losses within (or through) an optical system. Losses in transmitted light through spectrometers are due to absorption, reflection, scattering, and optical misalignment; the losses can vary with temperature and wavelength. The quantity of optical loss can be expressed as an attenuation rate in decibels (dB) of optical power per unit distance (cm).

The usual attenuation occurs through inefficiencies in source launch optics, lenses, mirrors, bandpass filters, slits and J-stop features, detector focus, fiber optic losses, and the like. Typical changes in the energy transmitted through a spectrometer over time result from laser emission power loss, launch optics misalignment, temperature variations resulting in mechanical alignment changes, inefficient optical couplings within an optical path, aging of mirrored surfaces, clouding/aging of lenses, and typical soiled optical surfaces (e.g., lenses, mirrors, windows).

The losses in energy transmitted through a spectrometer can be calculated by using Equation 3.5, which demonstrates the loss of power in watts (or other units of energy) over time. To compute this, one would substitute the following variables into the equation: A_{db} is the attenuated power in Watts, P_I is the power output at the detector over time in Watts, and P_{Io} is the initial power output at the detector over time in Watts. The initial value is measured at the time of construction of the new instrument.

3.1.7 *Raman Instrument Comparison and Evaluation Methods*

A number of standard measurements may be completed that allow detailed comparison of overall instrument performance. These are delineated in this section. The following series of tests is used to qualify instrument performance and to determine which issues are problematic due to deficiencies in instrument design features. These tests are related to alikeness in measurement performance between instruments and to overall accuracy, and precision (as repeatability and reproducibility). The techniques used for evaluation of spectrometer performance are described in detail within Chapter 6, entitled *Data Analysis and Presentation.*

Recommended tests for Raman instruments include ASTM E1683-02(2014)e1, "Standard Practice for Testing the Performance of Scanning Raman Spectrometers," a practice used to determine if the Raman spectrometer has changed significantly over time such that measurement quality or integrity would be affected.[26] This practice applies for single-, double-, or triple monochromator instruments, but not for array detection or Fourier transform (FT-Raman) instruments. Another important document is the ASTM E1840-96(2014) entitled, "Standard Guide for Raman Shift Standards for Spectrometer Calibration." This guide describes the process of wavenumber calibration using seven pure compounds and one liquid mixture with peak wavenumber positions given by seven calibrated laboratories. A list of typical materials, wavenumber positions, and wavenumber certainties is given in Table 3.5.

Table 3.5 Typical materials used for Raman wavenumber calibration.

Material type	Peak position (cm^{-1})	Peak uncertainty (cm^{-1})
Silicon	520.7	±0.5
O_2 (Air)	1555.0	±1.0
N_2 (Air)	2332.0	±1.0
Calcite ($CaCO_3$)[25]	156	—
	281	±1.0
	713	—
	1085	±2.0
Cyclohexane[27]	284.10	±0.78
	426.30	±0.41
	801.30	±0.96
	1028.30	±0.45
	1157.60	±0.45
	1266.34	±0.94
	1444.40	±0.3
	2664.40	±0.42
	2852.90	±0.32
	2923.80	±0.36
	2938.30	±0.51
Indene[1]	730.4	±0.5
	1018.3	±0.5
	1205.6	±0.5
	1552.7	±0.5
	1610.2	±0.5
	2892.2	±1.0
	3054.7	±1.0

Note that more details for calibrating Raman spectrometers are given in Chapter 6 of this volume.[27]

3.1.8 *Advanced Raman Measurement Techniques*

Advanced methods and instrumentation for these methods are discussed in Chapters 1, 2, and 4 of this volume. The techniques described include Stimulated Raman CARS (Coherent Anti-Stokes Raman Spectroscopy), RRS/RR (Resonance Raman Spectroscopy), SERS (Surface-Enhanced Raman Spectroscopy), SERRS (Surface-Enhanced Resonance Raman Spectroscopy), SRS (Spontaneous Raman Spectroscopy), SRS/SPSRS (Stimulated Raman Spectroscopy), SORS (Spatially Offset Raman Spectroscopy), TERS (Tip-Enhanced Raman Spectroscopy), and SPPERS (Surface Plasmon Polariton (SPP) Enhanced Raman Spectroscopy).

References

1. J.R. Ferraro and K. Nakamoto, *Introductory Raman Spectroscopy*, 1st edn. (Academic Press, Boston, 1994).
2. A.A. Michelson, The Relative Motion of the Earth and the Luminiferous Ether, *American Journal of Science* **22** (1881) 120–129.
3. A.A. Michelson and E. Morley, On the Relative Motion of the Earth and the Luminiferous Ether, *American Journal of Science* **34**(203) (1887) 333–345.
4. D.M. Livingston, *The Master of Light: A Biography of Albert A. Michelson*, 1st edn. (The University of Chicago Press, 1979).
5. A.A. Michelson, *Light Waves and Their Uses*, 1st edn. (The University of Chicago Press, 1902).
6. T. Hirschfeld and B. Chase, FT-Raman spectroscopy: development and justification, *Appl. Spectrosc.* **40**(2) (1986), 133–137.
7. Horiba: http://www.horiba.com/scientific/products/raman-spectroscopy/raman-academy/
8. R. Haisheng, A. Liu, R. Jones, O. Cohen, D. Hak, R. Nicolaescu, A. Fang, and M. Paniccia, An all-silicon Raman laser, *Nature* **433**(7023) (2005), 292–294.
9. R. Haisheng, S. Xu, O. Cohen, O. Raday, M. Lee, V. Sih, and M. Paniccia, A cascaded silicon Raman laser, *Nature Photonics* **2**(3) (2008), 170–174.
10. H. Chuan and T.H. Chyba, Solid-state barium nitrate Raman laser in the visible region, *Optics Communications* **135**(4) (1997), 273–278.
11. D.L. Gerrard and H.J. Bowley, *Instrumentation for Raman spectroscopy,* (Springer Berlin Heidelberg, 1989).
12. C.A. Murray and S.B. Dierker, Use of an unintensified charge-coupled device detector for low-light-level Raman spectroscopy, *JOSA A* **3**(12) (1986), 2151–2159.
13. M. Bowden, D.J. Gardiner, G. Rice, and D.L. Gerrard, Line-scanned micro Raman spectroscopy using a cooled CCD imaging detector, *Journal of Raman Spectroscopy* **21**(1) (1990), 37–41.
14. S. Raman, C. Yonezawa, H. Matsue, H. Iimura, and N. Shinohara, Efficiency calibration of a Ge detector in the 0.1–11.0 MeV region, *Nuclear Instruments and Methods in Physics Research Section A: Accelerators, Spectrometers, Detectors and Associated Equipment* **454**(2) (2000), 389–402.
15. R.C. Jones, The Ultimate Sensitivity of Radiation Detectors, *Journal of the Optical Society of America* **37**(11) (1947), 879–888.
16. R.C. Jones, A New Classification System for Radiation Detectors, *Journal of the Optical Society of America* **39**(5) (1949), 327–341.
17. R.C. Jones, Erratum: The Ultimate Sensitivity of Radiation Detectors, *Journal of the Optical Society of America* **39**(5) (1949), 343.
18. R.C. Jones, Factors of Merit for Radiation Detectors, *Journal of the Optical Society of America* **39**(5) (1949), 344–356.
19. Roy Cain, personal communications, *A User's Guide to Fourier Transform Infrared Spectrometers*, (ASTM, March 2006). Roy was an active leader in ASTM activities and was a recipient of the ASTM Fellow, Award of Merit in (1999).
20. ASTM (American Society for Testing and Materials) Method ASTM E1421-99(2009), Standard Practice for Describing and Measuring Performance of Fourier Transform Mid-Infrared (FT-MIR) Spectrometers: Level Zero and Level One Tests, ASTM International, 100 Barr Harbor Drive, PO Box C700, West Conshohocken, PA, 19428-2959 USA (2009).

21. P.R. Griffiths and J.A. de Hasseth, *Fourier Transform Infrared Spectrometry,* 2nd edn. (Wiley-Blackwell, 2007).
22. D.M. Buchla and W. McLachlan, *Applied Electronic Instrumentation and Measurement*, 1st edn. (Prentice Hall, 1991).
23. J. Turner and M. Hill, *Instrumentation for Engineers and Scientists* (Textbooks in Electrical & Electronic Engineering), Textbooks in Electrical & Electronic Engineering, Book 8, 1st edn. (Oxford University Press, 1999).
24. G.S. Sawhney, *Biomedical Electronics and Instrumentation Made Easy*, Paperback, 1st edn. (I. K. International Publishing House, 2011).
25. ASTM E2529-06(2014), Standard Guide for Testing the Resolution of a Raman Spectrometer, Book of Standards Volume: 03.06, ASTM International, 100 Barr Harbor Drive, PO Box C700, West Conshohocken, PA, 19428-2959 USA (2014).
26. ASTM E1683-02(2014)e1, Standard Practice for Testing the Performance of Scanning Raman Spectrometers, Book of Standards Volume: 03.06, ASTM International, 100 Barr Harbor Drive, PO Box C700, West Conshohocken, PA, 19428-2959 USA (2014).
27. ASTM E1840-96(2014), Standard Guide for Raman Shift Standards for Spectrometer Calibration, Book of Standards Volume: 03.06, ASTM International, 100 Barr Harbor Drive, PO Box C700, West Conshohocken, PA, 19428-2959 USA (2014).

Chapter 4

Sampling Considerations for Raman Spectroscopy

Contents

4.1 Introduction 100
4.2 General Considerations 100
4.3 Sample Type versus Sampling Method 101
4.3.1 Gases and Vapors 102
4.3.2 Liquids 103
4.3.2.1 General Liquid Sampling 103
4.3.2.2 Volatile Materials 104
4.3.2.3 Viscous Materials 105
4.3.2.4 Pastes, Emulsions, and Slurries 105
4.3.3 Solids 106
4.3.3.1 General 106
4.3.3.2 Powders 107
4.3.3.3 Low Levels of Analyte in Solution 108
4.3.3.4 Amorphous Materials (Organic), Continuous Sheets, and Polymeric Films 109
4.3.3.5 Lumps, Granules, or Pellets 110
4.3.3.6 Intractable Materials 110
4.4 Special Raman Sampling Techniques 111
4.4.1 Stimulated Raman 111
4.4.2 CARS (Coherent Anti-Stokes Raman Spectroscopy) 112
4.4.3 RRS/RR (Resonance Raman Spectroscopy) 112
4.4.4 SERS (Surface-Enhanced Raman Spectroscopy) 112
4.4.5 SERRS (Surface-Enhanced Resonance Raman Spectroscopy) 112
4.4.6 SRS (Spontaneous Raman Spectroscopy) 113
4.4.7 SRS/SPSRS (Stimulated Raman Spectroscopy) 114

4.4.8 SORS (Spatially Offset Raman Spectroscopy) 116
4.4.9 TERS (Tip-Enhanced Raman Spectroscopy) 116
4.4.10 SPPERS (Surface Plasmon Polariton (SPP) Enhanced Raman Spectroscopy) 116
4.4.11 Summary 116
References 118

4.1 Introduction

The quality of any spectrum, including Raman spectra, is strongly dependent on precise sample preparation and the specific optical interface between the sample and the Raman spectrometer. There are a variety of sampling accessories and techniques that allow optimization of the Raman signal obtained from any particular sample.

This chapter is intended to serve as a basic guide to provide instructions for the optimum method for Raman sampling given most sample types. For many applications, the basic measurement techniques should be sufficient. When measuring spectral data, it is essential to note the specific mathematical processing used and any numerical terms or factors that are applied. If spectral subtraction is used, it is strongly recommended that the original sample spectrum, and the spectrum used for subtraction are retained in a stored database.

4.2 General Considerations

Raman spectra are the spectra of molecular vibration covering the general spectral region of 3500 to 100 wavenumbers (cm^{-1}). Often Raman spectra are compared to infrared spectra for identification of molecular structures and for this purpose the overlap region for Raman and infrared is typically 4000 to 400 wavenumbers (cm^{-1}).

Raman interactions with matter are recorded as Raman intensity (in counts) versus wavenumbers (cm^{-1}), either as a direct or indirect measurement. Laser excitation is directed onto the surface of the sample where inelastic collisions occur between the laser excitation photons and the sample molecules and atoms. The Raman scatter is directed away from the sample and collected by a series of sampling optics directly from the sample surface or from a sample normally contained within a glass vial. The Raman scatter cross section of molecules is matrix dependent and as such a series of reference calibration samples are prepared when performing quantitative analysis. Qualitative analysis is accomplished by comparing a test spectrum to a library of Raman spectra that have been carefully measured.

Sections describing the algorithmic methods useful for qualitative analysis may be found in the main index for this volume set.

It is assumed that the reader has a basic knowledge of optical spectroscopy and the molecular theory of chemistry. Raman spectroscopy is one of the analytical techniques available for the measurement of scattered radiation as a consequence of collisions with vibrating molecular bonds. The reader is directed to a standard reference text on organic chemistry or spectroscopy[1,2] or to texts focused on the structure/spectra relationships of molecular compounds to understand the details of molecular vibrations.[3–10] There are several sections within this set of volumes specifically referring to this topic for Raman spectroscopy, such as Chapter 1, Section 1.4 and Chapter 7.

4.3 Sample Type versus Sampling Method

Raman spectroscopy may be applied to most sample types in any physical state, or matrix. In this chapter section, sampling will be discussed in terms of physical state and/or physical nature of the sample. One point that will be emphasized throughout is the importance of documentation of all experimental procedures, sampling methods and conditions, making note of any abnormalities observed during sample preparation procedures.

Aqueous solutions or slurries are more readily measured using Raman spectroscopy than by using infrared or near infrared techniques. This is because O-H stretching vibrations are poor scattering centers and give only a small Raman signal. Water gives a large infrared and near infrared absorption signal and as such measurements of aqueous samples are challenging for these other techniques.

Note if any sample is not homogenous and is comprised of more than one phase, it is useful to obtain spectra from each of the separated phases. Limited penetration of excitation energy requires that poorly mixed samples (i.e., non-homogeneous) must be well mixed or separated and the individual components measured in order to obtain a useful set of spectral measurements. Techniques such as filtration, decanting, and centrifuging may be used to ensure representative sampling of a particular phase. If separating samples, record the percentage of each phase of material as well as separate spectra for each phase. One may record the various structural layers of material, their appearance, and densities if needed to provide a more complete record of analysis for any multi-phase sample type. This procedure also applies to layered solids, such as multi-layered polymer films, multi-layer tablets, capsules, and layered biological samples. In this way a sort of manual mapping technique is provided, as well as a quantitative evaluation that is more accurate than a single measurement of the layered material.

4.3.1 *Gases and Vapors*

The sampling for gases and vapors is relatively straightforward, and the main issues in selecting the best sampling conditions are pressure (molecules per unit volume = concentration), temperature, and the relative concentrations of the analytes for mixed gases. For concentrated samples it is normal to use a standard transmission cell (or gas handling cell) with a relatively small diameter, typically in the range of 1–5 cm, the final choice being dependent on the concentration of the most important analytes.

For gases, it is important to record the pressure and partial pressures that are used, especially when performing a quantitative measurement. This also applies if a non-scattering diluent gas, such as helium, is used. Diatomic gases, such as O_2 and N_2 have strong peaks in Raman spectroscopy and are used as reference samples for wavenumber calibration. Figure 4.1 demonstrates a configuration for gas analysis using a gas cell and Raman spectrometer.

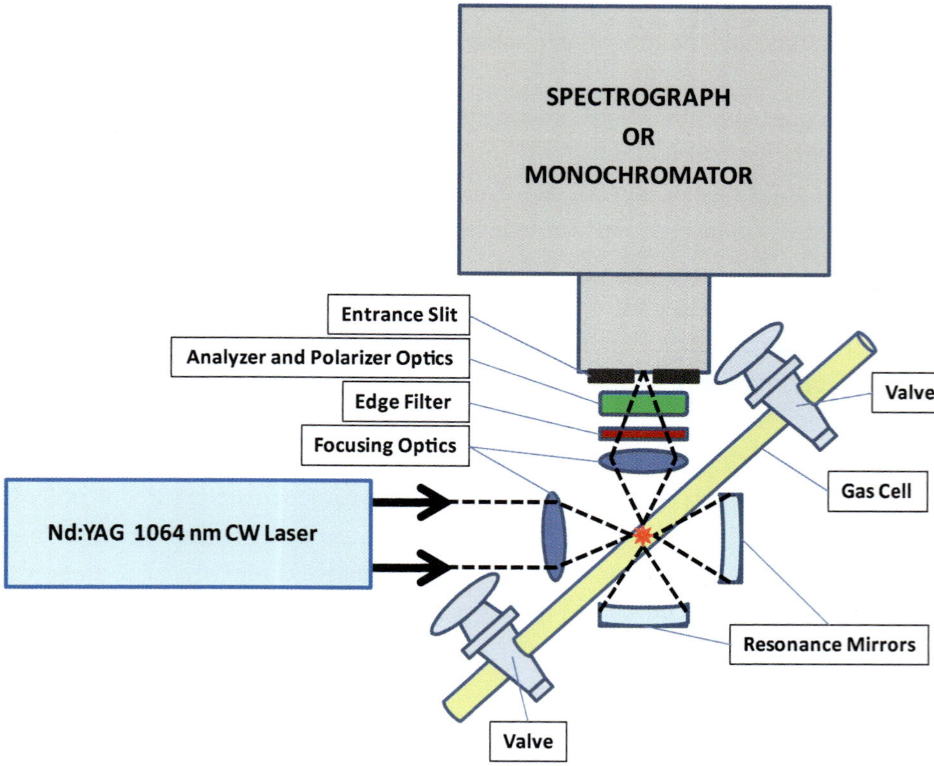

Fig. 4.1 Raman configuration for gas sample measurements.

4.3.2 *Liquids*

Raman is compatible with quartz, glass, and silicate cell windows. Raman scatter for O-H in glass and silica is negligible.

4.3.2.1 *General Liquid Sampling*

Traditionally, liquids have been considered to be relatively easy to sample and measure using Raman, except for volatile, low-molecular weight organic compounds, which evaporate quickly or have large Raman scatter intensities. Sampling these materials with Raman requires a small sample in a sealed glass vial of a few mm in diameter (1 to 10 mm), which typically contain a few mL (0.5 to 10 mL) of liquid (Figure 4.2). The exact pathlength is not critical, since Raman measurements rely on the scattering phenomena occurring only in the first few microns depth of the sample for most organic materials, as shown in Figure 4.3.

For liquids in general, it is very helpful to note if the sample is known to be a single material, a mixture, or a dilute solution. If the sample is a dilute solution, the major solvents, if known, should also have their spectra recorded, preferably during the same time frame. In this way, it is possible to differentiate the solute and solvent, especially with the aid of computer-based spectral subtraction techniques.

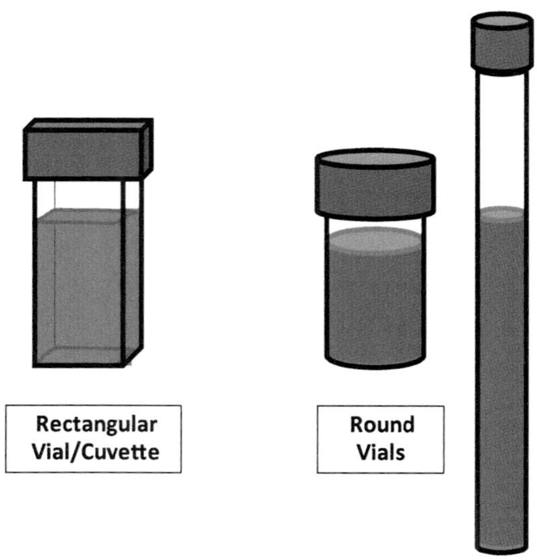

Fig. 4.2 Illustration of typical glass sampling vial for Raman measurements.

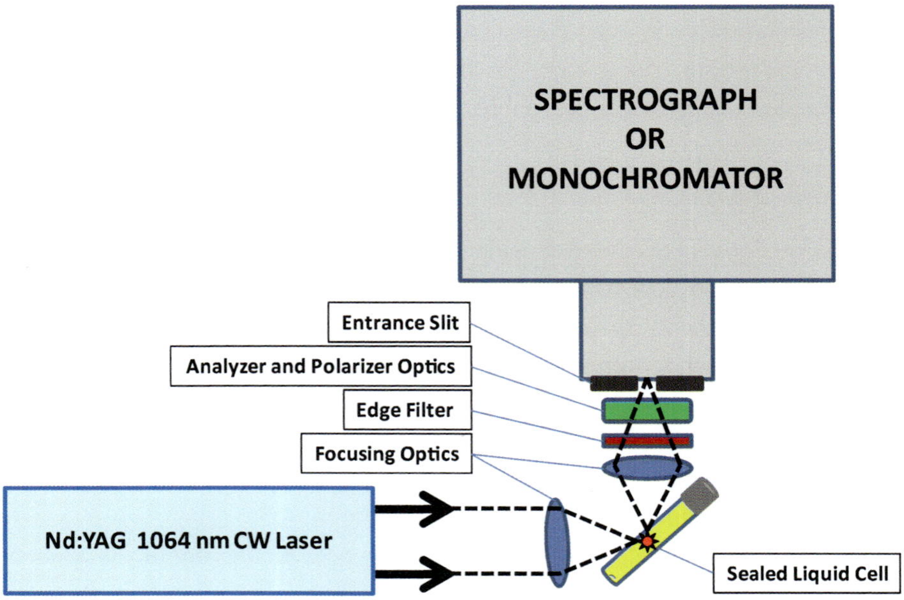

Fig. 4.3 Raman configuration for liquid sample measurements.

4.3.2.2 *Volatile Materials*

There are several important issues to consider with volatile liquids, such as the need to use a sealed glass cell, potential loss of sample during analysis, protection of the instrument (in particular sensitive optics), and general fire and safety precautions. The traditional approach to measuring volatile liquids and solutions (based on volatile solvents) is to use a sealed, vial. If the sample is extremely volatile, then thermal control of the vial (cooling) may be advised to prevent leakage or damage to the seals of the vial itself.

As commented upon earlier, it is possible to analyze the sample in the vapor phase by placing it within a heated gas cell. The latter suggestion is useful when the total material is volatile or when there is the need to obtain information about volatile components only. If a vapor-phase approach is used, the sample temperature in the gas cell must ideally be known because this will impact the appearance of the final spectrum. Note that the appearance of a vapor-phase spectrum may be significantly different than the corresponding spectrum in the condensed (i.e., liquid) phase. Raman spectral collections of common organic compounds in the condensed phase are available in both hardcopy and digital formats.[11–13]

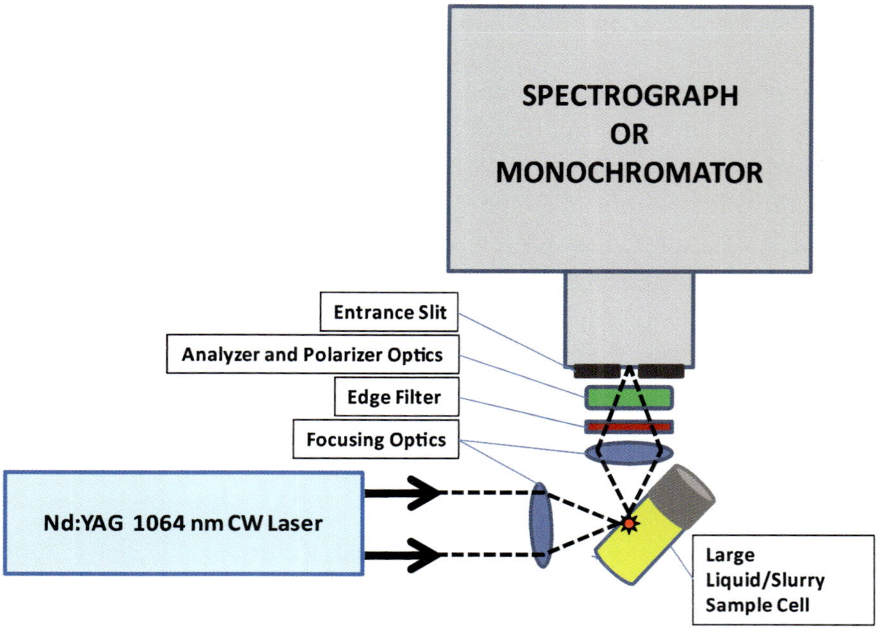

Fig. 4.4 Standard Raman measurement with large liquid/slurry vial.

4.3.2.3 *Viscous Materials*

A simple and convenient method for measuring viscous, nonvolatile materials is to pack the sample into a glass vial, the same as those used for liquid sampling with a larger diameter for ease of filling. Large sampling vials may be used for samples difficult to place into small diameter vials. The Raman measurement is independent of the physical pathlength of the sample, see Figure 4.4.

4.3.2.4 *Pastes, Emulsions, and Slurries*

Dependent upon the constituents, pastes and slurries generally may be handled in a similar manner as those proposed for viscous liquids. Always be aware of potential homogeneity problems. Also be cautious of the effect of any abrasive materials, retained in suspension, on optical window or crystal surfaces. This is particularly so for any system designed for continuous flow. Constant abrasion will result in light losses (due to increased light scattering) and eventually permanent damage to optical surfaces in contact with the abrasive.

Emulsions containing water are not a problem for Raman measurements. Make sure that separation does not occur within the sample vial or inaccurate spectra will be obtained. Sometimes the polar nature of the optical element will cause segregation to occur. Common examples are emulsions, such as mayonnaise and protein-containing materials; in this later case, the protein may actually form a coating on the optical surface. Also, some emulsions are stabilized by amino or ammonium compounds and these will be detected using Raman.

Liquid samples may also be measured using a rotating sampling device, which uses centrifugal force to maintain the sample at the window for the measurement cycle. Such a device provides for the sample to be measured without inordinate heating by the laser. Volatile samples, biological samples, photosensitive samples, and thermally sensitive samples may be safely measured using such a sampling device (Figure 4.5).

4.3.3 *Solids*

4.3.3.1 *General*

Of the three physical phases, the solid phase poses the greatest challenge for both infrared and Raman spectroscopy. The main issues for such samples are how to get effective interaction between the Raman excitation energy and the sample; how to minimize artifacts and distortions caused by refractive index, scattering,

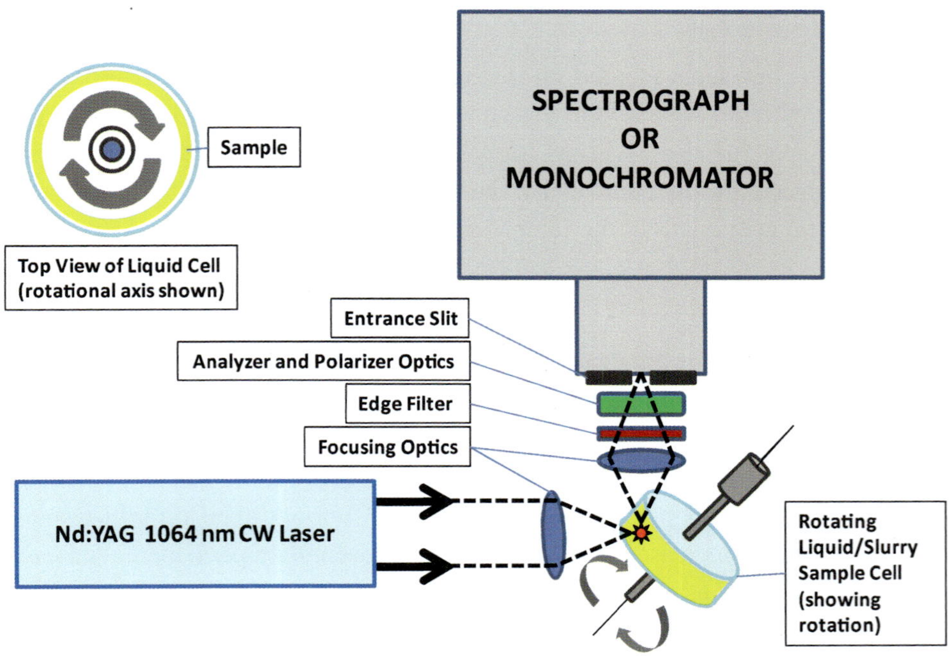

Fig. 4.5 Standard Raman measurement with rotating liquid/slurry cell.

and polarization effects; and how to measure the sample reproducibly and in a manner that is representative of the sample. A wide range of sampling techniques are available for solids, and their suitability is dependent on the nature and physical form of the sample. These include measuring the sample as a powder, diluted powder, pellet, or as a deposited film.

4.3.3.2 *Powders*

A standard approach for grindable powders is the preparation of a compressed potassium halide (KBr or KCl) pellet. The objective is to grind the sample, ideally, down to a sub-micrometer particle size and to disperse the finely ground material in a KBr or KCl matrix, which has a closely matched refractive index. This approach is intended to minimize or eliminate the impact of light scattering caused by individual particles from the sample. If the particle size is similar or larger than the wavelength of the excitation radiation, then a significant portion of the light may be subject to elastic scattering. This is scattered light with no information regarding the sample chemistry. Producing a smaller particle size than the wavelength of the excitation energy will reduce elastic scattering and allow better mixing with the alkali halide diluent.

In the case of compressed halide pellets, the sample is first ground to the desired nominal particle size and is then mixed (with partial grinding) with the halide salt (usually KBr but sometimes KCl). The mixture is then compressed in a special high-pressure die, with a nominal pressure of approximately 10 tons (for a standard 13-mm-diameter pellet). It is important to ensure that there is no ionic interaction between the sample and the potassium halide. Also, be aware of materials that undergo phase changes when subjected to the extreme pressures of the compressed pellet technique. Make sure that the potassium halide is kept in a clean, dry area, and if there is any doubt about its quality, run a blank pellet without the sample present and retain the spectrum for later comparison.

Coarse or hard powders are not well served by either the compressed pellet or mull techniques mainly because of difficulties associated with grinding these materials. In such situations, the best approaches require the use of an accessory, such as a direct sample measurement. Mixing the sample with a potassium halide may be beneficial to reduce anomalous optical behavior at the surface of the sample. This is especially helpful with inorganic compounds or other materials that have high absorptivities or have a high index of refraction. As a rule of thumb, organic compounds can be measured in about a 10% mixture and inorganics with approximately 1 or 2% mixture (in KBr or KCl). See Figure 4.6 for an illustration of solids measurements.

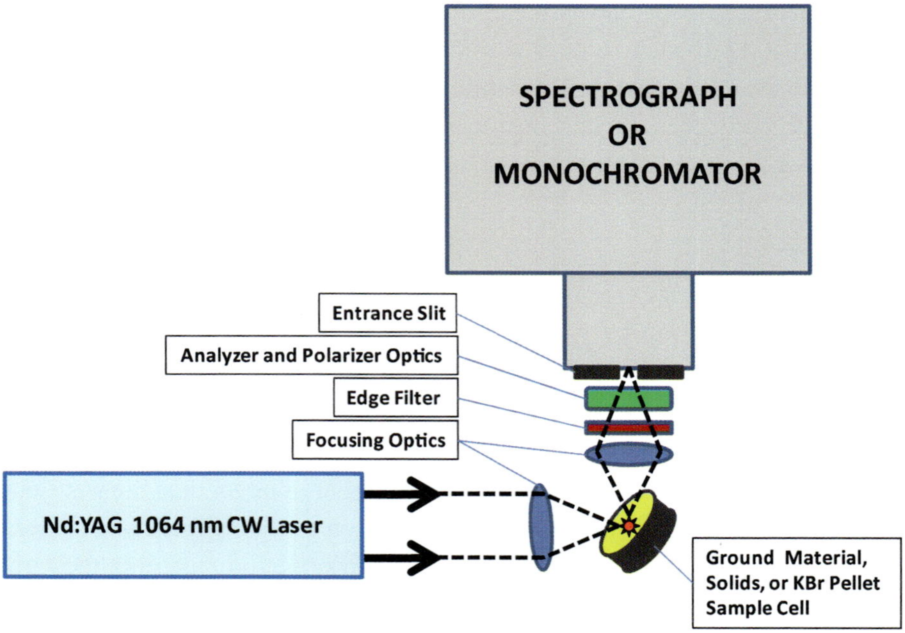

Fig. 4.6 Standard Raman with solids or KBr pellet cell.

Solid samples sensitive to heat or light degradation may also be measured using a rotating solid sample cell (ring shaped sample press). With this accessory the sample is pressed into a rotating disk such that the excitation light dwells only for a fraction of a second on the sample surface allowing an 'averaged' Raman spectrum for the sample while mitigating photo-bleaching or thermal damage to the sample. See Figure 4.7 to illustrate this sampling technique.

4.3.3.3 *Low Levels of Analyte in Solution*

Low levels of analyte in aqueous solution may be measured using a Teflon waveguide. The use of a capillary tube made from a Teflon tube allows the Raman sampling to occur along an extended pathlength of material such that low levels of analytes may be detected in an aqueous medium. This technique is most applicable to environmental analysis where low levels of solvents are to be analyzed in wastewater or naturally occurring lakes, streams, and oceans. This technique requires special launch optics and detection optics to excite the sample down the capillary tube and collect the signal at the opposite end of the tube, see Figure 4.8 for an illustration.

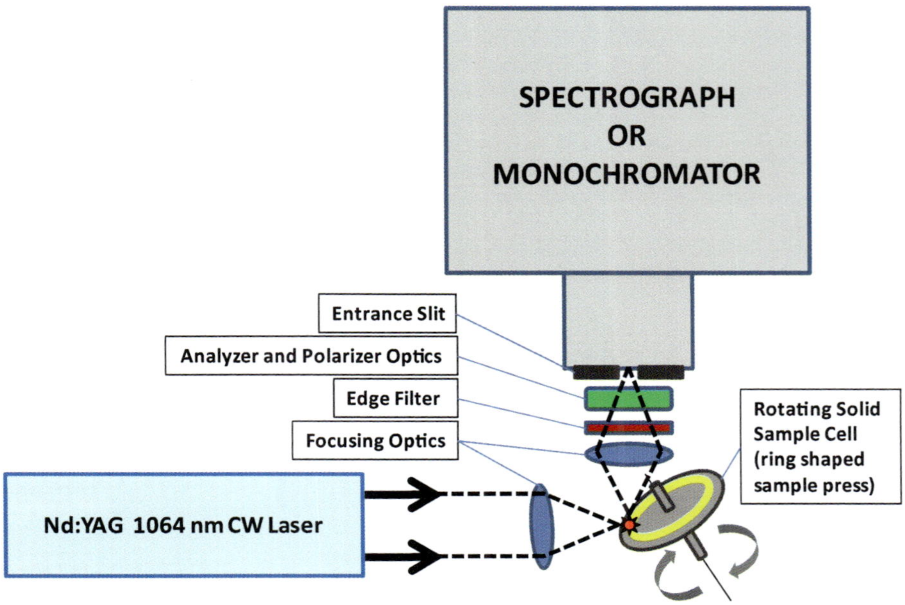

Fig. 4.7 Standard Raman with ring solids rotating cell.

4.3.3.4 *Amorphous Materials (Organic), Continuous Sheets, and Polymeric Films*

Moldable materials may be prepared as self-supporting films for direct Raman measurements, such as shown in Figure 4.6. Materials not pre-existing as films, such as polymer pellets, may be molded or hot-pressed; accessories are available for producing films with well-defined thickness. When dealing with a coating on a metal surface — a metal sheet or a foil — direct Raman measurements may be used.

Less common situations to consider are when studies of surface contamination are required and when the surface cannot be separated and brought to the instrument. In the first example, a practical solution is to grind a small amount of alkali halide powder (KBr or KCl) at the point of the contamination. This will result in the transfer of some of the contaminant to the alkali halide, and this in turn may be examined using the compressed pellet technique. A way to assist this surface extraction is to add a few drops of a solvent, such as methylene chloride, in the area of the contamination. Sampling may also be performed by gently abrading the surface with silicon carbide paper. This action will transfer a small amount of the surface to the abrasive paper, and this in turn may be examined by direct Raman measurement of the surface of the abrasive paper.

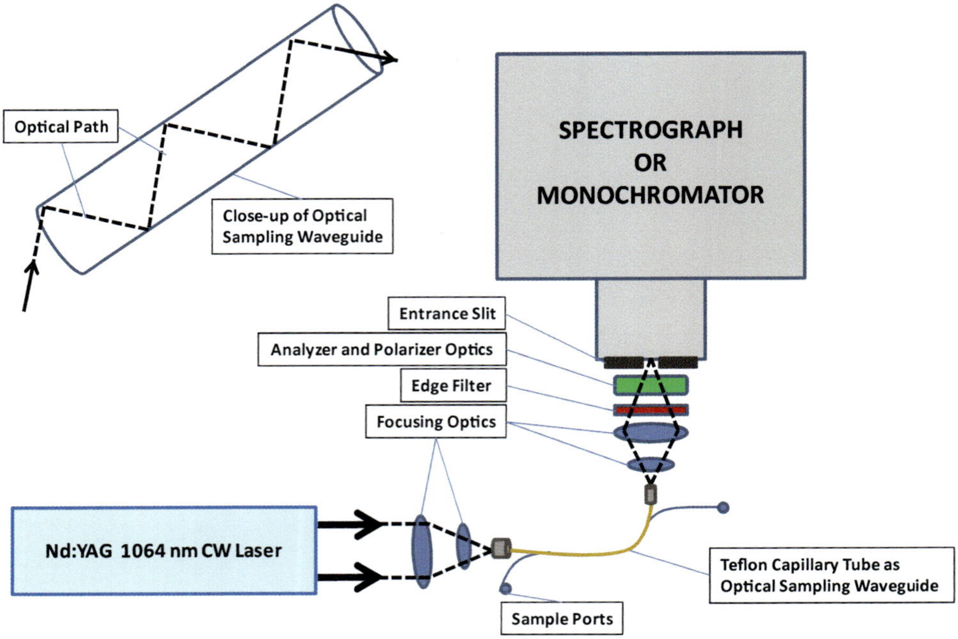

Fig. 4.8 Raman sampling with teflon liquid waveguide.

4.3.3.5 *Lumps, Granules, or Pellets*

This class of solids is an extension of the sample types already discussed, and many of the procedures already highlighted may be used here. If the material dissolves in a suitable solvent, then liquid sampling may be used. Moldable materials, such as polymer pellets, may be prepared as hot-pressed films, with care taken to ensure that material does not thermally degrade. Grindable materials can be handled as previously discussed for powders using the compressed halide pellet.

Elastomeric or rubber-like materials may be directly measured using Raman spectroscopy. Fibrous materials may be analyzed by a number of different methods, including the standard method for sample preparation — compressed halide pellets. In this latter case, normally the fiber structure is destroyed during sample preparation. Alternatively, one can consider using Raman microscopy, it is possible to study the orientation characteristics of the fiber material and to correlate this information to certain mechanical properties of the fiber.

4.3.3.6 *Intractable Materials*

Elastomeric materials filled with inorganics or carbon black can be problematic to obtain high quality spectra, and they often are considered to be intractable. Raman microscopy or an infrared method may be most applicable, as described in Volume 4, Chapter 4.

One other technique, previously described, is the abrasion of the surface sample with a silicon carbide paper. The abraded material may be collected as a dust and analyzed as a compressed alkali halide pellet. Alternatively, the Raman spectrum of the sample from the surface of the abrasive paper may be obtained directly.

4.4 Special Raman Sampling Techniques

This section provides information for special Raman sampling techniques and the specific studies where these techniques may be applied. The following illustrations delineate the specific purpose of each Raman technique used to enhance signal, isolate specific molecular structures, and to profile spatial and chemical information using imaging or mapping techniques. See Chapter 2 of this volume for references and greater details on these measurement techniques.

4.4.1 *Stimulated Raman*

Used for amplification of the Raman signal. When combined with polarization (s and p) the technique effectively measures molecular alignment and orientation properties of molecules (Figure 4.9).

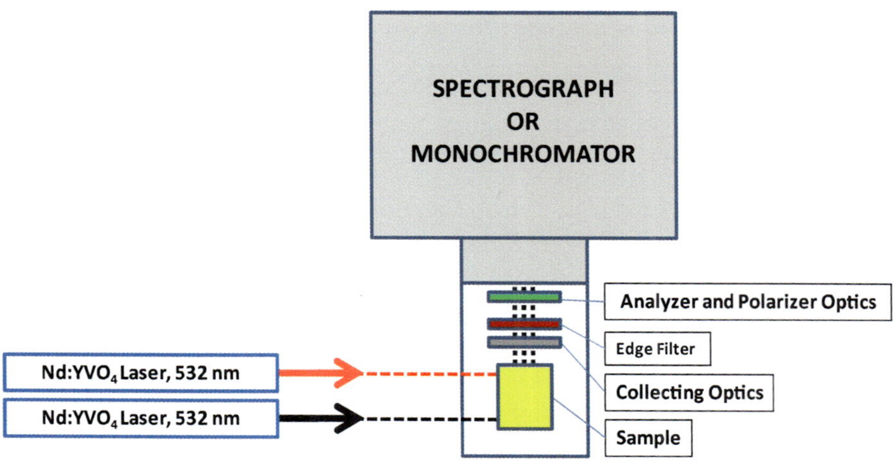

Fig. 4.9 Illustration of sampling for stimulated Raman (SRS).

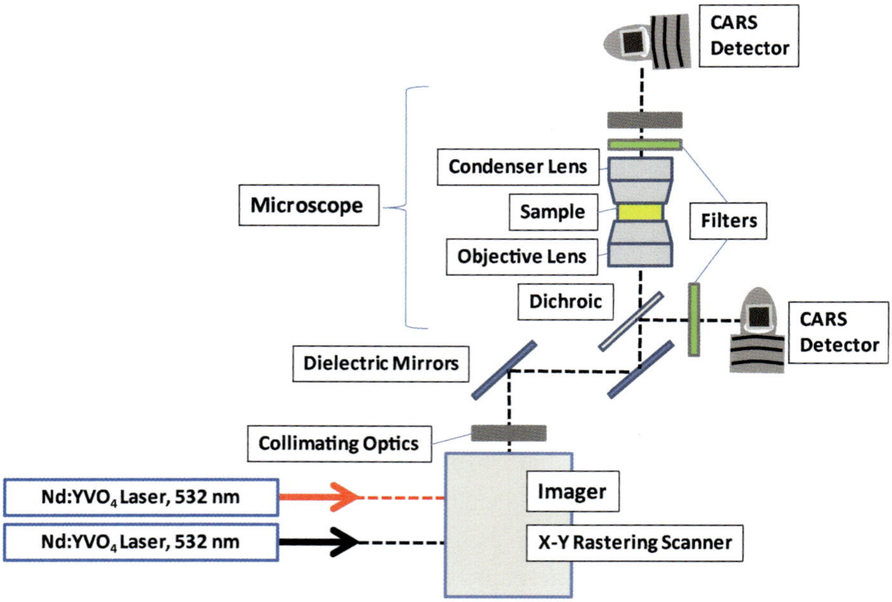

Fig. 4.10 Illustration of sampling for CARS (Coherent Anti-Stokes Raman Spectroscopy).

4.4.2 *CARS (Coherent Anti-Stokes Raman Spectroscopy)*

Used to enhance signal at specific spatial locations with a 10^5 signal enhancement over conventional Raman. CARS imaging is possible in real-time for mapping molecular vibrations at known locations for image reconstruction. The use of dyes is not required (Figure 4.10).

4.4.3 *RRS/RR (Resonance Raman Spectroscopy)*

This technique provides "isolation" of specific molecular vibrational modes of interest for the study of specific Raman active bands from large and complex molecules (Figure 4.11).

4.4.4 *SERS (Surface-Enhanced Raman Spectroscopy)*

This method enhances Raman signal by 10^{10} to 10^{11} for measurement of low levels of analyte (Figure 4.12).

4.4.5 *SERRS (Surface-Enhanced Resonance Raman Spectroscopy)*

This Raman technique enhances the signal of a specific molecular vibration under study (Figure 4.13).

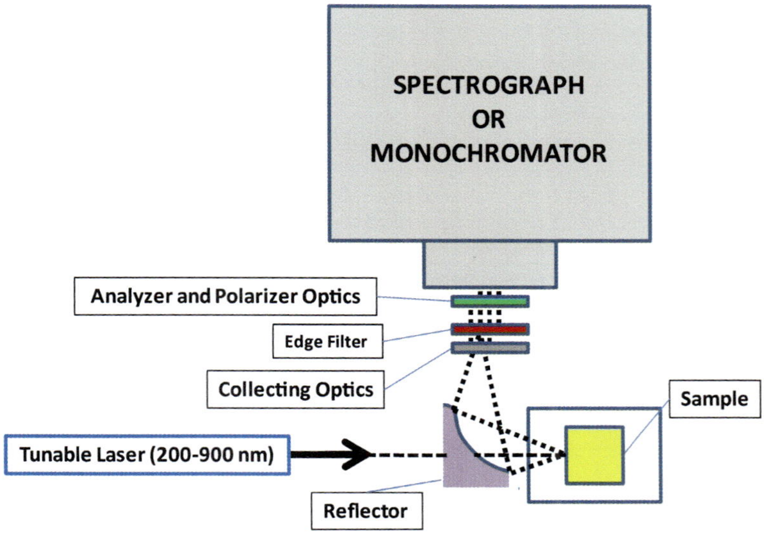

Fig. 4.11 Illustration of sampling for RRS/RR (resonance Raman spectroscopy).

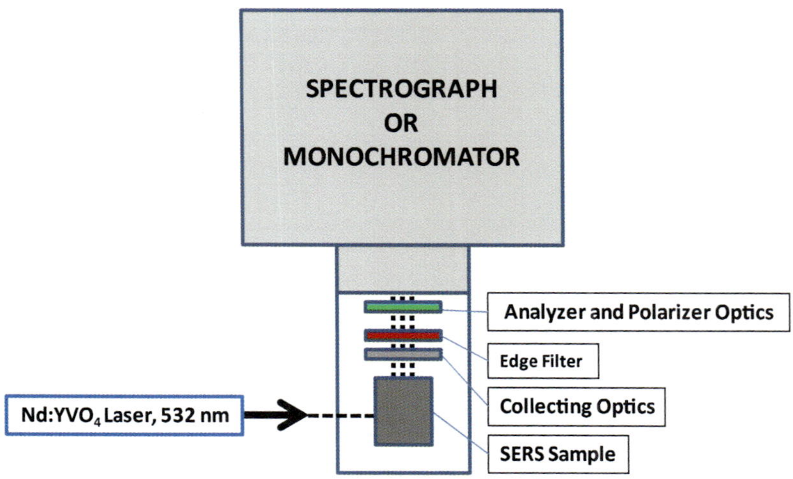

Fig. 4.12 Illustration of sampling for SERS (Surface-Enhanced Raman Spectroscopy).

4.4.6 *SRS (Spontaneous Raman Spectroscopy)*

The Raman scattering signal is enhanced to 40–50% of the excitation laser energy. The overall Raman signal is enhanced by 4×10^4 to 5×10^4 over traditional spontaneous Raman spectroscopy (Figure 4.14).

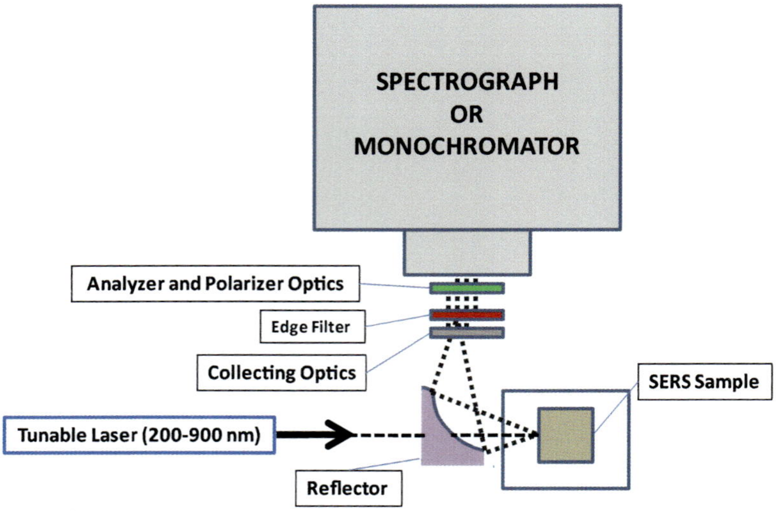

Fig. 4.13 Illustration of sampling for SERRS (Surface-Enhanced Resonance Raman Spectroscopy).

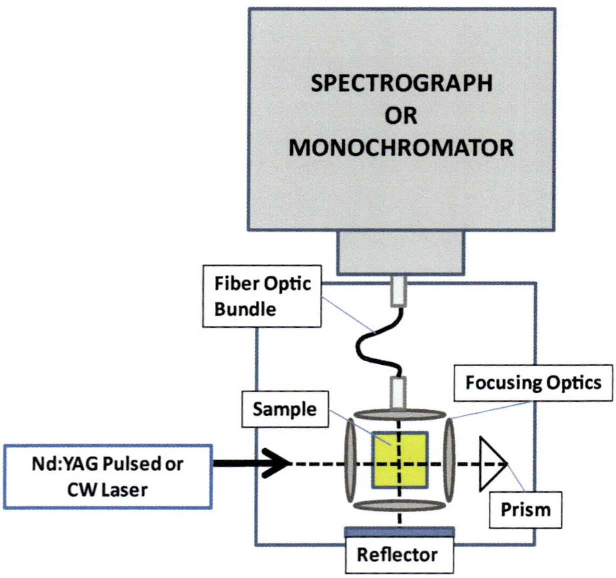

Fig. 4.14 Illustration of sampling for SRS (Spontaneous Raman Spectroscopy).

4.4.7 *SRS/SPSRS (Stimulated Raman Spectroscopy)*

The Raman scattering signal is enhanced to 40–50% of the excitation laser energy. The overall Raman signal is enhanced by 4×10^4 to 5×10^4 over traditional spontaneous Raman spectroscopy (Figure 4.15).

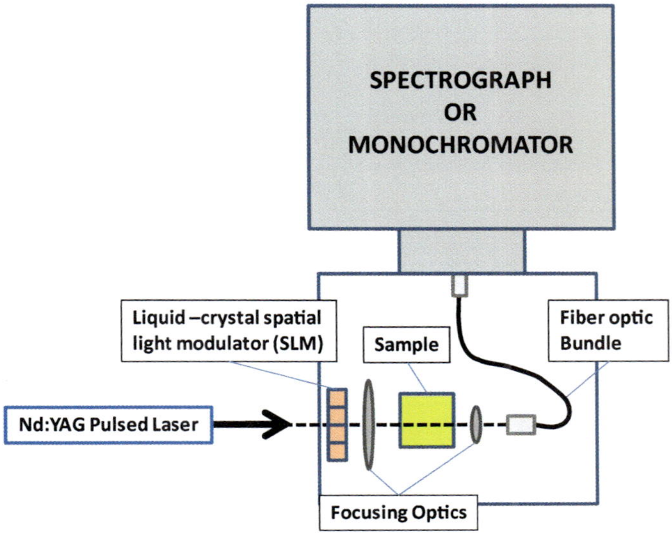

Fig. 4.15 Illustration of sampling for SRS/SPSRS (Stimulated Raman Spectroscopy).

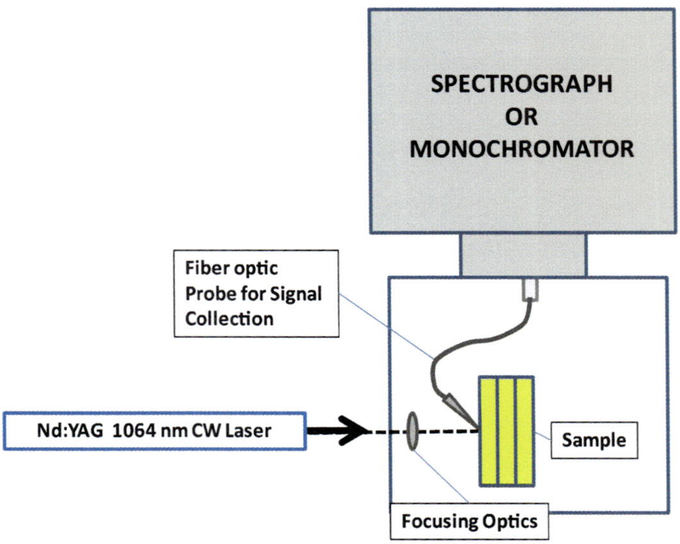

Fig. 4.16 Illustration of sampling for SORS (Spatially Offset Raman Spectroscopy) (zero position).

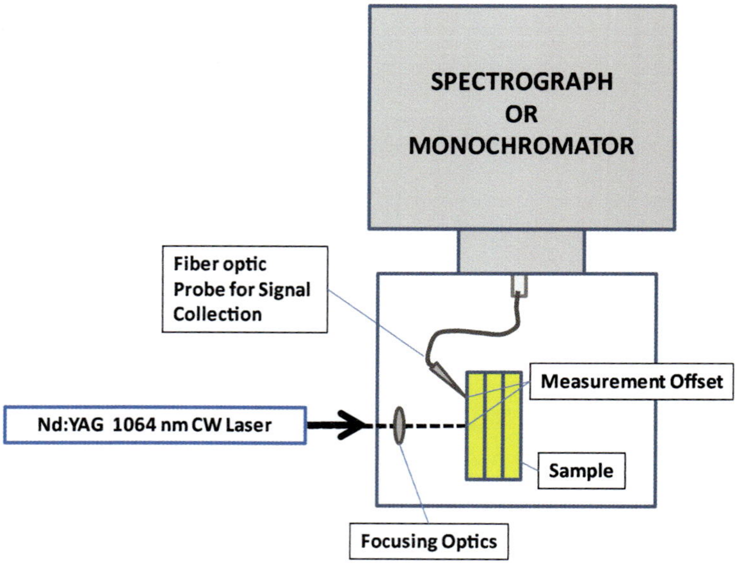

Fig. 4.17 Illustration of sampling for SORS (Spatially Offset Raman Spectroscopy) (offset position).

4.4.8 *SORS (Spatially Offset Raman Spectroscopy)*

This technique produces a Raman depth profile spectrum (Figures 4.16 and 4.17).

4.4.9 *TERS (Tip-Enhanced Raman Spectroscopy)*

An enhanced Raman signal is combined with AFM topography for a nano-scaled, three-dimensional image of the spatially related chemistry for the sample (Figure 4.18).

4.4.10 *SPPERS (Surface Plasmon Polariton (SPP) Enhanced Raman Spectroscopy)*

For measurement of sample areas as small as 100 microns in diameter, having the thickness of a molecular monolayer of sample material (Figure 4.19).

4.4.11 *Summary*

This chapter has focused on the issues of sampling as they pertain to routine and specialized Raman techniques for gathering spectra useful for both qualitative and quantitative analysis. There are many routine specialized approaches to sampling that may be adopted for research applications. These are typically experiment

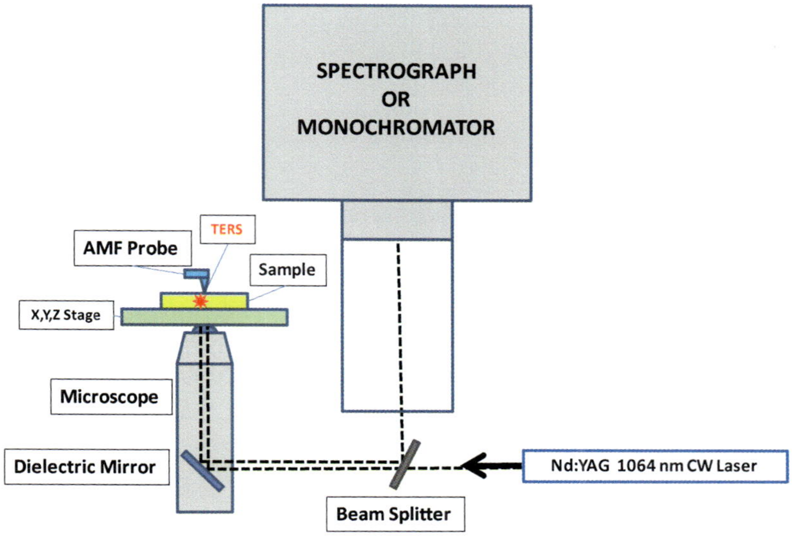

Fig. 4.18 Illustration of sampling for TERS (Tip-Enhanced Raman Spectroscopy) (bottom illumination).

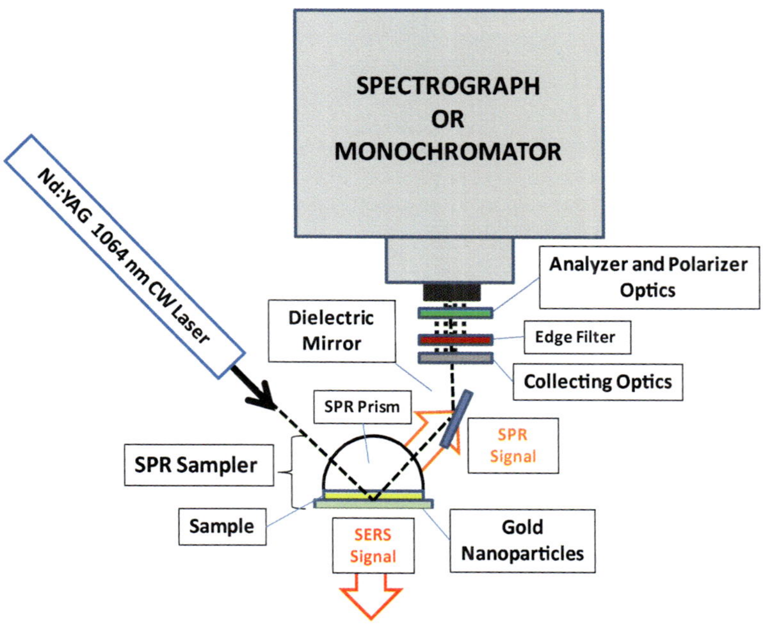

Fig. 4.19 Illustration of sampling for SPPERS (Surface Plasmon Polariton (SPP) Enhanced Raman Spectroscopy).

dependent, and their success may be somewhat limited to the application for which they are customized. The number of new sampling techniques is growing rapidly.

References

1. J.M. Hollas, *Modern Spectroscopy* (Wiley, Chichester, UK, 1996).
2. A.L. Smith, *Infrared Spectroscopy, Practical Handbook of Spectroscopy.* (J. W. Robinson, Ed.), (CRC Press, Boca Raton, FL, 1991).
3. N.B. Colthup, L.H. Daly, and S.E. Wiberley, *Introduction to Infrared and Raman Spectroscopy.* 3rd edn (Academic Press, Boston, 1990).
4. K. Feinstein (1995). *Guide to Spectroscopic Identification of Organic Compounds* (CRC Press, Boca Raton, FL, 1995).
5. R.M. Silverstein, G.C. Bassler, and T.C. Morrill, *Spectrometric Identification of Organic Compounds.* 5th edn (Wiley, New York, 1991).
6. J. R. Ferraro and K. Nakamoto, *Introductory Raman Spectroscopy*, 1st edn. (Academic Press, Boston, 1994).
7. G. Socrates, *Infrared and Raman Characteristic Group Frequencies: Tables and Charts.* 3rd edn. (John Wiley & Sons, 2004).
8. D. Lin-Vien, N.B. Colthup, W.G. Fateley, and J. G. Grasselli, *The Handbook of Infrared and Raman Characteristic Frequencies of Organic Molecules.* (Elsevier Publishers, Amsterdam, 1991).
9. K. Nakamoto, *Infrared and Raman Spectra of Inorganic and Coordination Compounds,* 1st edn. (John Wiley & Sons, Ltd, 1978).
10. R.A. Nyquist and R.O. Kagel, *Handbook of Infrared and Raman Spectra of Inorganic Compounds and Organic Salts: Infrared Spectra of Inorganic Compounds,* 1st edn. (Academic Press, Boston, 1972).
11. R.A. Nyquist, R.O. Kagel, and M.A. Leugers, *Handbook of Infrared and Raman Spectra of Inorganic Compounds and Organic Salts*, 1st edn. 4 Volumes (Academic Press, Boston, 1997).
12. A.H. Kuptsov and G.N. Zhizhin, *Handbook of Fourier Transform Raman and Infrared Spectra of Polymers*, 1st edn. (Elsevier, Amsterdam, 1998).
13. Aldrich Raman Condensed Phase Library (14,033 spectra). FT-Raman spectra, containing many common chemicals found in the *Aldrich Handbook of Fine Chemicals. Aldrich Chemical Company* (16 cm^{-1} resolution by 8-bit ordinate precision. Contain spectral data from 4,000 to 450 cm^{-1}). https://www.sigmaaldrich.com/catalog/product/aldrich/z276529?lang=en&region=US

Chapter 5

Data Acquisition for Raman Spectra

Contents

5.1 Recommended Spectrum Acquisition, Presentation Format and Data Manipulation Procedures 119
5.1.1 Recommended Spectrum Acquisition Parameters 119
5.1.2 Recommended Spectral Range 121
5.1.3 Recommended Presentation Formats 122
5.1.4 Spectral Data Manipulation 122
5.1.5 Data Formatting 123
5.1.6 Summary 123
References 124

5.1 Recommended Spectrum Acquisition, Presentation Format and Data Manipulation Procedures

5.1.1 *Recommended Spectrum Acquisition Parameters*

The following are basic guidelines for typical Raman data acquisition situations. Depending upon the resolution setting of the Raman spectrometer, the time for data acquisition will vary. Higher resolution requires more replicate spectra to be collected to attain the required signal-to-noise. Raman spectra are collected as Raman intensity (in counts) with respect to Raman shift in wavenumber (cm^{-1}). See the various Chapters 1 through 4 in this volume on Raman Methods, Instrumentation and Sampling, respectively, for detailed information regarding these topics.

Commercial Raman spectrometers may have resolution settings to as high as 2 cm^{-1} with a typical spectral coverage of from 3600–50 cm^{-1}. Note that as the resolution setting is decreased by a factor of 2 when using a spectrometer, the number of data points recorded doubles. For example, if a 4 cm^{-1} setting acquires 3451 data points, then a 2 cm^{-1} setting would acquire 6902 data points ($n \times 2$).

Noise-reduction using co-added scans follows the square root of n reduction where the n is the number of replicate co-added scans. Note that 4 scans yields a noise reduction of 2 and an increase in signal to noise of 2 as compared to a single scan. Likewise 100 scans would have 1/10 the noise and 10 times the signal to noise as a single scan. In this way the number of scans is estimated for each type of spectrum measured.

For most Raman measurements measuring typical samples the following set of parameters would be most common (Tables 5.1 and 5.2).

Note that increased replicate scanning for more than 1–3 minutes may approach a point of diminishing returns in terms of spectral quality and S/N improvement (limited by the square root of *n* law for random variation, or white noise where n is

Table 5.1 Typical settings for liquids and solids measurements using dispersive Raman spectroscopy.

Measurement parameter:	Typical setting:
Spectral Range:	3500 cm^{-1} to 50 cm^{-1} (Stokes) 6000 cm^{-1} to 50 cm^{-1} (extended Stokes)
Laser Excitation:	532 nm (<300 mW) 633 nm (<300 mW) 780 nm (<300 mW)
Detector Type:	InGaAs, Ge, Si, CCD
Resolution Setting:	2 cm^{-1} to 11 cm^{-1}
Sample Size:	50 to 100 mg

Table 5.2 Typical settings for liquids and solids measurements using FT-Raman spectroscopy.

Measurement parameter:	Typical setting:
Spectral Range:	4000 cm^{-1} to 50 cm^{-1} (Stokes)
Data Points:	1.0 to 0.5 wavenumber data resolution
Laser Excitation:	785 nm 1064 nm (500 mW to 2000 mW)
Detector Type:	InGaAs (20° C) Ge (Liquid N_2) Si (20° C)
Beamsplitter:	Quartz, CaF_2, KBr
Resolution Setting:	2 cm^{-1} (8 cm^{-1} typical)
Sampling Time:	~1 minute
Sample Size:	>50 mg

the number of replicate scans). Raman spectrometers may drift during a duration of several minutes or structured noise may enter the spectral signal; structured signal variation is not averaged by repetitive measurements. Structured noise, due to sampling or instrument perturbations, may become a part of spectra measured for extended periods. Note that for gas samples much higher resolution is a requirement and would generally demand additional scan times and replicate measurements to attain optimum signal to noise.

Irrespective of the final selection of data acquisition parameters, full documentation of the experimental details should be recorded with all spectra, especially if regulated procedures are followed. Also, as a general rule, it is recommended that good laboratory practice guidelines are adopted throughout, which include laboratory notebooks, either paper or electronic and recorded measurement parameters, analyst results, and corroborating signatures.[1]

5.1.2 *Recommended Spectral Range*

Unless special circumstances dictate otherwise, Raman spectra are recorded from 3500 cm^{-1} to 50 cm^{-1} (Stokes shift), or extended ranges for some dispersive instruments from 6000 cm^{-1} to 50 cm^{-1} (Stokes shift). If the Raman measurements are strictly to be compared to infrared measurements then a range of 4000 to 400 cm^{-1} is most appropriate (i.e., the ASTM International definition for the infrared region). A typical scan may include 3200 cm^{-1} to 200 cm^{-1} since classic Raman spectral libraries may have been measured using this data range, and some Raman spectral atlases cover the range of 4000 cm^{-1} to near 100 cm^{-1}.[2–4]

Typical research grade commercial Raman spectrometers will have a full scan range of 50 cm^{-1} to above 3500 cm^{-1}. This wavenumber restriction is due to the number of octaves that may be covered with various optical systems, detectors, and notch filters. Some very special research instruments may scan from below 50 cm^{-1} to above 6000 cm^{-1} yielding full scale molecular spectra. Note that specialized techniques also require special sample preparation or sample holders.

With modern instruments, it is normal to measure the entire individual spectral range with each measurement. When calibration or data transfer is required then data may be truncated (or windowed) so that one spectrometer collects the identical spectral range and data points as another spectrometer. Such data management allows easier compatibility of databases and calibrations developed using spectral search libraries for qualitative analysis. Note information on calibrating instruments may be found in the index or the special chapters on data analysis (Chapter 6 and others) within this text volume. When high resolution data is obtained it may be converted to lower resolution data but not vice versa (i.e., low to high resolution is forbidden).

5.1.3 *Recommended Presentation Formats*

The normal intensity scale obtained for spectral measurements are in counts for Raman Stokes scattering intensity. This may be normalized or scaled but generally falls within 0 to 70000 or more. If an abbreviated or expanded scale is used, it is important to ensure that the scale limits are presented, preferably with immediate scale markings shown. Note that for all numerical operations on spectral data, it is necessary to perform the operations on the final Raman intensity form of the spectrum, including baseline corrections, spectral subtractions, and any numerical scaling of data. An important exception is the numerical and/or digital treatment of noise. Noise is essentially linear over the entire Raman intensity scale. All processing that assumes a linear distribution of noise, such as digital smoothing, and digital signal processing must be applied to the Raman intensity form of the data prior to any additional mathematical processing that may convert data to exponential form. This is important for quantitative or qualitative analysis. Note that when using some detectors a cosmic ray correction may be essential since in array detection spurious 'spikes' may occur that seriously affect qualitative and quantitative use of spectra.

The standard wavenumber scale is normally used for Raman shift data, representing the Fundamental vibration group frequencies and fingerprint frequencies associated with molecular vibrations. However, it is entirely appropriate to represent Raman spectra in wavelength scale as microns (μm), or nanometers. However most often the wavenumber scale is used for Raman shift representation.

5.1.4 *Spectral Data Manipulation*

At all times, it is strongly recommended that the ratioed replicate spectral data are retained, along with the replicate spectra of the reference sample and dark background. Recall the formula representing the ratio of sample and reference spectra with a dark signal subtraction for each (Chapter 3). For Raman spectroscopic measurements this relationship is shown in the following description. A typical Raman sample measurement as (I_{RC}) or Raman Corrected Intensity is made by taking the ratio of the I_R (Sample) — I_R (Dark Signal) to the I_R (Blank Sample) — I_R (Dark Signal). This procedure compensates for the extraneous noise from the instrument operation, both mechanically and electronically. It also has an effect to mitigate sampling errors when a pure solvent is measured as the blank signal using the identical sampling procedure. Raman spectra may be measured using no correction or correction depending upon the experiment. However, the corrected Raman spectrum would then be as:

$$I_{RC} = \frac{I_R(Sample) - I_R(Dark\_Signal)}{I_R(Blank\_Sample) - I_R(Dark\_Signal)} \tag{5.1}$$

Retaining the individual replicate scan measurements allows improved diagnostics for determining the cause of error while taking repetitive measurements. For appearance purposes, some limited digital smoothing and nominal baseline corrections are acceptable if they improve overall quality of the final spectra for multivariate product calibration. If there are any doubts in terms of acceptability for data processing, it is recommended that original raw data are kept in storage to support any data that have been numerically manipulated (irrespective of the reasons for the manipulation). All documentation concerning data manipulation should be retained with the spectra in a digital format.

5.1.5 *Data Formatting*

A standard data format is essential for the utilization of Raman spectral data. Data from various instruments is routinely combined for quantitative calibration transfer and utilization of spectral libraries for qualitative searches for the purpose of identification.

In general, there has not been a universal adoption of standardization regimes for Raman and other spectroscopic data. In some ways, this has limited the potential of the technique for generalized methods of analysis. The goal for future spectrophotometry is to give users the ability to produce an exact and very reproducible standard spectrum for a given material by a prescribed method of sample preparation, independent of operator or instrument type, such as model or vendor. Users are encouraged to refer to the standard procedures for sample preparation as defined in texts, such as the ASTM International *Annual Book of Standards*.[5–6] Each instrument manufacturer may have from 1–10 (or more) data formats including older format versions in their software. A list of the most typical spectroscopy software formats is given in Table 5.3.

5.1.6 *Summary*

This chapter has focused on the issues of data sampling as they pertain to routine data collection for qualitative and quantitative Raman analysis. There are many

Table 5.3 Most common spectroscopic data file formats and extensions.

Data format name	Extension
AB SCIEX Data Explorer	(.pkm)
ASCII Table	(.csv)
Bruker	(.XML)
Excel	(.xls)
Finnegan	(.asc)
JCAMP-DX	(.dx)
MatLab	(.mat)
mzData	(.XML)
Perkin Elmer	(.sp)
Plain Text	(.txt)
Spectra-Calc (Archival)	(fixed point .spc)
Thermo Galactic	(.spc)
XLM Marked-up text	(.XML)

routine specialized data collection approaches that may be adopted for research applications. These are typically experiment dependent, and their success may be somewhat limited to the application for which they are customized. For this reason, such examples have been generally excluded from this text.

References

1. Guidance for Industry Good Laboratory Practices Questions and Answers, U.S. Department of Health and Human Services Food and Drug administration Office of Regulatory affairs, June 1981 (Minor editorial and formatting changes made December 1999 & July 2007).
2. R.A. Nyquist, R.O. Kagel, and M.A. Leugers, *Handbook of Infrared and Raman Spectra of Inorganic Compounds and Organic Salts*, 1st edn. 4 Volumes (Academic Press, Boston, 1997).
3. D. Lin-Vien, N.B. Colthup, W.G. Fateley, and J.G. Grasselli, *The Handbook of Infrared and Raman Characteristic Frequencies of Organic Molecules*. (Elsevier, 1991).
4. J. R. Ferraro and K. Nakamoto, *Introductory Raman Spectroscopy*, 1st edn. (Academic Press, Boston, 1994).
5. ASTM (American Society for Testing and Materials) Method E169-04(2009). "Standard Practices for General Techniques of Ultraviolet-Visible Quantitative Analysis", ASTM International, 100 Barr Harbor Drive, PO Box C700, West Conshohocken, PA (2009).
6. ASTM (American Society for Testing and Materials) Method E1655-05 (2012). "Standard Practices for Infrared Multivariate Quantitative Analysis", ASTM International, 100 Barr Harbor Drive, PO Box C700, West Conshohocken, PA (2012).

Chapter 6

Raman Data Analysis and Presentation

Contents

6.1 General Data Processing for Raman Spectroscopy 126
6.1.1 Introduction 126
6.1.2 Survey of Methods used for Raman Data Processing 126
6.2 Multivariate Regression Methods use for Calibration in Spectroscopy 127
6.2.1 Introduction 127
6.2.2 Multivariate Linear Regression (MLR) 127
6.2.3 Principal Component(s) Analysis (PCA) 129
6.2.4 Principal Component(s) Regression (PCR) 131
6.2.5 Partial Least Squares Singular Value Decomposition (PLS SVD) 133
6.2.6 Partial Least Squares Regression (PLSR) 134
6.2.7 Singular Value Decomposition (SVD) 135
6.3 Derivatives in Spectroscopy 139
6.3.1 Methods of Computing the Derivative for Spectroscopy 139
6.3.2 Limitations of the Savitzky-Golay (S-G) Method 143
6.3.3 Extensions to the S-G Derivative Method 145
6.4 The Statistics of Spectral Searches (Comparisons) 149
6.4.1 Introduction to Standard Approaches 149
6.4.2 Mahalanobis Distance Measurements 150
6.4.3 Euclidean Distance 151
6.4.4 Common Spectral Matching (Correlation or Dot Product) . . . 151
6.4.5 Computing the Probability that Two Sample Spectra are Statistically Identical 153
6.4.5.1 Example Problem 153

6.4.5.2 Geometric Interpretation of Cosign Similarity for Spectral Matching and use of 0-100 Scale for Hit Quality Index 156
6.5 Raman Spectral Resolution Test(s) 157
6.5.1 Introduction 157
6.5.2 Wavenumber repeatability or precision test 158
6.5.3 Measuring Spectral Resolution 162
6.6 Raman Spectrometer Calibration Basics 162
6.6.1 Introduction 162
6.6.2 Wavenumber/Wavelength Calibration 163
6.6.3 Intensity Calibration (Simplified Method) 165
6.6.4 Background Correction Modeling 169
References 169

6.1 General Data Processing for Raman Spectroscopy

6.1.1 *Introduction*

Data processing for Raman spectroscopy includes statistics and chemometrics. This chapter will deal more with the basic chemometric approaches to data processing. In spectroscopy, chemometrics is broadly concerned with digital signal processing (DSP), spectral data manipulation and transforms, pattern recognition, imaging algorithms, quantitative and qualitative analysis algorithms and statistical process control (SPC). Note to the reader that this chapter discusses regression methods, derivative use, Raman spectral search techniques, Raman spectral resolution testing, and Raman calibration methods for wavenumber and intensity scale calibration. Within each volume, Chapter 6 describes many techniques commonly used for general spectral data analysis. The reader is referred to these chapters as well as the index for each data processing term. Instrument performance testing and calibration methods for accuracy and precision are also discussed in these chapters and the basic concepts and equations may also be applied to performance testing for Raman spectrometers. (See specifically Volume 3, Section 6.10.2, as accuracy and precision calculations are very similar for most spectrometers.)

6.1.2 *Survey of Methods used for Raman Data Processing*

Following intensity and wavenumber calibration of each Raman spectrometer, the main methods for preprocessing Raman spectra include: autoscaling, normalization, mean centering, spectral subtraction, derivatives, and smoothing.[1–2]

Fluorescence correction may also be employed when using visible laser excitation in order to correct for, and measure, samples with native fluorescence.[3] Advanced wavelet-based noise reduction techniques have been demonstrated.[4] Data augmentation with other spectral regions is used to increase the data rank and thus the information content of spectroscopic data.[5] Raman may also include the use of internal standards for quantitative analysis where a specific Raman active component is added to a sample at a fixed concentration and where the Raman shift peak is different than the analyte. The correction is made by ratioing the analyte intensity to the internal standard intensity and computing a linear regression calibration model for predicting the unknown test sample concentration. The internal standard intensity is then used as a correction factor for the analyte intensity for quantitative analysis.[6–8] (See Section 1.3.2, Chapter 1 for more details.)

Regression methods, such as multiple linear regression (MLR),[9–10] principal components regression (PCR),[11–12] and partial least squares (PLS)[11–12] are often used for quantitative analysis.

6.2 Multivariate Regression Methods use for Calibration in Spectroscopy

6.2.1 *Introduction*

Note that the following sections discuss the matrix operations used for computations of the most common forms of: Multivariate Linear Regression (MLR), Principal Components Analysis (PCA), Principal Components Regression (PCR), Singular Value Decomposition (SVD), Partial Least Squares SVD, and Partial Least Squares Regression (PLSR). The computations for Simple Linear Regression (SLR) are given elsewhere in these volumes, for example, see Volume 2, Chapter 6 (Section 6.16.3.6); Volume 4, Chapter 6 (Section 6.5.4); and the index for detailed examples and demonstrations.

To begin this discussion a table demonstrating common symbols used for matrix notation is given in Table 6.1. In this chapter, matrix operations for linear equations are demonstrated for solving example problems.

6.2.2 *Multivariate Linear Regression (MLR)*

The following sections of this chapter describe multivariate linear regression (MLR), principal components regression (PCR), partial least squares regression (PLSR), and singular value decomposition (SVD). In all cases the same matrix notation is used and mathematical operations are described using MatLab (Matrix Laboratory)

Table 6.1 Common symbols used in matrix notation for regression methods.

Name of matrix variable or operation	Symbol
Matrix	[X] or X
Determinant	| X|
Vectors	x
Scalars	x
Parameters or matrix names	A, B, C, G, H, P, Q, R, S, U, V
Errors and residuals	D, E, F
Addition	+
Subtraction	−
Multiplication	× or ·
Division	÷ or /
Empty or null set	Ø
Inverse of a matrix	$[X]^{-1}$
Transpose of a matrix	[X]' or $[X]^T$
Generalized inverse of a matrix	$[X]^-$
Identity Matrix	[I] or [1]

software.[13–14] For more basic details the reader is referred to references showing the manual methods for calculating matrix algebra and regression methods covered within this section.[15–19]

To begin, if given a spectral data matrix denoted by A, representing a set of intensity or amplitude (Raman) or absorbance (i.e., absorption spectroscopy) values for three samples at two wavenumbers (as a rows × columns matrix), the example matrix is demonstrated in Equation 6.1. For this example each row represents a different sample spectrum, and each column represents a different data channel, as intensity or absorbance versus frequency, respectively; this is as shown in Equation 6.1.

$$A = \begin{bmatrix} A_{11} & A_{12} \\ A_{21} & A_{22} \\ A_{31} & A_{32} \end{bmatrix} = A_{I \times K} = \begin{bmatrix} 1 & 7 \\ 4 & 10 \\ 6 & 14 \end{bmatrix} \tag{6.1}$$

Thus, the integers 1 and 7 represent the instrument signals for two data channels (frequencies 1 and 2) for sample spectrum #1; 4 and 10 represent the instrument signals for the data channels (e.g., frequencies 1 and 2) for sample spectrum #2; and 6 and 14 represent the instrument signals for frequencies 1 and 2 for sample spectrum #3.

If the known or reference concentration c is a vector representing a single parameter, it may be displayed as a single column of numbers is as (Equation 6.2). Each row is the reference concentration for each sample #1 through #3; the concentration values being 4, 8, and 11, respectively.

$$c_{r\times c} = \begin{bmatrix} c_{11} \\ c_{21} \\ c_{31} \end{bmatrix} = c_{I\times 1} = \begin{bmatrix} 4 \\ 8 \\ 11 \end{bmatrix} \tag{6.2}$$

Now all the data necessary to calculate the matrix of multiple linear regression coefficients or b values (i.e., the regression vector) is given by Equation 6.3.

$$b = \begin{bmatrix} b_{11} \\ b_{21} \end{bmatrix} = (A'A)^{-1}A'c = A^{+}c = \hat{p} \tag{6.3}$$

This b vector (also known as $\hat{p}$ = the *prediction vector*) is often referred to as the *regression vector* or set of *regression coefficients*. Note that $(A'A)^{-1}A'$ is referred to as the *pseudoinverse* of A designated as A^{+}. Note that there is always one regression coefficient for each frequency (or data channel).

The matrix of predicted values is easily obtained as a dot product or Matrix A (the data matrix) × Vector b (the regression coefficients) = Vector c (the predicted values). This is shown in matrix notation as Equation 6.4.

$$A \times b = c \tag{6.4}$$

Now using a MatLab command line software, the solution may be demonstrated for the multivariate problem identified in (Equations 6.1 through 6.4). The operations for this computation are demonstrated using a series of simple matrix operations as shown in Table 6.2.

6.2.3 *Principal Component(s) Analysis (PCA)*

This section describes the matrix operations for a basic principal components regression (PCR) operation. A spectral data matrix is measured and denoted as A. A is used to represent a set of intensity (amplitude) or absorbance values for three samples as a rows × columns matrix. For this example each row represents a different sample spectrum, and each column a different data channel or frequency. A is designated as in Equation 6.5.

$$A = \begin{bmatrix} A_{11} & A_{12} \\ A_{21} & A_{22} \\ A_{31} & A_{32} \end{bmatrix} = A_{I\times K} = \begin{bmatrix} 1 & 7 \\ 4 & 10 \\ 6 & 14 \end{bmatrix} \tag{6.5}$$

For this discussion, 1 and 7 represent the instrument signal for two data channels (frequencies 1 and 2) for sample spectrum #1; 4 and 10 represent the instrument

Table 6.2 Matrix operations in MatLab to compute Equations 6.1 through 6.4.

Command line	Comments
≫ A = [1 7;4 10;6 14]	Enter the A matrix
A =	Display the A matrix
1 7	
4 10	
6 14	
≫ c = [4;8;11]	Enter the concentration vector c
c =	Display the concentration vector c
4	
8	
11	
≫ b = inv(A′*A)*A′*c	Calculate the regression vector [Note: The inverse applies only to (A′*A)]
b =	Display the regression vector b
0.7722	
0.4662	
≫ A*b	Predict the concentrations [Note: a residual (or difference) exists between the predicted and the actual concentrations 4, 8, and 11].
ans =	
4.0356	
7.7509	
11.1601	

signals for frequencies 1 and 2, for sample spectrum #2; and 6 and 14 represent the instrument signals for frequencies 1 and 2 for sample spectrum #3.

The data necessary to compute the *singular value decomposition* (SVD) for matrix A is present. The operation performed in SVD is sometimes referred to as *eigenanalysis*, *principal components analysis*, or *factor analysis*. The SVD operation is performed on the A matrix, with the result being 3 matrices, termed the left singular values (LSV) matrix or the U matrix; the singular values matrix (SVM) or the S matrix; and the right singular values matrix (RSV) or the V matrix.

The SVD operation provides enough information to find the Scores matrix and the Loadings matrix. First of all the Loadings matrix is simply the right singular values matrix or the V matrix; this matrix is often referred to as the P matrix in Principal Components Analysis (PCA) terminology. The Scores matrix is calculated as Equation 6.6.

$$\text{The data matrix A} \times \text{the Loadings matrix V} = \text{Scores matrix T} \qquad (6.6)$$

Note: the Scores matrix is often referred to as the T matrix in Principal Components Analysis (PCA) terminology. Table 6.3 demonstrates the set of operations for Principal Components Analysis (PCA) using MatLab.

Table 6.3 Matrix operations in MatLab to compute the SVD of data matrix A to obtain a PCA Analysis (as Equations 6.5 and 6.6).

Command line	Comments
≫ A = [1 7;4 10;6 14]	Enter the A matrix
A =	Display the A matrix
1 7 4 10 6 14	
≫ [U,S,V] = svd(A);	Perform SVD on the A matrix
≫ U	Display the U matrix or the left singular values (LSV) matrix
U =	
0.3468 0.9303 0.1193 0.5417 −0.0949 −0.8352 0.7656 −0.3543 0.5369	
≫ S	Display the S matrix or the singular values (SV) matrix
S =	
19.8785 0 0 1.6865 0 0	
≫ V	Display the V matrix or the right singular values (RSV) matrix (Note: this is also known as the P matrix or Loadings matrix)
V = 0.3576 −0.9339 0.9339 0.3576	
≫ T = A*V	
T =	
6.8948 1.5690 10.7691 −0.1600 15.2198 −0.5976	Calculate the Scores Matrix or the T matrix

6.2.4 *Principal Component(s) Regression (PCR)*

As in the example for Multiple Linear Regression, if the known or reference concentration c is a vector representing a single parameter, it may be displayed as a single column of numbers is as (Equation 6.7). Each row is the reference concentration for each sample #1 through #3; the values being 4, 8, and 11, respectively.

$$c_{r\times c} = \begin{bmatrix} c_{11} \\ c_{21} \\ c_{31} \end{bmatrix} = c_{I\times 1} = \begin{bmatrix} 4 \\ 8 \\ 11 \end{bmatrix} \tag{6.7}$$

The values of S, V, and T are used to calculate all results required for Principal Components Regression (PCR), as noted in the following Equations 6.8 through 6.10 and Table 6.4.

Table 6.4 Matrix operations in MatLab to compute PCR (as Equations 6.7 through 6.10).

Command line	Comments
≫ Aest = T*V′	Estimate the A data matrix
≫ Aest = 1.0000 7.0000 4.0000 10.0000 6.0000 14.0000	Display the estimate for A
≫ b = V(:,1:2)*inv(S(1:2,1:2))*U(:,1:2)′*c;	Calculate the regression vector [Note: The inverse operation refers only to the singular values matrix S]. The calculation to determine b can only be performed using two columns in each of the V, S, and U matrices; this number is equivalent to the number of latent variables (or principal components) used.
b = 0.7722 0.4662	Display the regression vector
≫ cest = (T*V′)*b	Predict the concentrations [Note: This computation is equivalent to (Aest × b)].
cest = 4.0356 7.7509 11.1601	Display the concentration vector [Note: For this example of PCR a residual (or difference) exists between the predicted and the actual concentrations 4, 8, and 11].

A reconstruction of the original data matrix A is computed by using the preselected number of principal components (i.e., columns in our T and V matrices) as Equation 6.8.

$$\text{A (estimated)} = \text{T} \times \text{V}' \tag{6.8}$$

The set of regression coefficients (i.e., the regression vector) is calculated as Equation 6.9.

$$\text{b (regression vector)} = \text{V} \times \text{S}^{-1} \times \text{U}' \times \text{c} \tag{6.9}$$

Finally, the predicted or estimated concentration values c are computed as

$$\text{c (estimated)} = (\text{T} \times \text{V}') \times \text{b} \tag{6.10}$$

When using MatLab software, the solution for PCR analysis is given in Table 6.4.

6.2.5 *Partial Least Squares Singular Value Decomposition (PLS SVD)*

This section demonstrates the matrix operations for computing Partial Least Squares (PLS) regression utilizing singular value decomposition (SVD). For PLS a special case of SVD (singular value decomposition) is used. The PLS form of SVD includes the use of the concentration vector c as well as the spectral data matrix A. The reader will note that the Scores and Loadings are determined using the concentration values for PLS SVD whereas only the data matrix A is used to perform SVD for Principal Components Analysis (PCA). All mathematical operations are completed using MatLab (Matrix Laboratory) software.[13–14]

For this example, the same simple matrix is used to represent a set of spectra, denoted as A. A is used to represent a set of intensity (amplitude) or absorbance values for three samples and (for this example) three data channels. The samples are represented as rows and the frequencies as columns. For this example each row represents a different sample spectrum, and each column a different data channel, or frequency. For this example, the spectral data matrix is given as Equation 6.11.

$$A_{r\times c} = \begin{bmatrix} A_{11} & A_{12} & A_{13} \\ A_{21} & A_{22} & A_{23} \\ A_{31} & A_{32} & A_{33} \end{bmatrix} = A_{I\times K} = \begin{bmatrix} 1 & 7 & 9 \\ 4 & 10 & 12 \\ 6 & 14 & 16 \end{bmatrix} \tag{6.11}$$

For this example, 1, 7, and 9 represent the instrument signal (as intensity or absorbance) for three data channels (as frequencies 1, 2, and 3). Spectrum #1 values are 1, 7, and 9 for frequencies 1, 2, and 3; for sample spectrum #2 the values are 4, 10, and 12 for frequencies 1, 2, and 3; and for spectrum #3 the values are 6, 14, and 16 for frequencies 1, 2, and 3, respectively.

The concentration c vector is given as Equation 6.12 representing a single component as a single column of numbers.

$$c_{r\times c} = \begin{bmatrix} c_{11} \\ c_{21} \\ c_{31} \end{bmatrix} = c_{I\times 1} = \begin{bmatrix} 4 \\ 8 \\ 11 \end{bmatrix} \tag{6.12}$$

Both the data matrix A and the concentration vector c are required to calculate PLS SVD. Both A and c are necessary to calculate the special case of *PLS singular value decomposition* (plsSVD). The operation performed in plsSVD is sometimes referred to as the pls form of *eigenanalysis*, or *factor analysis*. If plsSVD is performed on the A matrix with the c vector, the result is 3 matrices, termed the left singular values (LSV) matrix or the U matrix; the singular values matrix (SVM) or the S matrix; and the right singular values matrix (RSV) or the V matrix.

By computing the PLS SVD there is enough information to find the PLS Scores matrix and the PLS Loadings matrix. First of all, the PLS Loadings matrix is simply the right singular values matrix or the V matrix; this matrix is referred to as the P matrix in PLS or PCA terminology. The PLS Scores matrix is calculated as Equation 6.13.

$$\text{The data matrix A} \times \text{the PLS Loadings matrix V} = \text{PLS Scores matrix T} \tag{6.13}$$

Note: the PLS Scores matrix is referred to as the T matrix in PLS and PCA terminology. The PLS SVD calculations may be shown in MatLab as demonstrated in Table 6.5.

6.2.6 *Partial Least Squares Regression (PLSR)*

After computing the PLS SVD, the S, V, and T matrices are to calculate the PLS Regression solution. The reconstruction of the original data matrix A is computed by using the preselected number of factors (i.e., columns in our T and V matrices) as Equation 6.14.

$$\text{A (estimated)} = \text{T} \times \text{V}' \tag{6.14}$$

The set of regression coefficients (i.e., the regression vector) is calculated as Equation 6.15.

$$\text{b (regression vector)} = \text{V} \times \text{S}^{-1} \times \text{U}' \times \text{c} \tag{6.15}$$

The predicted or estimated values c for the PLS model are computed as Equation 6.16.

$$\text{c (estimated)} = (\text{T} \times \text{V}') \times \text{b} \tag{6.16}$$

This Equation 6.16 expression is equivalent to Equation 6.17.

$$\text{c (estimated)} = \text{A(estimated)} \times \text{b} = \text{A} \times \text{b} \tag{6.17}$$

The PLS solution may be used to predict a single sample spectrum (a) using the expression in Equation 6.18.

$$\text{c (estimated)} = \text{a (estimated)} \times \text{b} = \text{a} \times \text{b} \tag{6.18}$$

Table 6.7 demonstrates the MatLab operations for PLS Regression following PLS SVD.

Table 6.5 Matrix operations in MatLab to compute the PLS SVD calculations of data matrix A (as Equations 6.11 through 6.13).

Command line	Comments
≫ A = [1 7 9;4 10 12;6 14 16]	Enter the A matrix
A =	Display the A matrix
1 7 9	
4 10 12	
6 14 16	
≫ c = [4;8;11]	Enter the c vector
≫ c =	Display the c vector
4	
8	
11	
≫≫ [U,S,V] = svdpls(A,c,3);	Perform PLS SVD on the A matrix. This is a CPAC[34] version of the PLS SVD algorithm.
≫ U	Display the U matrix or the left singular values (LSV) matrix
U =	
0.3817 −0.9067 −0.1797	
0.5451 0.0638 0.8359	
0.7465 0.4170 −0.5186	
≫ S	Display the S matrix or the singular values (SV) matrix
S =	
29.5796 −0.2076 0.0000	
0.0000 1.9904 −0.0367	
0.0000 0.0000 0.2038	
≫ V	Display the PLS V matrix or the right singular values (RSV) matrix (Note: this is also known as the P matrix or PLS Loadings matrix)
V =	
0.2446 0.9345 0.2588	
0.6283 0.0506 −0.7764	
0.7386 −0.3525 0.5747	
≫ T = A*V	Calculate the PLS Scores Matrix or the T matrix
T =	
11.2894 −1.8839 −0.0034	
16.1236 0.0138 0.1680	
22.0801 0.6750 −0.1210	

6.2.7 *Singular Value Decomposition (SVD)*

Calculations for MLR, PCA, PCR, SVD, and PLSR have been demonstrated in the previous sections on regression mathematics. The computations of PCA, PCR, and PLS depend upon the use of SVD. This section describes more details related to the basic calculations and mathematical relationships for computing Singular Value Decomposition or SVD.

Table 6.6 Matrix operations in MatLab to compute (Equations 6.14 to 6.18).

Command line	Comments
≫ Aest = T*V′	Estimate the A data matrix
≫ Aest =	Display the estimate for A
1.0000 7.0000 9.0000	
4.0000 10.0000 12.0000	
6.0000 14.0000 716.0000	
≫ b = V*inv(S)*U′*c	Calculate the PLS regression vector [Note: The inverse operation refers only to the singular values matrix S. The calculation to determine b is performed using three columns in each of the V, S, and U matrices; this number is equivalent to the number of latent variables (or PLS factors) used.
b =	Display the regression vector
1.1667	
−0.6667	
0.8333	
≫ cest = (T*V′)*b	Predict the concentrations [Note: This computation is equivalent to (Aest × b)].
cest =	
4.0000	
8.0000	Display the concentration vector
11.0000	
	[Note: For this simple example of PLS no residual (or difference) exists between the predicted and the actual concentrations 4, 8, and 11].

SVD is used for modern computations of Principal Components Regression (PCR) and Partial Least Squares Regression (PLSR), although slightly different forms of SVD are used for each set of computations. For PCR, SVD is used, and for PLS a special case of SVD termed PLS SVD was used. The PLS form of SVD includes the use of the concentration vector c as well as the data matrix A. The reader will note that the scores (T) and loadings (V) are determined using the concentration values for PLS SVD whereas only the data matrix A is used to perform SVD for Principal Components Analysis.

For this example, a representative spectral data matrix is denoted by A. A is used to represent a set of intensity or absorbance values for three samples and three data channels, as a rows × columns matrix. For this example each row represents a different sample spectrum, and each column a different data channel (i.e., frequency).

Table 6.7 Simple SVD calculation performed on matrix A using MatLab; other matrix relationships are also shown (see Equations 6.19 through 6.22).

Command line	Comments
≫ A = [1 7 9;4 10 12;6 14 16]	Enter the A matrix
A =	Display the A matrix
1 7 9 4 10 12 6 14 16	
≫[U,S,V] = svd(A)	Calculate the SVD of A
U = 0.3821 0.9061 −0.1814 0.5451 −0.0624 0.8361 0.7463 −0.4183 −0.5178	Display the U matrix, also known as the left singular values matrix, and rarely referred to as the scores matrix. The scores matrix is most often denoted as U × S or A × V which as it turns out are exactly the same.
S =	Display the S matrix or the singular values matrix. This diagonal matrix contains the variance described by each principal component. Note: the squares of the singular values are termed the *eigenvalues*.
29.5803 0 0 0 1.9907 0 0 0 0.2038	
V =	Display the V matrix or the right singular values matrix; this is also known as the loadings matrix. Note: this matrix is the *eigenvectors* corresponding to the positive eigenvalues.
0.2380 −0.9312 0.2762 0.6279 −0.0694 −0.7752 0.7410 0.3579 0.5681	
≫ U*S*V′ =	U*S*V′ is equivalent to the original data matrix A derived using the SVD computation
ans = 1.0000 7.0000 9.0000 4.0000 10.0000 12.0000 6.0000 14.0000 16.0000	
≫ T = A*V	The scores matrix (often designated as T) can be calculated as A × V
T = 11.3024 1.8038 −0.0370 16.1231 −0.1243 0.1704 22.0748 −0.8328 −0.1055	
≫ U*S	As mentioned in the text of this section, the scores matrix T can also be calculated as U × S.
ans = 11.3024 1.8038 −0.0370 16.1231 −0.1243 0.1704 22.0748 −0.8328 −0.1055	

(*Continued*)

Table 6.7 (*Continued*).

Command line	Comments
≫ T*V′	The original data matrix A can be estimated as the scores matrix (T) × the transpose of the loadings matrix (V′) as shown.
ans = 1.0000 7.0000 9.0000 4.0000 10.0000 12.0000 6.0000 14.0000 16.0000	
≫ A*V*V′	Just another way to estimate the original data matrix A. In this case, V times the transpose of V (itself) is a diagonal matrix with a value of ones (1) along the diagonal, such as shown below. Note: this matrix of ones along the diagonal is called an identity matrix or (I).
ans = 1.0000 7.0000 9.0000 4.0000 10.0000 12.0000 6.0000 14.0000 16.0000	 1.0000 0.0000 0.0000 0.0000 1.0000 0.0000 0.0000 0.0000 1.0000

We arbitrarily designate A for our example as Equation 6.19.

$$A_{r\times c} = \begin{bmatrix} A_{11} & A_{12} & A_{13} \\ A_{21} & A_{22} & A_{23} \\ A_{31} & A_{32} & A_{33} \end{bmatrix} = A_{I\times K} = \begin{bmatrix} 1 & 7 & 9 \\ 4 & 10 & 12 \\ 6 & 14 & 16 \end{bmatrix} \tag{6.19}$$

Thus, 1, 7, and 9 represent the instrument signals (as intensity or absorbance) for three data channels (frequencies 1, 2, and 3) for sample spectrum #1; 4, 10, and 12 represent the instrument signals for frequencies 1, 2, and 3 for sample spectrum #2; and 6, 14, and 16 represent the instrument signals for frequencies 1, 2, and 3 for sample spectrum #3.

Given any data matrix A of arbitrary size (as rows × columns) the matrix A can be written or defined using the computation of Singular Value Decomposition (SVD) as Equation 6.20.

$$A = USV' = U \times S \times V' \tag{6.20}$$

Where U is the left singular values matrix, V is the loadings matrix, and S is the diagonal matrix containing information on the variance described by each principal component (as the S matrix columns). It is important to note when reviewing the use of SVD in the literature that many references define the scores matrix (T) as U × S. Keep in mind that the scores can be calculated as Equation 6.21.

$$\mathrm{U} \times \mathrm{S} = \mathrm{A} \times \mathrm{V} = \mathrm{T} \tag{6.21}$$

and it holds that the original data matrix A may then be reconstructed as Equation 6.22.

$$U \times S \times V' = T \times V' = A \times V \times V' = A \times I = A \qquad (6.22)$$

The interrelationships between the different matrices resulting from the SVD calculations may be practically demonstrated by the use of MatLab as shown in Table 6.7.

By studying the relationships between the various matrices resulting from the computation of SVD one can observe that there are several ways to compute the same final results, making it somewhat challenging to follow multiple variations within the chemometrics literature. However knowing these inner mathematical relationships can help clarify one's understanding of the different chemometric nomenclature.[20–22]

6.3 Derivatives in Spectroscopy

6.3.1 *Methods of Computing the Derivative for Spectroscopy*

More detailed discussions of the use of derivatives for spectroscopy are given throughout these volumes, see for example, Volume 3, Chapter 6. This derivative discussion is a complementary section on the use of derivatives.[46–49] Note that the index for this set of volumes lists multiple sections of additional material covering this subject.

The derivative of any set of data points representing a spectrum may be estimated by calculating the difference between intensity (or absorbance values) of two data points spaced some wavenumber (or wavelength) distance apart and dividing by that $\Delta\lambda$ (i.e, ΔX spacing). (A detailed set of equations for this method are given for first- through fourth-order derivatives in Volume 3, Equations 6.23 through 6.29). This ratio method is the most elementary form of derivative computation available. This method causes a decrease in the signal-to-noise ratio as compared with the underlying Raman intensity (or infrared absorbance band intensity), and this decrease in signal to noise has two main sources. The lesser source of error is the increase in the noise level due to the addition of variances that occur when numbers are added or subtracted, and thus precision is lost. The far greater effect is due to the fact that the derivatives are much smaller than the signal values, and the second derivative is much smaller (by an order of magnitude) than the first derivative, and each successive derivative order is smaller still. The net result is the closer the theoretical approximation to the true derivative is, the noisier the actual computed derivative becomes.

Several methods have been devised to circumvent this characteristic of the process of computing derivatives. One of the very common methods is to reduce the initial noise of the spectrum by computing averages; averaging the spectral data over some number of wavenumbers (or wavelengths) before estimating the derivative by calculating the difference between the resulting averages. This process is sometimes called "smoothing" since it smoothes out the noise of the spectrum. Sections on smoothing are dispersed within these volumes (see, for example, Volume 3, Chapter 6, Section 6.7.2).

The next common method of computing derivatives is the use of Savitzky-Golay convolution functions (see also Volume 3, Chapter 6, Section 6.8). The application to spectroscopy is based on what is one of the most often cited papers in spectroscopy computing literature.[23] This classic paper, presents the concept underlying the most used method for computing derivatives (including the zero-order derivative, which reduces to what is basically a smoothing operation); Figure 6.1 shows this diagrammatically. The assumption is that the mathematical nature of the underlying spectral curve is unknown, but can be represented over some finite region by a polynomial; "polynomial" in this sense in general and includes straight lines. If the

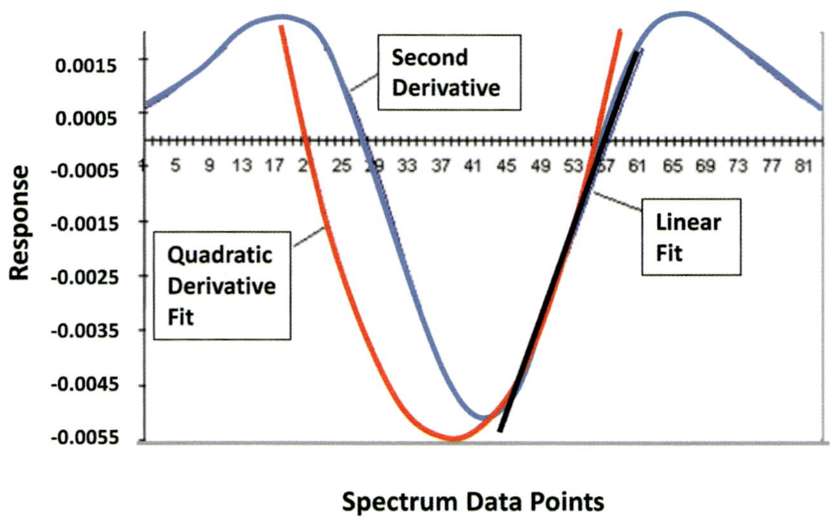

Fig. 6.1 An illustration that polynomials best fit may be used to approximate a derivative curve at the point of interest. The Savitzky-Golay method of computing derivatives is based on a least-squares fit of a polynomial to the data of interest. This figure illustrates both linear and quadratic fits to a Normal spectral curve.

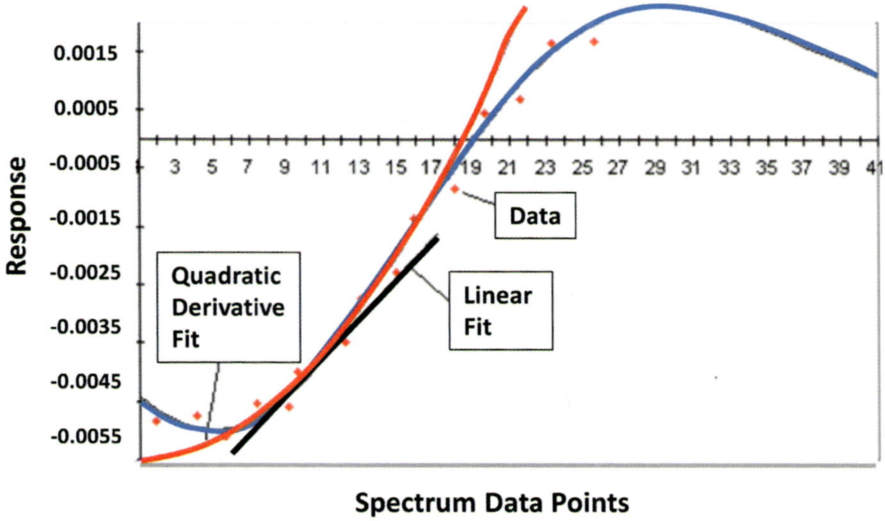

Fig. 6.2 An expansion of Figure 6.1 demonstrating how the polynomials are determined using a least-squares fit to the actual data in the region where the derivative is computed, even when the data is contaminated with noise. (Note that red dots represent the actual data.) This figure illustrates that when the derivative curve is based on data that has error, the polynomials can be validly computed using a least-squares fit to the data. At the point for which the derivative is computed, the lines in the figure are tangent to one another. The Savitzky-Golay approach allows the use of varying numbers of data points to be used in the computation of the fitting polynomial.

equation for the polynomial is known, then the derivative of the spectrum can be calculated from the properties of the fitted polynomial.

The polynomial best fit function may be calculated from the spectral data, by performing a least-squares fit of the polynomial to the data in the region of interest, as shown in Figure 6.2. Figure 6.1 shows that various polynomials may be used to approximate the derivative curve at the point of interest, and Figure 6.2 shows that when the derivative curve is based on data that has error, the polynomials can be computed using a least-squares fit to the data. At the point for which the derivative is computed, the lines in Figure 6.2 are tangent to one another. The Savitzky-Golay approach provides for the use of varying numbers of data points to be used in the computation of the fitting polynomial. Figure 6.3 illustrates the difference between the ordinates of two points equally spaced around $\mu + \sigma$ as a function of the spacing. In this figure the underlying absorbance curve has a bandwidth of 20 nm.

The steps that Savitzky and Golay[23] used to create their classic method were: (1) Fit a polynomial curve of the desired order to the spectral data, using least-squares curve fitting; (2) Compute the desired order of derivative of that polynomial; (3) Evaluate the expression for the derivative of that polynomial at

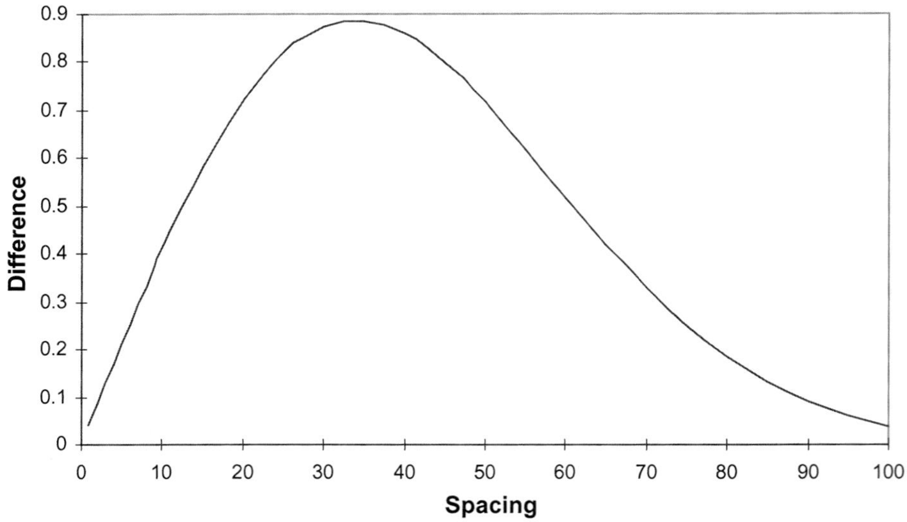

Fig. 6.3 The difference between the ordinates of two points equally spaced around $\mu + \sigma$ as a function of the spacing is illustrated in this figure (The curve has a bandwidth of 20 nm).

the point for which the derivative is to be computed. In the Savitzky-Golay paper, this is the central point of the set used to fit the data. In general this need not be the case, although doing so simplifies computations; (4) Convert those formulas into a set of coefficients that can be used to multiply the data spectrum in order to produce the value of the derivative according the specified polynomial fit, at the point of the center of the set of data. However, their paper overlooks some key points. And finally, while this work was of theoretical interest, Savitzky and Golay took one more step that turned the theory into a form that could be easily put to practical use; for step (5) they provided recipes for a good number of sets of derivative orders, fitting polynomials and numbers of data points, they calculated and printed in their paper tables of the coefficients needed for each of the derivative cases considered. Thus the practicing chemist could easily make use of the results produced.

Unfortunately there are also several caveats that have to go along with the use of the Savitzky-Golay results. The most important and also the best-known caveat is that there are errors in the tables in their paper. This was pointed out by Steinier in a paper that is invariably cited along with the original Savitzky-Golay paper, and which should be considered a "must read" along with the original paper by anyone taking an interest in the Savitzky-Golay approach to computation of derivatives.[24]

The Savitzky-Golay coefficients provide a simplified form of computation for the derivative of the desired order at a single point. To produce a derivative spectrum the coefficients must be applied successively to sets of spectral data, each set offset from the previous one by a single wavelength increment. This is known as the convolution of the two functions. Having done that, the result of all the theoretical development and computation is that the derivative spectrum so produced simultaneously is based on a smoothed version of the spectrum. The amount of smoothing depends on the number of data points used to compute the least-squares fit of the polynomial to the data, use of more data points is equivalent to performing more smoothing. Using higher-degree polynomials as the fitting function, on the other hand, is equivalent to using less smoothing, since high-order polynomials have more degrees of freedom to follow the details of the data curves more readily.

6.3.2 *Limitations of the Savitzky-Golay (S-G) Method*

The publication of the Savitzky-Golay paper (augmented by the Steinier paper)[24] was a major breakthrough in data analysis of chemical and spectroscopic data. Nevertheless it does have some limitations, and some more caveats that need to be considered when using this approach.

One limitation is that the method as originally described, is applicable only to computations using odd numbers of data points. This was implied earlier when it was noted that a derivative (of any order) is computed at the central point (i.e., wavenumber or wavelength) of the data set used.

Another limitation is that because of the computation being applicable to the central data point, there is an "end effect" to using the Savitzky-Golay approach; it does not provide for the computation of derivatives that are "too close" to the end of the spectrum. The reason is that at the end of the spectrum there is no spectral data to match up to the coefficients on one side or the other of the central point of the set of coefficients, therefore the computation at or near the ends of the spectrum cannot be logically performed.

Of course, an inherent limitation of the method is the fact that only those combinations of parameters (viz., derivative order, polynomial degree and number of data points) that are listed in the Savitzky-Golay/Steinier tables are available for use. While those cover what are likely to be the most common needs, anyone wanting to use a set of different parameters beyond those has no coefficients or solutions to apply.

A caveat to the use of the Savitzky-Golay tables is that, even after Steinier's corrections, they apply only to a special case of data, and do not, in general, produce the correct value for the true derivative. The reason for this is that applying the Savitzky-Golay coefficients to a set of spectral data is equivalent to assuming that the data is separated by unit X distance, and therefore is equivalent to computing only the numerator term of a finite difference computation, without taking into account the ΔX (or frequency/wavelength spacing) to which the computed ΔY (intensity or absorbance value) corresponds. Therefore, in order to compute the Savitzky-Golay estimate of a true derivative, the value computed using the Savitzky-Golay coefficients must be divided by $(\Delta X)^n$, where n is the order of the derivative.

Another limitation is perhaps not so much a limitation as, perhaps, a strange characteristic, albeit one that can catch the unwary. To demonstrate, consider the simplest S-G derivative function, that for the first derivative using a 5-point quadratic fitting function, the convolution coefficients (after including the normalization factor) are:

$$-0.2, \quad -0.1, \quad 0, \quad 0.1, \quad 0.2$$

If a second derivative is computed by applying this first derivative function twice, the effect is easily demonstrated to be equivalent to applying the S-G convolution coefficients, as follows:

$$0.04, \quad 0.04, \quad 0.01, \quad -0.04, \quad -0.1, \quad -0.04, \quad 0.01, \quad 0.04, \quad 0.04$$

This is a collection of nine coefficients that produces a second derivative, based on the S-G first derivative coefficients. However, this collection of convolution coefficients appears nowhere in the S-G tables. The nine-point S-G second derivative with a Quadratic or Cubic polynomial fit has the coefficients:

$$0.0606, \quad 0.0152, \quad -0.0173, \quad -0.0368, \quad -0.0433, \quad -0.0368, \quad -0.0173, \quad 0.0152, \quad 0.0606$$

And the nine-point S-G second derivative with a Quartic or Quintic polynomial fit has the coefficients:

$$-0.8811, \quad 2.5944, \quad 1.0559, \quad -1.4755, \quad -2.5874, \quad -1.4755, \quad 1.0559, \quad 2.5944, \quad -0.8811$$

The original S-G paper[23] describes how to compute other S-G convolution coefficients from given ones; these other coefficients are also functions that follow the basic concepts of the S-G procedure, the derivative of a least-square, best-fitting polynomial function. Since they do not produce the convolution coefficients generated by applying the S-G first derivative coefficients twice, one is forced to the conclusion that even though the coefficients for the first derivative follow the S-G concepts, applying them twice (or multiple) times in succession does not produce a set of convolution coefficients that is part of the S-G collection of convolution functions. This seems to be generally true for the S-G convolution coefficients as a whole.

6.3.3 *Extensions to the S-G Derivative Method*

Several extensions have been developed to the original concept. There are extensions that do not change the fundamental structure of the Savitzky-Golay approach, but simply make it easier to use. The main development along this line is the elimination of the tables. On the one hand, tables of coefficients are easy to deal with conceptually, because they can be applied mechanically — just copy down the entries and use them to multiply the data. However, this approach has the inherent limitation of including only those conditions that are listed in the Savitzky-Golay tables, extensions to the derivative order, polynomial degree, or number of data point used are excluded.

Therefore an extension of this idea was presented in a paper by Hannibal Madden.[25] Instead of presenting the already-worked-out numbers, Madden derived formulas from which the coefficients could be computed, and presented a table of those formulas in his paper. This is definitely a step up, since it confers several advantages, such as: (1) By using these formulas, the Savitzky-Golay convolution coefficients could be computed for a convolution function using any odd number of data points for the convolution; and (2) Since the coefficients are being calculated by the computer, there is no chance for typographical errors occurring in the coefficients.

Madden's paper, however, also has limitations: (1) the paper contains formulas for only those derivative orders and degrees of polynomials that are contained in the original Savitzky-Golay paper, and is therefore still limited to those derivative order and polynomial degrees; (2) the coefficients produced still contain the implicit assumption that the value of $\Delta X = 1$. Therefore, to produce correct derivatives it will still be necessary to divide the results from the formulas by $(\Delta X)^n$, as above. Further limitations include: (3) the formulas are at least as complicated, difficult and tedious to enter as the tables they replace, and as fraught with the possibility of typos during their entry. This is exacerbated by the fact that, being a formula in a computer program, everything must be perfectly reproduced, and all the parentheses, etc., must be in the correct places, which, for formulas as complicated as those are, is not easy to do. Nevertheless, as with the tables, once it is done correctly it need not be completed again; and finally (4) there is an error in one of the formulas! There is a problem with one of the formulas, however, the one for the third derivative using a quintic (fifth degree) polynomial fitting function. The coefficients produced are thus incorrect.

The solution to this problem has been identified by Premek Lubal, one of the members of the International Chemometrics Society.[26] In the formula for the case involved, the quintic fitting function for the third derivative, the term (50 * m) has the wrong sign. The sign in the printed formula is negative (−), and it should be positive (+). After changing the sign of that term, the correct coefficients are the result.

So now the question presents itself, is there a more general method of computing coefficients for any arbitrary set of combinations of derivative order, polynomial degree and number of data points to fit? That is, is there an automated method for computing Madden's formulas, or at least the Savitzky-Golay convolution coefficients? The answer turns out to be Yes. Chris Brown pointed out some pertinent literature citations[27–28] and summarized them into a general solution.[29] Is the solution as "simple" as the tables in Savitzky-Golay/Steinier or the formulas in Madden? This is a matter of perception. If this general solution was presented to the chemical/spectroscopic community in 1964 (at the time of the original Savitzky-Golay paper) it would have been considered far beyond what a "mere" chemist would be expected to know, and would never have gained the popularity it currently enjoys.

Wentzell, *et al.*[28] present this scheme in a very clear way. A matrix is created based on the index of coefficients that are to be ultimately produced. Savitzky and Golay labeled the coefficients in relation to the central data point of the convolution, therefore the coefficients, a three-term set of coefficients are labeled −1, 0, 1. A five-term set is labeled −2, −1, 0, 1, 2 and so forth.

The matrix (**M**) is set up like this table (this, of course, is only one example, for demonstration purposes) as Equation 6.23.

$$\mathbf{M} = \begin{vmatrix} 1 & -3 & 9 & -27 \\ 1 & -2 & 4 & -8 \\ 1 & -1 & 1 & -1 \\ 1 & 0 & 0 & 0 \\ 1 & 1 & 1 & 1 \\ 1 & 2 & 4 & 8 \\ 1 & 3 & 9 & 27 \end{vmatrix} \tag{6.23}$$

The key characteristics of this matrix, are as follows. The first one is that each column contains the set of index numbers raised to the n − 1 power, where n is the column number in the table. Thus the first column contains the zeroth power, which is all 1's, the second column contains the first power, which is the set of index numbers themselves, and the rest of the columns are the second and third powers of the index numbers.

The number of rows is determined by the number of coefficients that are to be calculated; in this example, a set of seven coefficients. The number of columns is determined by the degree of the polynomial that will be used as the fitting function. The number of columns also determines the maximum order of derivative that can be computed. In this example, a third-power fitting function is used and one can produce up to a third derivative. Note that coefficients for lower-order derivatives are also computed simultaneously.

The matrix **M** is then used as the argument for the following matrix as Equation 6.24.

$$\text{Coefficients} = (\mathbf{M}'\mathbf{M})^{-1}\mathbf{M}' \tag{6.24}$$

Where, by convention the boldface **M** refers to the matrix produced, the superscript $'$ refers to the transpose of the matrix and the superscript -1 denotes the matrix inverse of the argument.

The matrix **M** is given above, as Equation 6.23. The transpose (**M'**), then, is as Equation 6.25.

$$\mathbf{M}' = \begin{vmatrix} 1 & 1 & 1 & 1 & 1 & 1 & 1 \\ -3 & -2 & -1 & 0 & 1 & 2 & 3 \\ 9 & 4 & 1 & 0 & 1 & 4 & 9 \\ -27 & -8 & -1 & 0 & 1 & 8 & 27 \end{vmatrix} \tag{6.25}$$

To multiply these two matrices together to form **M'M** as in Equation 6.26 (rules for matrix multiplication are given in many books, including reference 15).

$$\mathbf{M}'\mathbf{M} = \begin{vmatrix} 7 & 0 & 28 & 0 \\ 0 & 28 & 0 & 196 \\ 28 & 0 & 196 & 0 \\ 0 & 196 & 0 & 1588 \end{vmatrix} \tag{6.26}$$

The matrix inverse $(\mathbf{M}'\mathbf{M})^{-1}$ of Equation 6.26 is as Equation 6.27.

$$(\mathbf{M}'\mathbf{M})^{-} = \begin{vmatrix} 0.33333 & 0 & -0.04762 & 0 \\ 0 & -0.26256 & 0 & -0.03241 \\ -0.04762 & 0 & 0.011905 & 0 \\ 0 & -0.03241 & 0 & 0.00463 \end{vmatrix} \tag{6.27}$$

Equation 6.28 contains scaled coefficients for the zeroth through third derivative convolution functions, using a third degree polynomial fitting function. The first row of Equation 6.27 contains the coefficients for smoothing, the second row contains the coefficients for the first derivative, and so forth.

Equation 6.28 gives the coefficients, but there is a scaling factor missing. Therefore there is one more final computation that needs to be performed to create the correct coefficients; each row must be multiplied by the scaling factor. The scaling factor is (p − 1)! where p is the row number. Therefore the scaling factors for the first two rows are unity, since 0! and 1! are both unity, the scaling factor for the third row is two and for the fourth row is six. The final set of coefficients, therefore is given in Equation 6.29; this gives $(\mathbf{M}'\mathbf{M})^{-1}\mathbf{M}'$ (corrected for scaling).

Finally, for those who are facile with the matrix mathematics, Bialkowski[27] also shows how the "end effect" can be obviated, as well as allowing the use of even numbers of data points, but the advanced considerations involved are beyond the scope of this treatment of the subject.

Finally, multiplying Equation 6.27 by Equation 6.25 gives Equation 6.28; and 6.28 is corrected for scaling as shown in Equation 6.29.

$$(\mathbf{M}'\mathbf{M})^{-1}\mathbf{M}' = \begin{vmatrix} -0.09523 & 0.142857 & 0.285714 & 0.333333 & 0.285714 & 0.142857 & -0.09523 \\ 0.087301 & -0.26587 & -0.23016 & 0 & 0.230158 & 0.265873 & -0.0873 \\ 0.059524 & 0 & & & -0.03571 & 0 & 0.059524 \\ -0.02778 & 0.027778 & & & -0.02778 & -0.02778 & 0.027778 \end{vmatrix} \tag{6.28}$$

$$(\mathbf{M}'\mathbf{M})^{-1}\mathbf{M}' = \begin{vmatrix} -0.09523 & 0.142857 & 0.285714 & 0.333333 & 0.285714 & 0.142857 & -0.09523 \\ 0.087301 & -0.26587 & -0.23016 & 0 & 0.230158 & 0.26583 & -0.0873 \\ 0.119047 & 0 & -0.07143 & -0.09524 & -0.07143 & 0 & 0.119048 \\ -0.16667 & 0.166666 & 0.166666 & 0 & -0.16667 & -0.16667 & 0.166666 \end{vmatrix} \tag{6.29}$$

The magnitudes of the various derivatives, and especially the relative magnitudes of derivatives of different orders, depend on the units used, particularly the units used to describe the X-axis. In the case of measured spectra, if spectra are measured in microns as the units for the X-axis, the same spectrum would have a calculated value for the first derivative that was 1000 times what would be calculated for a "nm-based" derivative. In that case, the first derivative for a 10 nm wide band (as 0.01 micron) would be 100 times greater than the maximum spectral value, rather than being 1/10 of it, as the value computed using nanometers for the X-scale.

The Savitzky-Golay method of computing derivatives is based on a least-squares fit of a polynomial to the data of interest. Figure 6.1 illustrates both linear and quadratic fits to a Normal (Gaussian) spectral curve. Figure 6.2 is an expansion of the Figure 6.1 demonstrating how the polynomials are determined using a least-squares fit to the actual data in the region where the derivative is computed, even when the data is contaminated with noise. (Note the red dots represent the actual data.) The difference between the ordinates of two points equally spaced around $\mu + \sigma$ as a function of the spacing is illustrated in Figure 6.3. In this figure the underlying absorbance curve has a bandwidth of 20 nm.

This discussion of derivatives is intended to supplement other coverage of this topic within the volumes of this set. Please refer to the main or master index for more information on the subjects of derivatives, including smoothing and Savitzky-Golay.

6.4 The Statistics of Spectral Searches (Comparisons)

6.4.1 *Introduction to Standard Approaches*

There are a variety of mathematical techniques used for determining the matching index (or agreement) between an unknown test spectrum (or signal pattern) and a set of known or reference spectra (or multiple signal patterns).[30–45] The set of known spectra are often referred to as a *reference spectral library*. In general, high match scores or high similarity is indicative of greater "alikeness" between an unknown test spectrum and single or multiple known reference spectra contained within a reference library.

A basic list of the techniques used to compare an unknown test spectrum to a set of known library spectra for qualification or identification purposes is found in Table 6.8. Some of the mathematical approaches used will be described in greater detail within this chapter. Other volumes within this set provide more information and descriptive examples, see the index for "spectral searches" or "spectra matching", or in general Chapters 6 of each volume, particularly Volume 4. Note that the variable names may be different for each description and are delineated along with each equation.

The ASTM (American Society for Testing and Materials) International has published a "Standard Practice for General Techniques for Qualitative Analysis" (Method E 1252 from Committee E13). The method describes techniques useful for qualitative evaluation of liquids, solids, and gases using the spectral measurement region of 4000 to 50 cm^{-1} (above 2500 nm).[30]

Table 6.8 A listing of classic spectral search algorithms and terminology.

1. Visual overlap *test spectrum* (t) and *reference spectrum (r)* to compare spectral shapes for similarity for simple difference
2. Search and identify (compare individual peaks or sets of spectral peaks)
3. Compare physical data or chemical measurements between samples
4. Use mathematical methods, such as Hit quality index (HQI) value (or 'similarity' value vs. library reference sample). An example list of such HQI methods includes:
   a. Euclidean distance (d) algorithm for

$$\mathrm{d} = \left\{ \sum_{i=1}^{n} (t_i - r_i)^2 \right\}^{\frac{1}{2}} \tag{6.30}$$

   b. First derivative (δ) Euclidean distance algorithm for

$$1\mathrm{Dd} = \left\{ \sum_{i=1}^{n} (\delta_1 t_i - \delta_1 r_i)^2 \right\}^{\frac{1}{2}} \tag{6.31}$$

   c. Sum of differences (sd) as

$$\mathrm{sd} = \sum_{i=1}^{n} (t_i - r_i) \tag{6.32}$$

   d. Correlation [row matrix (row vector) dot product] as r

$$\mathbf{r} = (T \cdot R) = \sum_{i=1}^{n} T_i R_i \tag{6.33}$$

5. Other approaches: Hamming networks, pattern recognition, wavelets, genetic algorithms, and neural network learning systems. Recent published descriptions of these methods for infrared and Raman are included in references 42–45.

6.4.2 *Mahalanobis Distance Measurements*

The Mahalanobis Distance statistic (or more correctly the square of the Mahalanobis Distance as D^2) is a scalar measure of where the spectral vector **a** lies within a set of samples and their multivariate parameter space used for a calibration model.[32–33] The Mahalanobis distance is used for spectral matching, for detecting egregious outliers during calibration or prediction, or for detecting extrapolation of the model during analyses. Various commercial software packages may use D instead of D^2, or may use other related statistics as an indication of high leverage outliers, or may refer to the Mahalanobis Distance by another term. The use of D^2 is preferred as a distance measure since it is more easily related to the number of samples and variables.

For validation purposes it is important to verify exactly what distance method is being used. Both mean-centered and not mean-centered definitions for Mahalanobis Distance exist, with the mean-centered approach being preferred. Regardless of whether mean-centering of data is performed, the statistic designated by D^2 has valid utility for qualitative calculations or spectral matching. The smaller the distance from the test spectrum to a library spectrum, the closer is the match.

If t is the spectrum of a test sample (i.e., a spectral vector (dimension f by 1)) and **R** is the matrix of reference spectra (of dimension n by f), then the Mahalanobis Distance is defined as Equation 6.34:

$$D^2 = \mathbf{t}'(\mathbf{RR}')^{+}\mathbf{t} \tag{6.34}$$

This equation may be used to compute the distance between the test sample spectrum and a reference set of sample spectra.

6.4.3 *Euclidean Distance*

The Euclidean Distance approach (Equation 6.35) may be used to compare a test spectrum (t) to a reference spectrum (r) or a library of reference spectra. The smaller the distance the greater the match between the test (t) and reference (r) spectra. A distance of near 0 is a perfect match. The equation is as:

$$\mathbf{d} = \{(\mathbf{t}_{11}-\mathbf{r}_{11})^2 + (\mathbf{t}_{12}-\mathbf{r}_{12})^2 + (\mathbf{t}_{13}-\mathbf{r}_{13})^2 + \cdots + (\mathbf{t}_{1i}-\mathbf{r}_{1i})^2\}^{1/2} \tag{6.35}$$

where t_{ki} are data points from the test spectrum, and r_{ki} are data points from the reference (or known comparison spectrum). For this nomenclature, k is the spectrum replicate or sample number and i is the data point number within each spectrum. The distance is calculated at each data point (from 1 to i), with a comparison distance between each test spectrum and each reference spectrum. The distance from a reference spectrum to the test spectrum is calculated as the Euclidean distance; small differences are the best match.

6.4.4 *Common Spectral Matching (Correlation or Dot Product)*

Another general method for comparing two spectra (test versus reference), where the reference is a known compound; or the mean spectrum of a set of known spectra, is given as the M.I. (Match Index). The M.I. is computed by comparing the vector dot products between the test and reference spectra. This computation is referred to as the correlation, dot product, or cosine similarity. The theoretical values for these dot products range from -1.000 to $+1.000$, where -1.000 is a perfect negative (inverse) correlation, and $+1.000$ is a perfect match. Since for near infrared spectroscopy only positive absorbance values are used to compute the dot products, the values for the match index must fall within the 0.0 to +1.0 range. The mathematics is straightforward and are demonstrated below.

The match index (M.I.), i.e., cosine similarity, is equal to the cosine of the angle (designated as α) between two row vectors (the test and reference spectra) projected onto a two-dimensional plane, and is equivalent to the correlation between the two spectra (as row vectors) as Equation 6.36. Note that in the following equations, r represents the reference spectrum and not the correlation.

$$M.I. = \cos\alpha = \left(\frac{t \cdot r}{|t||r|}\right) \tag{6.36}$$

Where t is the test spectrum row matrix, and r is the reference spectrum row matrix.

Note the following Equation 6.37:

$$(T \cdot R) = \sum_{i=1}^{n} t_i r_i \tag{6.37}$$

where t_i represents the individual data points for the test spectrum (designated as the intensity values of spectrum t from wavenumbers i = 1 through n), and r_i represents the individual data points for the reference spectrum (designated as the intensity values of spectrum r from wavenumbers i = 1 through n).

$$\text{And where: } |t||r| = \left(\sum_{i=1}^{n} t_i^2\right)\left(\sum_{i=1}^{n} r_i^2\right) \tag{6.38}$$

Note, the angle (α), in degrees, between two vectors can be determined from the match index (M.I.) or cosine similarity using Equation 6.39.

$$\alpha = \cos^{-1}(M.I.) \tag{6.39}$$

The "alikeness" of one test spectrum (t), or series of test spectra, to a reference spectrum (r) can be determined by calculating a point-by-point correlation between absorbance data for each test and reference spectrum. The correlation matching can be accomplished for all data points available or for a pre-selected set only. The more alike the test and reference spectra are, the higher (closer to 1.000) are the various r (correlation coefficients) and R^2 (coefficient of determination) values. A perfect match of the two spectra would produce r or R^2 values of 1.0000000.

The sensitivity of the technique can be increased by pre-treating the spectra as first to higher order derivatives and then calculating the correlation between test and reference derivative spectra. Full spectral data can also be truncated (i.e., windowed or reduced) to include only spectral regions of particular importance, a practice which will further improve matching sensitivity for a particular spectral feature of interest.

Sample selection using this technique involves selecting samples most different from the mean population spectrum for the full sample set. Those samples with correlations of the lowest absolute values (including negative correlations) are selected first and then samples of second lowest correlation are selected (and so on) until the single sample of highest correlation is found. The distribution of spectra about the mean is assumed to follow a normal distribution with a computable standard deviation. This assumption indicates that a uniformly distributed sample set can be selected based on the correlation between test spectra and the mean spectrum of a population of spectra.

6.4.5 *Computing the Probability that Two Sample Spectra are Statistically Identical*

6.4.5.1 *Example Problem*

For converting correlation between two spectra (as shown in Equations 6.36 through 6.39) to a parametric statistic one may use the Fisher's z-transform, also known as the r-to-z′ transformation (i.e., Fisher's z′ Transform) from the correlation (r).[50–52]

The r-to-z′ transformation statistic is used to compare two correlation coefficients to see if they are significantly different, such as another correlation possibly obtained in the same or a similar experiment. As an example, if the null hypothesis test is $r = 1.000$, it will indicate if the measured correlation is significantly different from 1.000. This example is a test to see if the spectral correlation of a reference sample (r) and test sample (t) are the same.

For comparing two correlation coefficients using parametric statistics, the r-to-z transformation is used. This refers to the "r-to-z′ transformation" for comparing two correlation coefficients, particularly when at least one r-value does not equal zero. For this example case the null (H_0) and test (H_A) hypotheses, respectively, are as H_0: $r = 1.00$, and H_A: $r < 1.00$ (or could be stated as r is not equal to 1.00).

For this question Equation 6.40 below is used in conjunction with the r to z′ transformation as Table 6.9 (this table gives the z′ transform for any correlation value, as Equation 6.41). The critical values for the t distribution are found in Table 6.10 (this gives the critical t values for any degrees of freedom for different significance levels). For a hypothetical example, given a correlation coefficient of $r_{Reference} = 0.999$ (as representing the correlation coefficient obtained by regressing the reference sample spectrum to itself for a number of replicate

Table 6.9 For r to z′ transformation.

r (correlation computed)	z transform (z′)
0.9999999999	11.859
0.9999999	8.406
0.999999	7.254
0.99999	6.103
0.9999	4.952
0.999	3.800
0.998	3.453
0.997	3.250
0.996	3.106
0.995	2.994
0.994	2.903
0.993	2.826
0.992	2.759
0.991	2.700
0.990	2.647
0.985	2.443
0.980	2.298
0.975	2.185
0.970	2.092
0.965	2.014
0.960	1.946
0.955	1.886
0.950	1.832
0.945	1.783
0.940	1.738
0.935	1.697
0.930	1.658
0.925	1.623
0.920	1.589
0.915	1.557
0.910	1.528
0.905	1.499
0.900	1.472

sample measurements, or for a reference sample set measured at time 1 (designated as X) versus time 2 (designated as Y), say 1 hour apart.

For this example, compute the correlation (r-value) and report it as (r_{Ref}). Next the correlation is computed for the test sample spectrum (r_{Test}) as compared to the reference sample spectrum. (Compute the r-value and report it as r_{Test}). Note if the r_{Test} is greater than the r_{Ref} the sample is considered a match.

The z observed statistic is then computed for this correlation comparison to determine whether it is significantly different from 0.999 (the Reference r-value or $r_{Ref.}$). To continue, if r_{Test} is 0.992, the test will be to compare 0.992 to 0.999 to see if they are statistically the same. For this example, note the test sample set is 20 replicate sample spectra with an average correlation value computed; for this example $n_{Ref} = 20$ and $n_{Test} = 20$.

$$Z_{obs} = \frac{|Z_{Ref} - Z_{Test}|}{\sqrt{\frac{1}{n_{Ref}-3} + \frac{1}{n_{Test}-3}}} \tag{6.40}$$

For this case, calculate the r to z′ transformation for both reference and test r-values as: 0.999 and 0.992, respectively. An r to z′ table (Table 6.9 or Equation 6.41) may be used to compute the r to z′ transformation. For this example, the $z_{Ref.} = 3.800$ and the $z_{Test} = 2.759$. One may now compute the $Z_{obs} = 3.04$. (Compare this value to the Critical Values for the t Distribution, Significance level of 0.025 for a one-tailed test and the critical value is 2.13, see Table 6.10 or the t-distribution table for any sample number (n)).

Conclusion for this example: Since 3.04 > 2.13 one rejects the null hypothesis and accepts the alternate hypothesis, thus r_{Ref} is not the same as r_{Test} and $r < 1$ for the Test sample comparison at 95% confidence. The correlation values are not the same for reference spectra and test spectra, as they have significantly different correlation values.

Note the equation for computing the r to z′ transformation; the sampling distribution of Pearson's correlation (r-value) is not normally distributed. Fisher developed a transformation now called "Fisher's z′ transformation" which converts the Pearson's r-value to the normally distributed variable z′. The r to z′ transformation equation is as Equation 6.41. This statistic is normally distributed with a standard error of $1/\{n\text{-}3\}^{1/2}$.

$$z' = 0.5[\ln(1 + r) - \ln(1 - r)] \tag{6.41}$$

Table 6.10 Critical values for the t distribution (20 test samples).

Confidence level (%)	Significance level (α), one-tailed test	Critical value of t
90	0.05	1.75
95	0.025	2.13
98	0.01	2.60
99	0.005	2.95
99.9	0.0005	4.07

Table 6.11 Showing the various parameters associated with correlation between vectors.

corr (r)	cod (r^2)	exp. Variance	(α) rad	(α) degrees	MI = cos(α)	HQI	z'
1.000	1.000	100.000	0.000	0.000	1.000	100.000	∞
0.999	0.998	99.800	0.045	2.563	0.999	99.900	3.800
0.990	0.980	98.010	0.142	8.110	0.990	99.000	2.647
0.985	0.970	97.023	0.173	9.936	0.985	98.500	2.443
0.980	0.960	96.040	0.200	11.478	0.980	98.000	2.298
0.975	0.951	95.063	0.224	12.839	0.975	97.500	2.185
0.970	0.941	94.090	0.246	14.070	0.970	97.000	2.092
0.965	0.931	93.123	0.265	15.204	0.965	96.500	2.014
0.960	0.922	92.160	0.284	16.260	0.960	96.000	1.946
0.955	0.912	91.203	0.301	17.254	0.955	95.500	1.886
0.950	0.903	90.250	0.318	18.195	0.950	95.000	1.832
0.940	0.884	88.360	0.348	19.948	0.940	94.000	1.738
0.930	0.865	86.490	0.376	21.565	0.930	93.000	1.658
0.920	0.846	84.640	0.403	23.074	0.920	92.000	1.589
0.910	0.828	82.810	0.428	24.495	0.910	91.000	1.528
0.900	0.810	81.000	0.451	25.842	0.900	90.000	1.472
0.750	0.563	56.250	0.723	41.410	0.750	75.000	0.973
0.500	0.250	25.000	1.047	60.000	0.500	50.000	0.549
0.250	0.063	6.250	1.318	75.522	0.250	25.000	0.255
0.100	0.010	1.000	1.471	84.261	0.100	10.000	0.100
0.000	0.000	0.000	1.571	90.000	0.000	0.000	0.000

6.4.5.2 *Geometric Interpretation of Cosign Similarity for Spectral Matching and use of 0-100 Scale for Hit Quality Index*

The relationship between two vectors, as reference and test spectra, may be expressed in multiple ways as convenient for interpreting the *cosine similarity* between the vectors as Table 6.11. This table demonstrates the following parameters: Column 1 is the correlation between ref and test spectrum, column 2 is the coefficient of determination (r^2), column 3 is the explained variance, column 4 is the geometric angle (α) between the spectra in radians, column 5 is the geometric angle (α) between the spectra in degrees, column 6 is the Match Index between the spectra (as cos(α)), and column 7 is the Hit Quality Index from 0 to 100 as (MI * 100, also equal to | r*100|). A column 8 may be used to indicate the z transform such that the values may be compared to a standard t table for hypothesis significance testing (see Table 6.12).

Final Evaluation: If the correlation is greater for the test sample than the reference sample then the sample is identified as identical to the reference material. If the test correlation is less than the reference correlation then the z-transform method is used to compute the probability of the test sample being the "same"

Table 6.12 Critical values for the t distribution (greater than 120 datapoints).

Confidence level (%)	Significance level (α), one-tailed test	Critical value of t
90	0.05	1.65
95	0.025	1.96
98	0.01	2.33
99	0.005	2.58
99.9	0.0005	3.29

as the reference sample material. Thus, IF r test $\geq$ r reference THEN the sample is identified as the same material. IF the r test $\leq$ r reference THEN the z-transform statistic is used to compute the probability that the samples are identical.

6.5 Raman Spectral Resolution Test(s)

6.5.1 *Introduction*

Spectral resolution for Raman spectrometers is complex due to the different types of instruments and various methods used for measuring resolution. There are dispersive and interferometric design spectrometers used with reference samples and emission lines, using different procedures to complicate the issues. Spectral resolution is a measure of the minimum distance between the centers of adjacent spectral bands that may be partially or fully separated such that each band is clearly observed. The concept of distance in this discussion refers to the separation in Raman shift spectral space (X-axis) in units of wavenumbers (cm^{-1}).

Spatial resolution is not the same as spectral resolution. Spatial resolution is an ability of an optical spectrometer to distinguish individual features on a surface or image to partially or completely separate the individual features for observation. The units of measure for such spatial resolution are generally in units of microns or millimeters. For the concept of resolution, the lesser the number, the greater the resolution. For example, 4 cm^{-1} spectral resolution is twice that of 8 cm^{-1} spectral resolution. Likewise a 50 micron spatial resolution is 4 times greater than a 200 micron spatial resolution.

As the resolution increases the overall signal to noise will decrease for the same spectrometer. This is due to the fact that higher (i.e., narrower) resolution decreases the energy throughput (i.e., etendue) of an optical system over the same system used for a low (broader) resolution measurement. Higher resolution requires smaller slit widths for dispersive spectrometers, and a smaller J-stop aperture for interferometers. The reduced energy throughput reduces overall signal intensity and thus with equivalent noise levels in mechanical and electronic operation, the overall noise is a greater percentage of the total signal. Selecting the optimum resolution is a trade-off between the bands of interest the user would like to resolve, the signal to noise requirements, and the speed of measurement. For example, a lower resolution setting for a spectrometer provides faster measurement for improved signal to noise.

When band shifts occur for certain sample types in Raman measurements, and these band shifts are to be monitored for analysis, it is often assumed that high spectral resolution is essential for these measurements. However, the important factor for measuring band shifts in samples over time is the wavenumber precision (as repeatability and reproducibility). For determining the precision of a Raman spectrometer in wavenumber space one may compute the standard deviation or root mean square (RMS) deviation to determine the theoretical limit for measuring band shifts. Chapter 6 in Volumes 2, 3, and 4 of this set describe the various methods for computing wavelength or wavenumber precision for a spectrometer. One may make these measurements by using the following equations and by taking 3 to 5 repeat measurements of a wavenumber reference standard to compute the RMS of the peak positions over various times and measurement conditions.

6.5.2 *Wavenumber repeatability or precision test*

Verify the wavelength repeatability of the spectrophotometer using a suitable reference material standard, for example: silicon, O_2, N_2, calcite or cyclohexane (See index and Table 3.6, Chapter 3 of this volume, for listings of band positions for these materials). The standard deviation of the selected wavenumber must be consistent with the required performance specifications for the application in use.

For this test, repeat measurements of the same reference material standard by placing it in the sample beam and not mechanically moving the sample over a normal measurement cycle for the instrument. For example, a 20 second measurement period, and 20 second reference spectrum may be typical. Then calculate the first derivative of each of the background replicate spectra — compute the inflection or zero-crossing positions for the center band at the expected reference peak in cm^{-1} for each scan (i.e., scan-to-scan for each sample) and the mean spectrum wavenumber position ($\bar{x}_i$) for the measured (x_{ij}). Do not smooth the data. The standard deviation

(RMS) is calculated as:

$$s_i = \sqrt{\frac{\sum_{j=1}^{n} (\nu_{ij} - \bar{\nu}_i)^2}{n - 1}} \quad (6.42)$$

where (s_i) is the standard deviation for the scan-to-scan wavelength precision or repeatability for scan-to-scan measurements (i.e., RMS), and where (ν_{ij}) are individual wavenumber shifts of the zero-crossover for sample i and scan-to-scan number j; $(\bar{\nu}_i)$ is the average value for the scan-to-scan set; and n is the number of replicate measurements (pool all scan-to-scan data). See other methods for peak selection by referring to the index of this set of volumes for details.

The mean spectrum wavenumber position $(\bar{\nu}_i)$ is calculated as:

$$\bar{\nu}_i = \frac{\sum_i^n \nu_i}{n} \quad (6.43)$$

The results are reported as: Wavenumber (cm^{-1}) Repeatability (RMS)

	Mean wavenumber $(\bar{\nu}_i)$, Eq. 6.43	Precision/Repeatability (s_i), Eq. 6.42
Scan-to-Scan	—	—

One may use a factor of three (3) as a multiplier term for the RMS value to represent a 99.7% confidence that a true deviation in wavenumber for a sample greater than 3 × RMS may be detected by the spectrometer. This may be stated in Equation 6.44 form as:

$$\text{Minimum Detection Limit for Band Shift Change} = 3 \times \text{RMS (in cm}^{-1}) \quad (6.44)$$

The most common method for determination of spectral resolution in Raman spectrometers is ASTM International Method E2529-06, which will be discussed in more detail later in this section. Figures 6.4 through 6.7 show the results of different resolution power for the same doublet or double band. Figure 6.4 shows a single band with lowest spectral resolution; Figure 6.5 illustrates slightly higher resolution; Figure 6.6 illustrates higher resolution to begin to resolve the two bands more clearly; and Figure 6.7 shows high spectral resolution with Raman shift bands fully resolved.

Spectral resolution is affected by line density of the diffraction grating in terms of lines/mm; the F number of the monochromator or spectrograph; the pixel spacing dimensions for a CCD or array detector; the slit width dimensions for a monochromator; and the J-stop and interferometer moving mirror velocity for an interferometer. These are listed in Table 6.13.

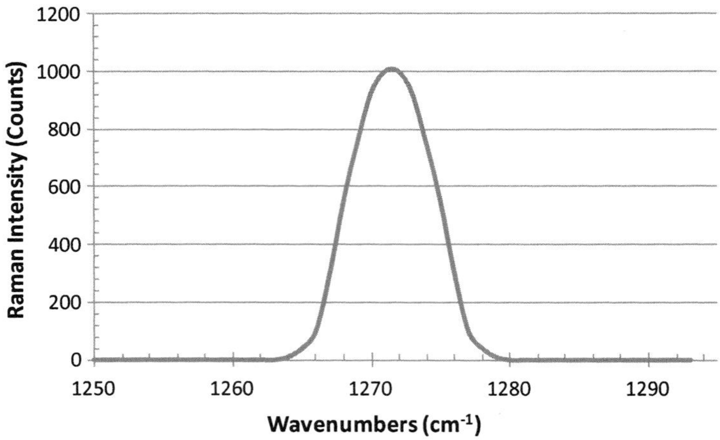

Fig. 6.4 Lowest spectral resolution of two bands (i.e., unresolved single band observed).

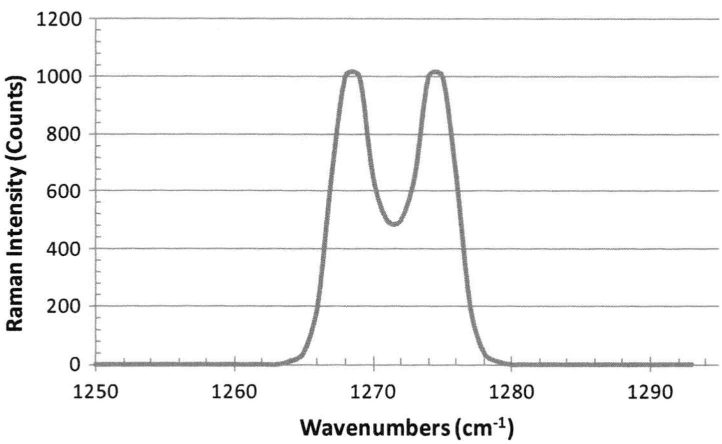

Fig. 6.5 Increase in resolution of two bands as compared to Figure 6.4.

Since resolution is often measured in different ways, it is important to first find agreement in the method that will be used to test resolution. Note that spectral resolution varies over the spectral region of Raman spectrometers where a standard design is employed; with spectral resolution increasing with increasing Raman shift (in cm^{-1}). Recall that dispersive (i.e., monochromator and CCD/array spectrograph) spectrometers have a relatively constant or uniform resolution in wavelength (nm or micron) space, but do not have constant resolution in wavenumber (cm^{-1}) space. In contrast, interferometers exhibit uniform resolution in wavenumber (cm^{-1}) space

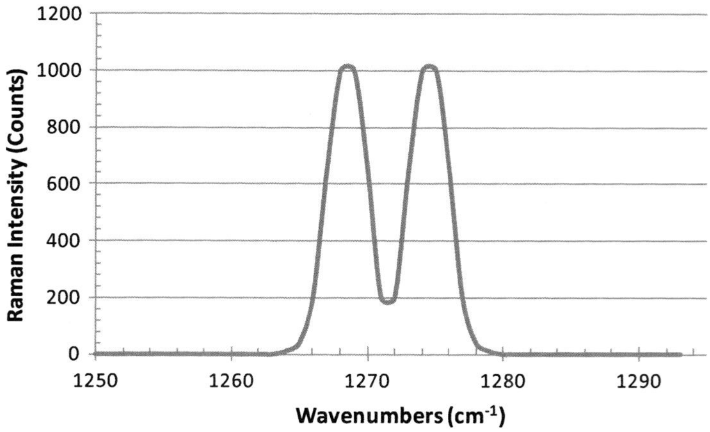

Fig. 6.6 Intermediate resolution of two bands, higher spectral resolution than Figs. 6.4 and 6.5.

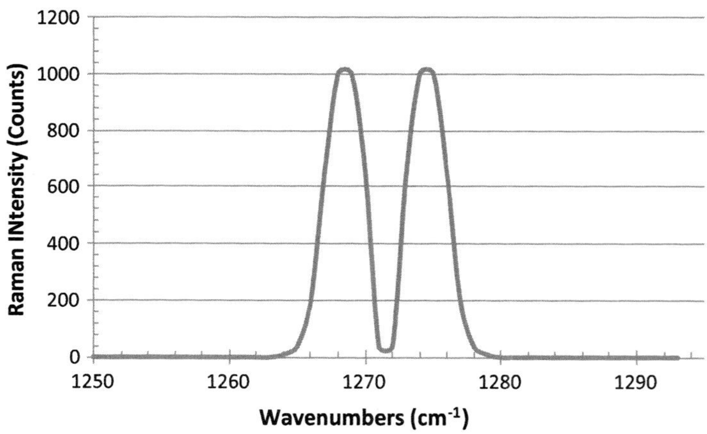

Fig. 6.7 Highest spectral resolution of two bands, higher than Figs. 6.4 through 6.6.

but not in wavelength space. Note the following equations showing a conversion between nanometers and wavenumbers.

Wavelength (in nanometers) to wavenumbers (cm^{-1}):

$$\bar{\nu}_{cm^{-1}} = \frac{10^7}{\lambda_{nm}} \quad (6.45)$$

Wavenumbers (cm^{-1}) to wavelength (in nanometers):

$$\lambda_{nm} = \frac{10^7}{\bar{\nu}_{cm^{-1}}} \quad (6.46)$$

Table 6.13 Factors affecting spectral resolution for Raman spectrometers.

Spectrometer type	Spectrometer parameter	High resolution setting/design
Interferometer	J-stop setting	Smaller aperture
	Moving mirror velocity	Faster mirror rate
Spectrograph	Line density for diffraction grating	More lines per millimeter (lines/mm)
	Pixel dimensions for CCD/array	More pixels, closer together
	Focal length	Longer focal length (F number)
Monochromator	Line density for diffraction grating	More lines per millimeter (lines/mm)
	Slit width	Smaller
	Focal length	Longer focal length (F number)

Note that spectral resolution also changes with the wavelength of the excitation laser. For example, a 780 nm laser may exhibit twice the resolution at a Raman shift of 3500 cm^{-1} when compared to the 100 cm^{-1} Raman shift.

6.5.3 *Measuring Spectral Resolution*

ASTM International E2529-06 uses a narrow band of calcite ($CaCO_3$) to determine the Full Width at Half Maximum (FWHM) for the Raman spectrometer at that specific band position (i.e., 1085 cm^{-1}).[53] A comparative neon emission source is used where the 588.2 nm neon line corresponds to a Raman shift band at 1796 cm^{-1}. A simulation of the calcite band is given in Figure 6.8, illustrating a resolution as FWHM of approximately 7 cm^{-1} and 3 cm^{-1} (broad and narrow bands, respectively).

6.6 Raman Spectrometer Calibration Basics

6.6.1 *Introduction*

Raman spectroscopy requires both wavenumber (wavelength, x-axis) and intensity (y-axis) calibration in order for the instrumentation to perform properly for qualitative and quantitative analysis, as well as spectral database building. This section discusses basic calibration methods, noting that additional detailed method descriptions are found in Chapter 8. Neon or argon emission lines are often used for wavelength calibration. A National Metrology Institute (NMI, i.e. such as NIST) traceable calibrated irradiance source, also referred to as a laboratory standard white-light source, may be used for intensity calibration.[54]

After the Raman spectrometer is calibrated, a reference standard material (such as those given in this volume, Chapter 3, Table 3.5; and as described in Chapter 8) may be used for final qualification of the spectrometer. Calibration spectra are collected

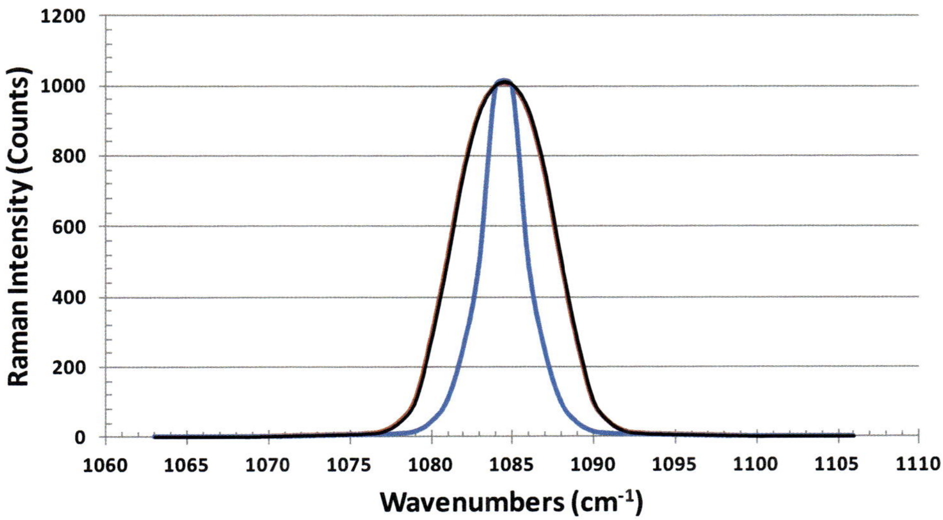

Fig. 6.8 Illustration of calcite spectral resolution test band at 1085 cm^{-1}; broad darker band is 7 cm^{-1} spectral resolution and narrow band is at 3 cm^{-1} resolution, respectively. Resolution is determined by width in wavenumbers at the half height (FWHM) position (in this case approximately 500 counts). For the broad band at 500 counts the computation is 1088 cm^{-1} − 1081 cm^{-1} = 7 cm^{-1}. For the narrow band it is 1086 cm^{-1} − 1083 cm^{-1} = 3 cm^{-1}.

and maintained over time for difference (i.e., repeatability and reproducibility) comparisons. The spectrometer may be accurately maintained using such practices in a conscientious manner.

6.6.2 *Wavenumber/Wavelength Calibration*

For calibrating dispersive spectrometers for wavelength or wavenumber accuracy, a reference standard material or emission lamp with known bandshape, linewidth, and peak position is measured and compared to the uncorrected spectrometer values. The measured peaks are compared to the known reference peaks and the dispersion coefficients are used to correct the grating to disperse light at the appropriate wavenumber through the spectrograph or monochromator. The fundamental grating equation is given by:

$$\sin\alpha + \sin\beta = 10^{-6}\,\mathrm{k}\cdot\mathrm{n}\cdot\lambda \tag{6.47}$$

In most monochromators, the location of the entrance and exit slits are fixed and the grating rotates around a plane through the center of the face. The angle, θ_V, is a constant determined by:

$$\theta_V = \beta - \alpha \tag{6.48}$$

If the value of α and β is to be determined for a given wavelength (λ), the grating equation may be expressed as:

$$kn\lambda \cdot 10^{-6} = 2\sin\left[\frac{\beta+\alpha}{2}\right]\cos\left[\frac{\beta-\alpha}{2}\right] \quad (6.49)$$

To calculate α in a Monochromator use:

$$\alpha = \sin^{-1}\left[\frac{10^{-6}kn\lambda}{2\cos\left(\frac{\theta_V}{2}\right)}\right] - \frac{\theta_V}{2} \quad (6.50)$$

The symbols for the grating equation are as follows. For the symbols in the grating equation note that α is the angle of incidence (in degrees); β is the angle of diffraction (in degrees); k is the diffraction order number (i.e., integers of 0, 1, 2, etc.); n is the groove density (in grooves per mm); θ_V is the angle of deviation or dispersion (in degrees); and λ is the wavelength of light (in nm). (Volume 3, Chapter 3 gives more details with respect to the use and theory of diffraction gratings.)

Once the appropriate grating equation is derived for the instrument using reference standards with known wavelength or wavenumber peak positions, the remaining differences between reference and measured peaks may be corrected using a second or third order polynomial as given in Equations 6.51 and 6.52, respectively.[55]

$$\lambda_{Ri} = b_0 + b_1\lambda_{Mi} + b_2\lambda_{Mi}^2 \quad (6.51)$$

$$\lambda_{Ri} = b_0 + b_1\lambda_{Mi} + b_2\lambda_{Mi}^2 + b_3\lambda_{Mi}^3 \quad (6.52)$$

Where λ_{Ri} is the reference or known wavelength for a standard reference sample or emission lamp; b_0 is the intercept for the polynomial estimation; b_1, b_2, and b_3 are polynomial regression coefficients; and λ_{Mi} is the measured value for wavelength (or wavenumber) for the λ_{Ri} reference value.

The following virtual example serves to illustrate the value of using a polynomial correction to further align the wavelength or wavenumber axis of a spectrometer using a reference standard sample. The units could be in nanometers or wavenumbers, depending on reference peak values stipulated.

Table 6.14 gives a virtual example showing the measured peak values for a reference material in column 1; and the standard reference peak values (as the measurands) in column 2. Column 3 represents the measured minus the reference peak differences. Column 4 represents the corrected peak positions using a second order polynomial, with column 5 showing the reference minus the predicted peak value differences for the second order polynomial corrected values. Column 6 represents the corrected peak positions using a third order polynomial, with column 7

Table 6.14 Polynomial correction for x-axis differences in a spectrometer.

Measured (λ_{Mi})	Reference (λ_{Ri}) *Measurand*	Meas. - Ref.	2O PRED	Ref.-Pred.	3O PRED	Ref.-Pred
451	450	1	450.4795	−0.4795	450.1038	−0.10381
902	900	2	900.3281	−0.3281	900.3541	−0.3541
1203	1200	3	1200.333	−0.33329	1200.58	−0.57967
1404	1400	4	1400.568	−0.56808	1401.008	−1.00766
1705	1700	5	1700.271	−0.27107	1701.154	−1.15386
1906	1900	6	1900.304	−0.30406	1901.635	−1.63522

showing the reference minus the predicted peak value differences for the second order polynomial corrected values.

The average difference for the uncorrected x-axis registration (column 3 values) is 3.5; note that this spectrometer was corrected for the dispersion constants, as Equations 6.47 through 6.50, but had no further higher order correction. The second order polynomial differences (column 5 values) average −0.38068, much improved over no correction. Lastly, the third order polynomial correction yields a result not as close as the second order correction; the third order correction averages −0.80572 (i.e., column 7 values). Note that the third order correction is much improved over no correction. The results and polynomial equations are graphically shown for the second and third order correction models in Figures 6.9 and 6.10, respectively.

6.6.3 *Intensity Calibration (Simplified Method)*

The process of calibrating a Raman spectrometer for intensity involves the measurement of a relatively flat (i.e., broad featured) white light spectrum of known and traceable intensity.[54] For this purpose, a National Metrology Institute (NMI, i.e. such as NIST) traceable calibrated irradiance source, also referred to as a laboratory standard white-light source, may be used for intensity calibration. The irradiance source is transmitted through the spectrometer or spectrograph and the full spectral region is measured and recorded. A correction factor (as a vector) is then computed as the ratio of the reference values for the white spectrum irradiance source, divided by the spectrometer measured response to the reference irradiance source. This is defined as Equations 6.53 and 6.54. So the simplified intensity correction vector (I_C) is computed as Equation 6.53.

$$I_C = \frac{r}{m} \tag{6.53}$$

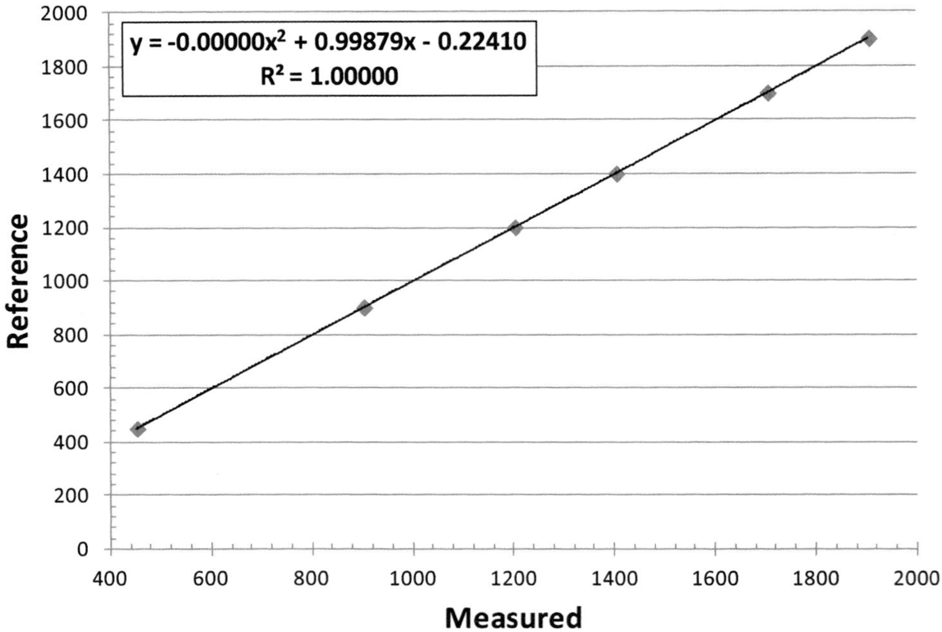

Fig. 6.9 Second order x-axis correction for wavelength or wavenumber alignment. Note in the equation nomenclature shown, the y is the reference peak value and the x is the measured peak value.

where I_C is the corrected intensity vector; r are the standard spectrum values for the NIST traceable lamp; and m are the measured spectrum intensity values for the NIST traceable lamp. An illustration of the reference and measured intensities is given in Figure 6.11 and the I_C correction vector for this example is given in Figure 6.12. A more complex and finer NIST intensity correction method for use with an actual measured sample, using polynomial fitting to a standard reference material and a standard intensity lamp spectrum, is described in Chapter 8 of this volume.

The corrected intensity spectrum at any time 1 is computed using Equation 6.54 as the dot product of the measured spectrum (as a vector) at time 1, times the I_C ratio computed at the time of calibration, designated as time 0.

$$\text{Corrected Intensity Spectrum} = (\text{Spectrum Measured at Time 1}) \times I_C \text{ ratio} \tag{6.54}$$

Thus, when the dot product is computed it will correct the intensity spectrum at any time based upon the initial intensity correction made at the time of calibration, when the I_C ratio was last computed. The official method for intensity calibration as recommended by NIST is given in Chapter 8 of this volume.

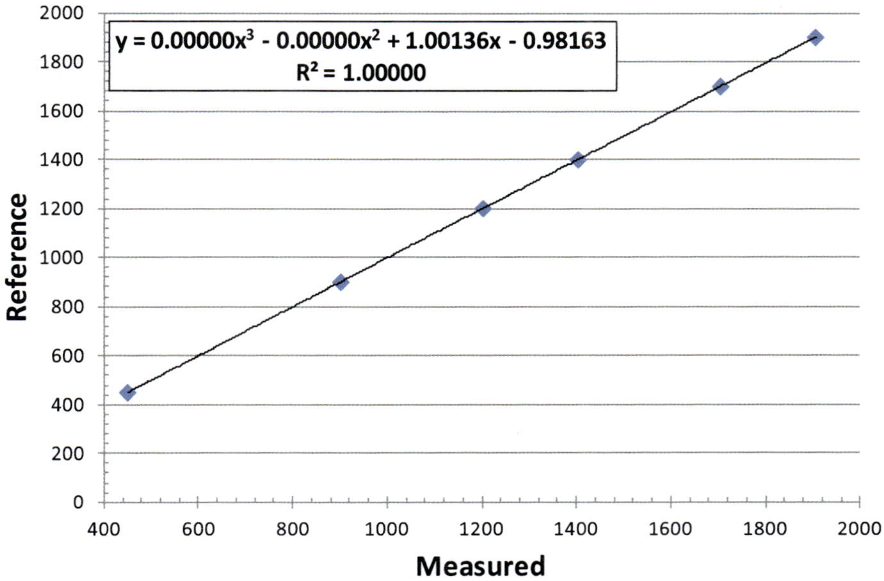

Fig. 6.10 Third order x-axis correction for wavelength or wavenumber alignment. Note in the equation nomenclature shown, the y is the reference peak value and the x is the measured peak value.

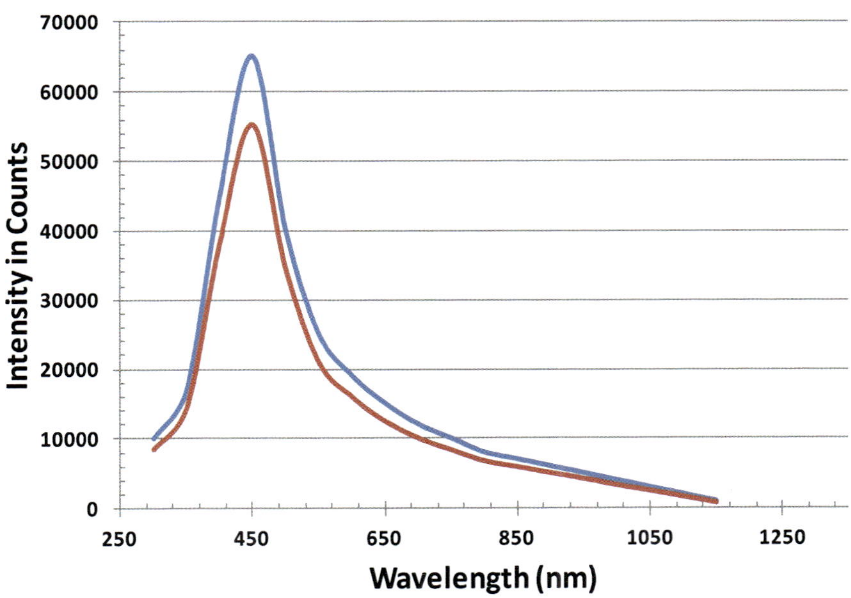

Fig. 6.11 Illustration of example spectrum showing calibrated irradiance white-light source (top spectrum), versus measured values for test spectrometer (bottom spectrum).

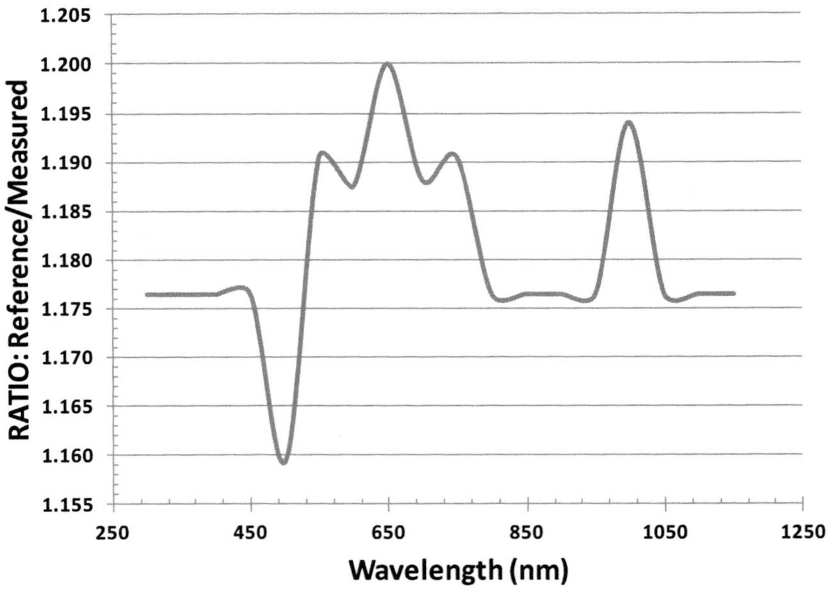

Fig. 6.12 The I_C correction vector ratio is shown in this figure and is computed as the ratio of (reference spectrum/measured spectrum), as in Equation 6.53. This is the ratio of the spectra from Figure 6.11 for the NIST traceable calibrated irradiance source (top spectrum) divided by the measured values for the irradiance source for a test spectrometer during calibration (bottom spectrum). Note: Figure 6.12 is comprised of the ratio of the two spectra from Figure 6.11.

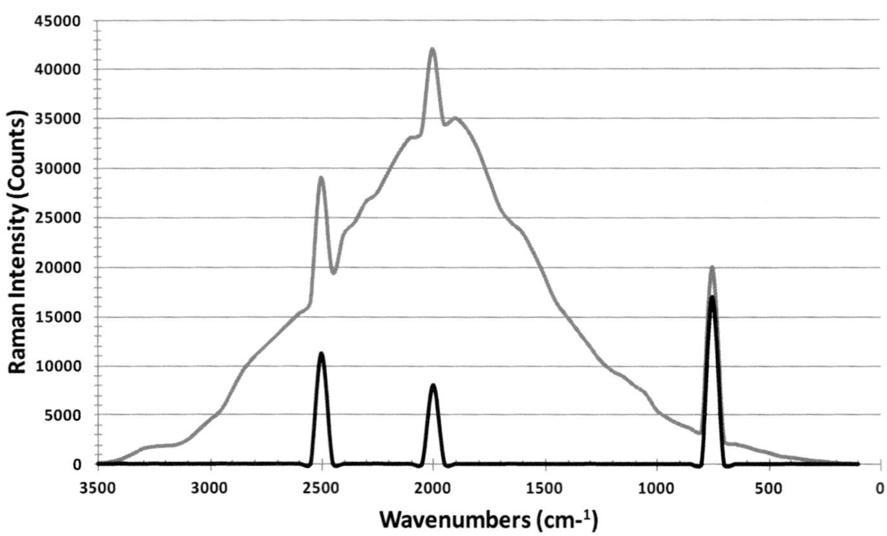

Fig. 6.13 Illustration of a polynomial background correction. Original Raman spectrum (top) and corrected spectrum (bottom). The corrected spectrum is computed by completing a polynomial background fitting to the original spectrum shape (excluding peaks), followed by spectral subtraction of the polynomial shape fit from the original spectrum. The remaining spectrum (bottom, dark) shows the majority of the background fluorescence signal removed and the Raman signal peaks remaining.

6.6.4 *Background Correction Modeling*

If fluorescence or other background interference is present in a Raman spectrum, a polynomial background correction is made by fitting a high order polynomial to the background curve and subtracting this from the original spectrum. This leaves a difference spectrum showing mostly the Raman spectral features with the absence of the background interference. This is demonstrated in Figure 6.13.

References

1. N.K. Afseth, V.H. Segtnan, and J.P. Wold, Raman spectra of biological samples: A study of preprocessing methods, *Applied Spectroscopy* **60**(12) (2006) 1358–1367.
2. M. Dyrby, S.B. Engelsen, L. Nørgaard, M. Bruhn, and L. Lundsberg-Nielsen, Chemometric quantitation of the active substance (containing C≡ N) in a pharmaceutical tablet using near-infrared (NIR) transmittance and NIR FT-Raman spectra, *Applied Spectroscopy* **56**(5) (2002) 579–585.
3. Z.-M. Zhang, S. Chen, Y.-Z. Liang, Z.-X. Liu, Q.-M. Zhang, L.-X. Ding, F. Ye, and H. Zhou, An intelligent background-correction algorithm for highly fluorescent samples in Raman spectroscopy, *Journal of Raman Spectroscopy* **41**(6) (2010) 659–669.
4. Y.-P. Wang, Y. Wang, and P. Spencer, Fuzzy clustering of Raman spectral imaging data with a wavelet-based noise-reduction approach, *Applied Spectroscopy* **60**(7) (2006) 826–832.
5. J.J. Workman Jr, Quantification of LDPE [low density poly (ethylene)], LLDPE [linear low density poly (ethylene)], and HDPE [high density poly (ethylene)] in polymer film mixtures "as received" using multivariate modeling with data augmentation (data fusion) and infrared, Raman, and near-infrared spectroscopy, *Spectroscopy Letters* **32**(6) (1999) 1057–1071.
6. D.R. Miller, O.H. Han, and P.W. Bohn, Quantitative Raman spectroscopy of homogeneous molecular profiles in optical waveguides using internal standards, *Applied Spectroscopy* **41**(2) (1987) 249–255.
7. A. März, K.R. Ackermann, D. Malsch, T. Bocklitz, T. Henkel, and J. Popp, Towards a quantitative SERS approach–online monitoring of analytes in a microfluidic system with isotope-edited internal standards, *Journal of Biophotonics* **2**(4) (2009) 232–242.
8. S. Song and S.A. Asher, Internal intensity standards for heme protein UV resonance Raman studies: excitation profiles of cacodylic acid and sodium selenate, *Biochemistry* **30**(5) (1991) 1199–1205.
9. D. Pratiwi, J.P. Fawcett, K.C. Gordon, and T. Rades, Quantitative analysis of polymorphic mixtures of ranitidine hydrochloride by Raman spectroscopy and principal components analysis, *European Journal of Pharmaceutics and Biopharmaceutics* **54**(3) (2002) 337–341.
10. M.B. Seasholtz, D.D. Archibald, A. Lorber, and B.R. Kowalski, Quantitative analysis of liquid fuel mixtures with the use of Fourier transform near-IR Raman spectroscopy, *Applied Spectroscopy* **43**(6) (1989) 1067–1072.
11. F. Estienne and D.L. Massart, Multivariate calibration with Raman data using fast principal component regression and partial least squares methods, *Analytica Chimica Acta* **450**(1) (2001) 123–129.

12. F. CC Oliveira, C. RR Brandao, H.F. Ramalho, L. AF da Costa, P. AZ Suarez, and J.C. Rubim, Adulteration of diesel/biodiesel blends by vegetable oil as determined by Fourier transform (FT) near infrared spectrometry and FT-Raman spectroscopy, *Analytica Chimica Acta* **587**(2) (2007) 194–199.
13. MATLAB software for Windows from The MathWorks, Inc., 24 Prime Park Way, Natick, Mass. 01760–1500. Internet: info@mathworks.com
14. T.C. O'Haver, Teaching and learning chemometrics with Matlab, *Chemometrics and Intelligent Laboratory Systems* **6**(2) (1989) 95–103.
15. H. Mark and J. Workman, *Chemometrics in Spectroscopy,* 1st edn. (Academic Press, Elsevier, Boston and Amsterdam, 2010).
16. J. Workman and H. Mark, Chemometrics in Spectroscopy — Part 21 — Calculating the Solution for Regression Techniques — Part I: Multivariate Regression Made Simple, *Spectroscopy* **12**(5) (1997) 32–36.
17. J. Workman and H. Mark, Chemometrics in Spectroscopy — Part 22 — Calculating the Solution for Regression Techniques — Part II: Principal Component(s) Regression Made Simple, *Spectroscopy* **12**(6), (1997) 32–33.
18. J. Workman and H. Mark, Chemometrics in Spectroscopy — Part 23 — Calculating the Solution for Regression Techniques - Part III: Partial Least Squares Regression Made Simple, *Spectroscopy* **12**(9) (1997) 36–37.
19. J. Workman and H. Mark, Chemometrics in Spectroscopy — Part 26 — Calculating the Solution for Regression Techniques — Part IV: Singular Value Decomposition, *Spectroscopy* **13**(5) (1998) 10–11.
20. J. Mandel, Use of the singular value decomposition in regression analysis, *The American Statistician* **36**(1) (1982): 15–24.
21. G.H. Golub and C.F. Van Loan, *Matrix Computations*, 3rd edn. (JHU Press, Baltimore, 2012).
22. S.R. Searle, *Matrix Algebra Useful for Statistics,* (Wiley, *New York* 1982).
23. A. Savitzky and M.JE. Golay, Smoothing and differentiation of data by simplified least squares, *Analytical Chemistry* **36**(8) (1964) 1627–1639.
24. J. Steinier, Y. Termonia, and J. Deltour, Smoothing and differentiation of data by simplified least square procedure, *Analytical Chemistry* **44**(11) (1972) 1906–1909.
25. H.H. Madden, Comments on the Savitzky-Golay convolution method for least-squares-fit smoothing and differentiation of digital data, *Analytical Chemistry* **50**(9) (1978) 1383–1386.
26. P.A. Gorry, General least-squares smoothing and differentiation by the convolution (Savitzky-Golay) method, *Analytical Chemistry* **62**(6) (1990) 570–573.
27. S.E. Bialkowski, Generalized digital smoothing filters made easy by matrix calculations, *Analytical Chemistry* **61**(11) (1989) 1308–1310.
28. P.D. Wentzell and C.D. Brown, Signal Processing in Analytical Chemistry, *In Encyclopedia of Analytical Chemistry*, R.A. Meyers (Ed). (John Wiley & Sons, Chichester, 2000).
29. C.D. Brown and P.D. Wentzell, Hazards of digital smoothing filters as a preprocessing tool in multivariate calibration, *Journal of Chemometrics* **13**(2) (1999) 133–152.
30. ASTM E1252-98(2013)e1, Standard Practice for General Techniques for Obtaining Infrared Spectra for Qualitative Analysis, ASTM Committee E13, 100 Barr Harbor Drive, PO Box C700, West Conshohocken, PA 19428–2959 (2013).
31. ASTM E1790-04(2010), Standard Practice for Near Infrared Qualitative Analysis, ASTM Committee E13, 100 Barr Harbor Drive, PO Box C700, West Conshohocken, PA 19428–2959 (2013).

32. P.C. Mahalanobis, On generalized distance in statistics, *Proc. Natl. Inst. Sci. India*, **12**(1936) 49–55.
33. H.L. Mark and D. Tunnell, Qualitative near-infrared reflectance analysis using Mahalanobis distances, *Analytical Chemistry* **57**(7) (1985) 1449–1456.
34. J. Workman, P. Mobley, B. Kowalski and R. Bro, Review of chemometrics applied to spectroscopy: 1985–95, Part I, *Applied Spectroscopy Reviews* **31**(1–2) (1996) 73–124.
35. R.G. Whitfield, M.E. Gerber, and R.L. Sharp, Near-infrared Spectrum Qualification via Mahalanobis Distance Determination, *Applied Spectroscopy* **41**(7) (1987) 1204–1213.
36. R. De Maesschalck, D. Jouan-Rimbaud, and D.L. Massart, The Mahalanobis distance, *Chemometrics and Intelligent Laboratory Systems* **50**(1) (2000) 1–18.
37. J.C. Reid and E.C. Wong, Data-Reduction and-Search System for Digital Absorbance Spectra, *Applied Spectroscopy* **20**(5) (1966) 320–325.
38. P.M. Owens and T.L Isenhour, Infrared spectra compression procedure for resolution independent search systems, *Analytical Chemistry* **55**(9) (1983) 1548–1553.
39. K. Tanabe and S. Saeki, Computer retrieval of infrared spectra by a correlation coefficient method, *Analytical Chemistry* **47**(1) (1975) 118–122.
40. L.V. Azarraga, R.R. Williams and J.A. de Haseth, Fourier encoded data searching of infrared spectra (FEDS/IRS), *Applied Spectroscopy* **35**(5) (1981) 466–469.
41. J.A. de Haseth and L.V. Azarraga, Interferogram-based infrared search system, *Analytical Chemistry* **53**(14) (1981) 2292–2296.
42. B.K. Lavine, K. Nuguru, N. Mirjankar, and J. Workman, Pattern recognition assisted infrared library searching, *Applied Spectroscopy* **66**(8) (2012) 917–925.
43. J. Karasinski, S. Andreescu, O.A. Sadik, B. Lavine, and M.N. Vora, Multiarray sensors with pattern recognition for the detection, classification, and differentiation of bacteria at subspecies and strain levels, *Analytical Chemistry* **77**(24) (2005) 7941–7949.
44. B.K. Lavine, K. Nuguru, N. Mirjankar, and J. Workman, Development of carboxylic acid search prefilters for spectral library matching, *Microchemical Journal* **103** (2012) 21–36.
45. B.K. Lavine, C.E. Davidson, and A.J. Moores, Genetic algorithms for spectral pattern recognition, *Vibrational Spectroscopy* **28**(1) (2002) 83–95.
46. H. Mark and J. Workman, Chemometrics in Spectroscopy — Part 54 — Derivatives in Spectroscopy, Part 1: The Behavior of Derivatives, *Spectroscopy*; **18**(4) (2003) 32–37.
47. H. Mark and J. Workman, Chemometrics in Spectroscopy — Part 55 — Derivatives in Spectroscopy, Part 2: The True Derivative, *Spectroscopy* **18**(9) (2003) 25–28.
48. H. Mark and J. Workman, Statistics in Spectroscopy — Part 11 — More and Different Statistics, *Spectroscopy* **3**(8) (1988) 13–15.
49. H. Mark and J. Workman, Chemometrics in Spectroscopy — Part 56 — Derivatives in Spectroscopy, Part 3: Computing the Derivative, *Spectroscopy* **18**(12) (2003) 106–111.
50. Automatic z′ transform calculator, see for example: http://vassarstats.net/rdiff.html
51. Online z′ transform calculator at: http://onlinestatbook.com/calculators/fisher_z.html
52. J.H. Steiger, Tests for comparing elements of a correlation matrix, *Psychological Bulletin* **87**(2) (1980) 245–251.
53. ASTM E2529-06(2014), Standard Guide for Testing the Resolution of a Raman Spectrometer, ASTM International, 100 Barr Harbor Drive, PO Box C700, West Conshohocken, PA, 19428–2959 USA (2014).

54. ASTM E2911-13, Standard Guide for Relative Intensity Correction of Raman Spectrometers, ASTM International, 100 Barr Harbor Drive, PO Box C700, West Conshohocken, PA, 19428–2959 USA (2013).
55. M.P. Ruffoni and R.F. Pettifer, Calibration of spectra from dispersive XAS beamlines, *Journal of Synchrotron Radiation* **13**(6) (2006) 489–493.

Chapter 7

Applications and Spectral Examples for Raman Spectroscopy

Contents

7.1 Raman Measurement History and Applications 174
7.1.1 Introduction 174
7.1.2 Characteristics of Raman Measurements 174
7.1.3 Agricultural Sciences 183
7.1.4 Animal Sciences 185
7.1.5 Biotechnology 192
7.1.6 Chemical Production 205
7.1.7 Cosmetics 205
7.1.8 Earth Sciences and Mineralogy 205
7.1.9 Environmental Sciences 206
7.1.10 Food and Beverages 207
7.1.11 Forensic Science 208
7.1.12 Gas Phase Analysis 208
7.1.13 Instrument Physics 209
7.1.14 Medicine and Clinical Chemistry 209
7.1.15 Military, Explosives, and Propellants 210
7.1.16 Petroleum, Natural Gas, and Fuels Research 211
7.1.17 Pharmaceutical Industry 211
7.1.18 Plant Sciences, Paper and Pulp 212
7.1.19 Polymer Science 215
7.1.20 Process Control Applications 215
7.1.21 Semiconductors, Electronics 216
7.1.22 Surface Analysis 216
7.1.23 Textiles 217
7.2 Raman Spectra-Structure Correlation Charts 217

7.3 Published Raman Standard Reference Methods and Practices 220
7.3.1 European Pharmacopoeia Commission 220
7.3.2 United States Pharmacopeia (USP) Methods[73–75] 220
7.3.3 International Organization for Standardization (ISO) Standards 221
7.3.4 Association of Official Analytical Chemists (AOAC) International Method 221
7.3.5 ASTM International Standards 221
References 222

7.1 Raman Measurement History and Applications

7.1.1 *Introduction*

Raman measurements exhibit high signal-to-noise and a reasonable cost for instrumentation. The dominant Raman spectral features are: acetylenic C≡C stretching, olefinic C=C stretch at 1680–1630 cm^{-1}, R-N=N-R′ (azo-) stretching, S-H (thiol-) stretching, C=S stretching, C-S stretching, S-S stretching bands, and CH_2 wagging and twisting bands. Raman spectra also contain such molecular vibrational information as: CH_2 twist and wagging, carbonyl C=O stretch associated with esters, acetates, and amides; C-Cl (halogenated hydrocarbons) stretching, and $-NO_2$ (nitro-/nitrite) stretching. In addition, Raman yields information content of phenyl-containing compounds at 1000 cm^{-1}. Organic and inorganic materials generally exhibit unique and identifiable Raman spectra. Bands that are strong in infrared but weak in Raman include: O-H, N-H, and C-F stretching, as well as C-H out-of-plane bending. See Chapter 1 of this volume for detailed information related to Raman bands and band identification.

Table 7.1 and Figure 7.1 demonstrate the major Raman functional groups,[1–4] locations, and band characteristics. Additional detailed figures and tables identifying spectra-structure correlation for Raman bands are also included within this chapter, and Chapter 1. For example, refer to Figures 7.2. through 7.23, and Tables 7.2 and 7.3.

7.1.2 *Characteristics of Raman Measurements*

This Raman application section outlines classic papers, with some recent citations, on the use of Raman measurements for a variety of sample types and measurement requirements. There are thousands of published papers on Raman techniques and

Table 7.1 Summary of Raman band positions and strengths for major functional groups.

Functional group	Location (cm^{-1})	Band character	Comments
O-H	3546–3448	v, w, b	Broad due to H bonding
N-H Stretching	3350–3320	m	Asymmetric Stretching
S-H (thiol-) Stretching	2597–2551	w	Thiol Stretching
C=S Stretching	1200–1050	s	Stretching
C-S Stretching	660	m	Stretching
S-S Stretching	519–486		Stretching
N-H Stretching	3350	m	Stretching
N-H Bending	1650	s	Bending
N-H Scissoring-bending	1550–1630	m	Amide II Band
N=N (azo-) Stretching	1629–1575	v	Azo Stretching
Ar-H	3030	v	Aromatic C-H Stretching
CH_3 Stretching	2973–2959	m-s	Asymmetric Stretching
CH_2 Stretching	2936–2915	m-s	Asymmetric Stretching
CH_3 Stretching	2864–2843	m-s	Symmetric Stretching
CH_2 Stretching	2864–2843	m-s	Symmetric Stretching
CH_3 Bending	1379–1370	m	Bending
CH_3 Bending	1470–1429	s	Bending
C=O Stretching	1779–1680		Exact position depends on type of carbonyl
C=O Stretching	1630–1690	s, sharp	Amide I Band
C-O Stretching	1100	s	Ethers and Alcohols
C=C Stretching	1680–1630	s, aromatic	Alkenes
C≡C Stretching	2150	m, sharp	Alkynes
C≡N Stretching	2250	m, sharp	Nitriles

Note: The following abbreviations for Band Intensities: w = weak, m = medium, m-s = medium-strong, s = strong, *υ* = variable, b = broad.

so this section represents a cross-section giving the reader a sampling or overview of the types of work being conducted using Raman spectrometers.

For this section on Raman applications the reader is reminded of a few unique characteristics of Raman spectroscopy that are somewhat different than absorption based methods such as ultraviolet, visible, near infrared and infrared spectroscopy.

- Raman shift is independent of laser type (i.e., wavelength)
- Fluorescence background may be an issue when using visible lasers (i.e., fluorescence increases with frequency of excitation laser)
- Excitation laser choice is based upon sample characteristics
- Rapid analysis is possible
- Sample may be measured through BK7 glass or most plastics/polymers (these are weak scatterers)
- Sample may consist in aqueous solutions

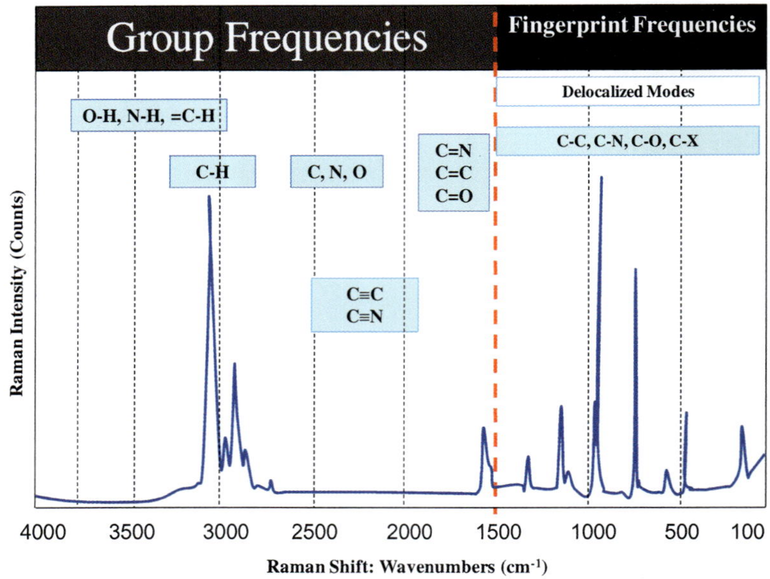

Fig. 7.1 Raman spectral chart: Raman fundamental group frequency and fingerprint frequency locations, showing a Raman intensity (in counts) spectrum with typical functional groups and band locations labeled (Raman shift of 4000 to 100 cm^{-1}).

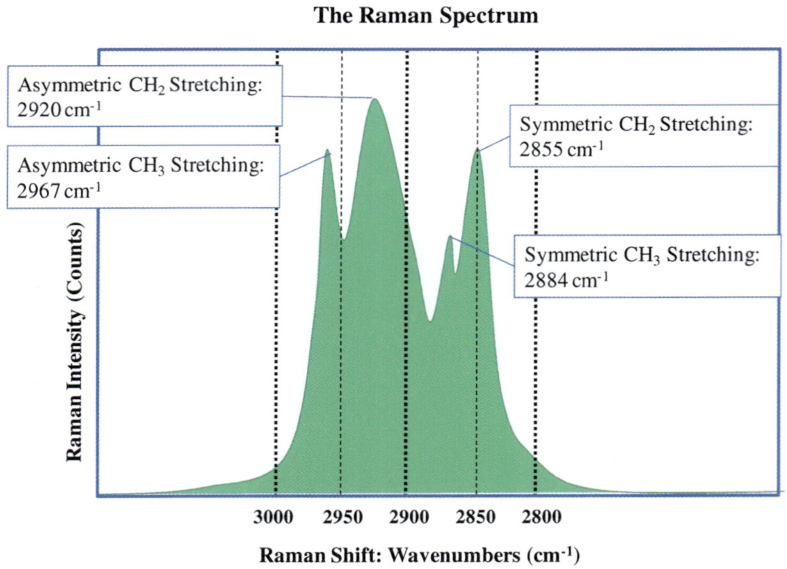

Fig. 7.2 Raman spectral chart: Raman fundamental group frequency locations, showing a Raman intensity (in counts) spectrum for n-alkanes with typical functional groups and band locations labeled (Raman shift of 3000 to 2800 cm^{-1} is shown). Relative peak heights are shown.

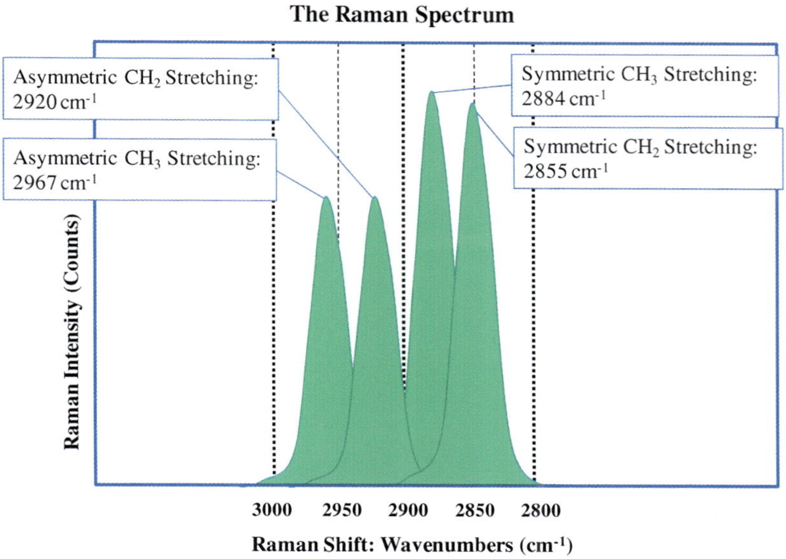

Fig. 7.3 Raman spectral chart: Raman fundamental group frequency locations, showing a Raman intensity (in counts) spectrum for n-alkanes C-H stretching for methyl (CH_3) and methylene (CH_2) functional group bands (Raman shift of 3000 to 2800 cm^{-1} is shown). Ideal peak locations shown.

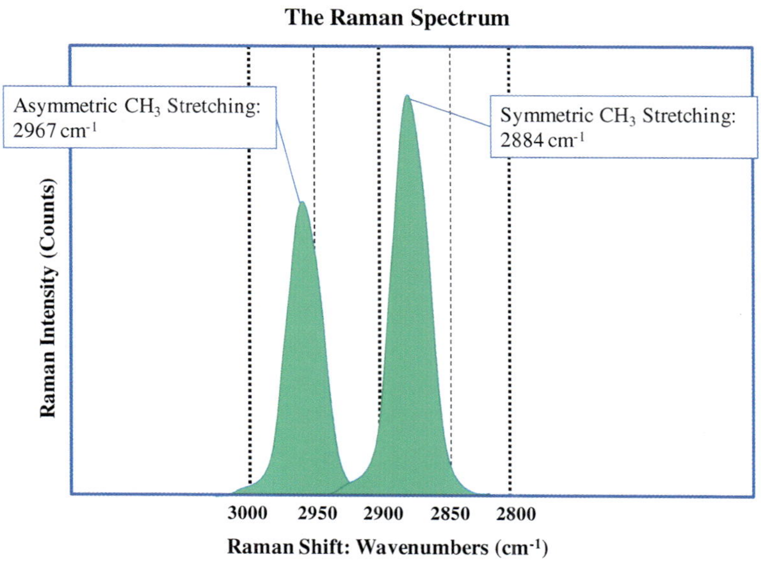

Fig. 7.4 Raman spectral chart: Raman fundamental group frequency locations, showing a Raman intensity (in counts) spectrum for n-alkanes asymmetric and symmetric C-H stretching for methyl (CH_3) functional group bands (Raman shift of 3000 to 2800 cm^{-1} is shown).

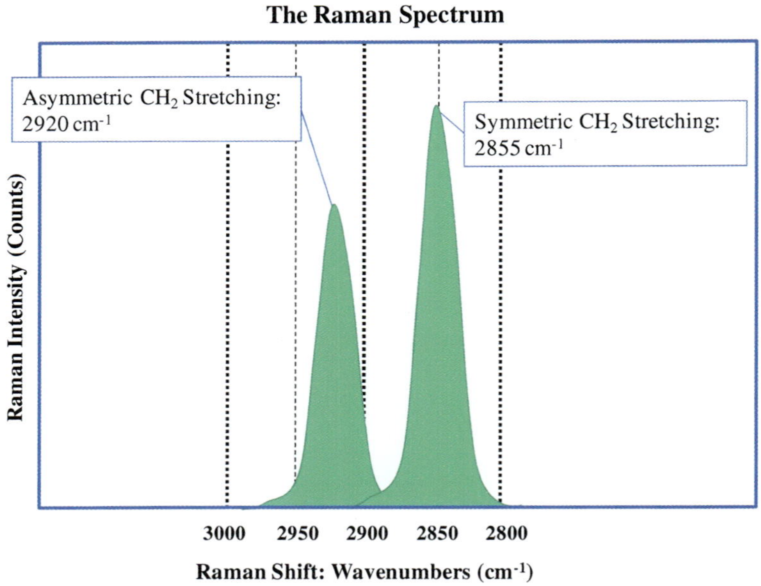

Fig. 7.5 Raman spectral chart: Raman fundamental group frequency locations, showing a Raman intensity (in counts) spectrum for n-alkanes asymmetric and symmetric C-H stretching for methylene (CH_2) functional group bands (Raman shift of 3000 to 2800 cm^{-1} is shown).

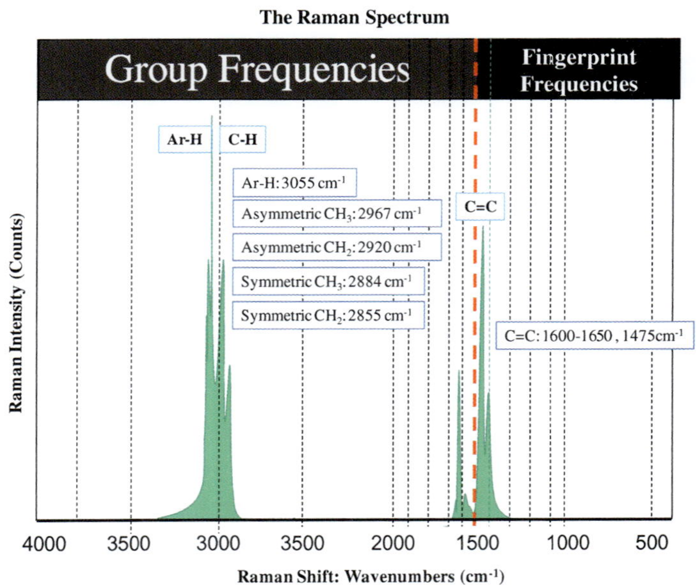

Fig. 7.6 Raman spectral chart: Raman fundamental group frequency and fingerprint frequency locations, showing a Raman intensity (in counts) spectrum for C-H and C=C functional group bands. These include aromatic =C-H, aliphatic C-H, and C=C groups (Raman shift of 4000 to 500 cm^{-1} is shown). Relative peak heights are shown.

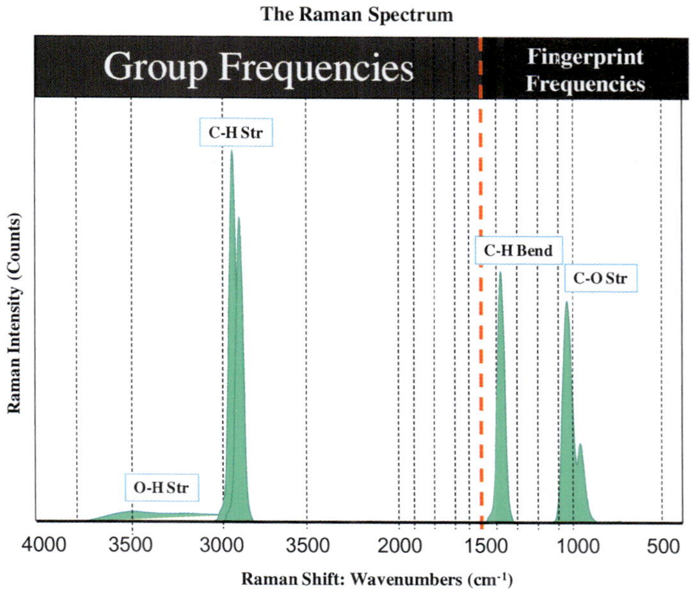

Fig. 7.7 Raman spectral chart: Raman fundamental group frequency and fingerprint frequency locations, showing a Raman intensity (in counts) spectrum for O-H and C-H stretching, C-H bending, and C-O stretching functional group bands. (Raman shift of 4000 to 500 cm^{-1} is shown). Relative peak heights are shown.

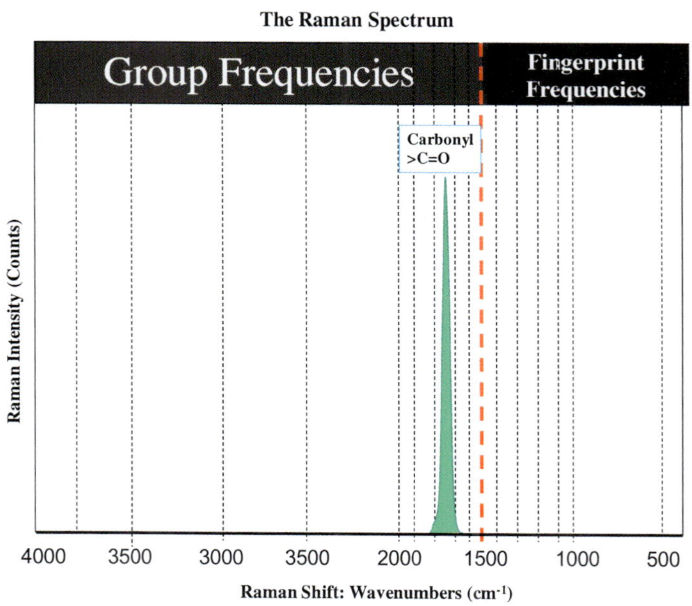

Fig. 7.8 Raman spectral chart: Raman fundamental group frequency and fingerprint frequency locations, showing a Raman intensity (in counts) spectrum for >C=O carbonyl stretching functional group band. (Raman shift of 4000 to 500 cm^{-1} is shown).

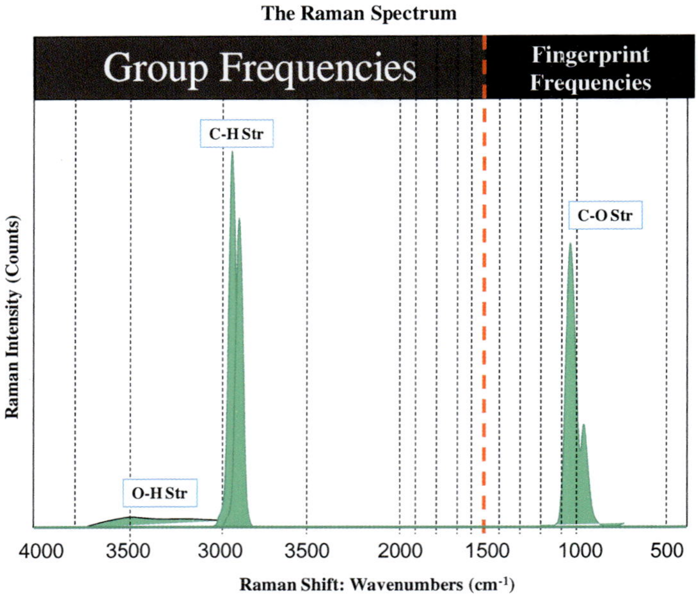

Fig. 7.9 Raman spectral chart: Raman fundamental group frequency and fingerprint frequency locations, showing a Raman intensity (in counts) spectrum for O-H and C-H stretching, and C-O stretching functional group bands. (Raman shift of 4000 to 500 cm^{-1} is shown). Relative peak heights are shown.

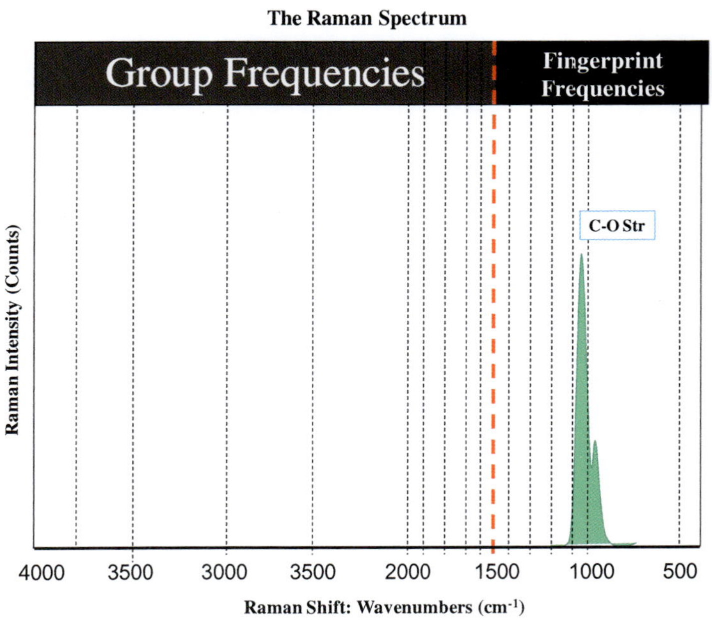

Fig. 7.10 Raman spectral chart: Raman fundamental group frequency and fingerprint frequency locations, showing a Raman intensity (in counts) spectrum for C-O stretching functional group bands. (Raman shift of 4000 to 500 cm^{-1} is shown).

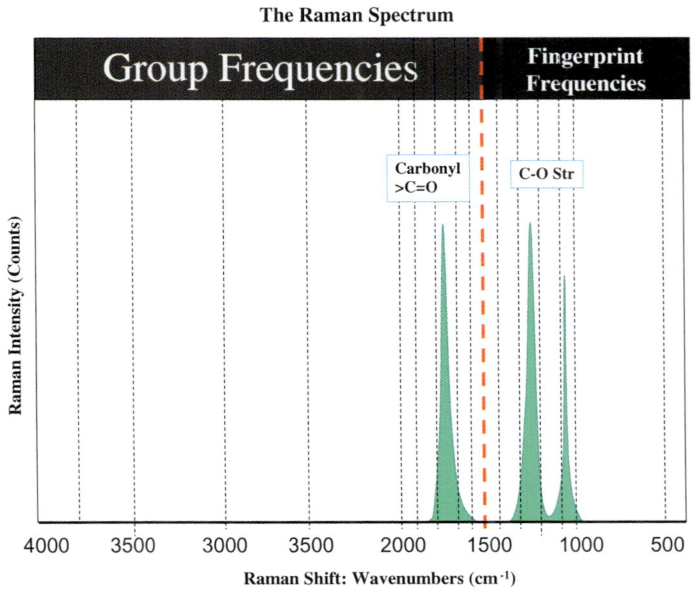

Fig. 7.11 Raman spectral chart: Raman fundamental group frequency and fingerprint frequency locations, showing a Raman intensity (in counts) spectrum for >C=O carbonyl and C-O stretching functional group bands. (Raman shift of 4000 to 500 cm^{-1} is shown). Relative peak heights are shown.

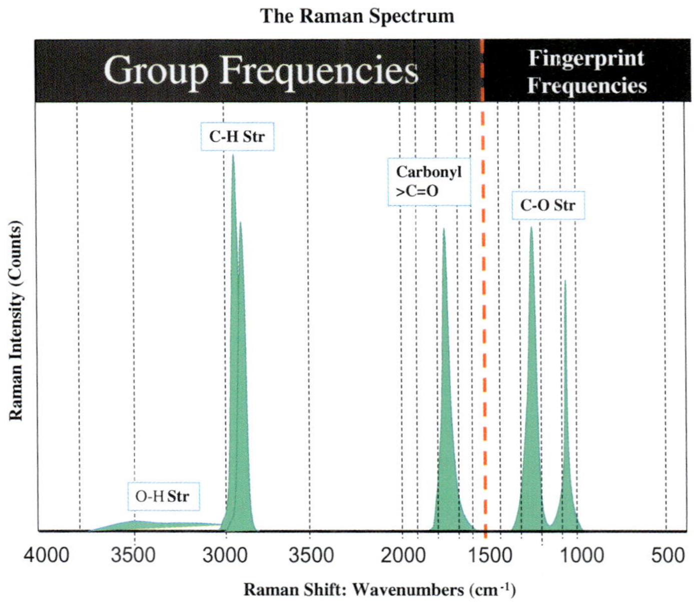

Fig. 7.12 Raman spectral chart: Raman fundamental group frequency and fingerprint frequency locations, showing a Raman intensity (in counts) spectrum for O-H and C-H stretching, as well as >C=O carbonyl and C-O stretching functional group bands. (Raman shift of 4000 to 500 cm^{-1} is shown). Relative peak heights are shown.

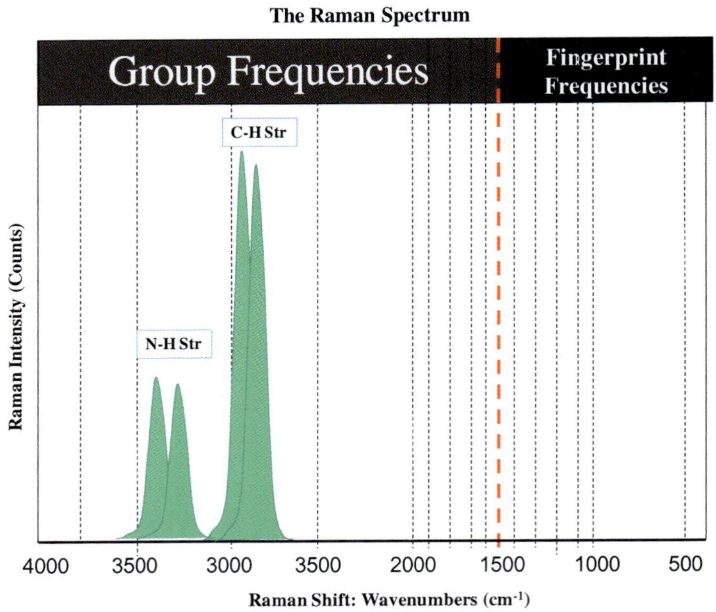

Fig. 7.13 Raman spectral chart: Raman fundamental group frequency and fingerprint frequency locations, showing a Raman intensity (in counts) spectrum for C-H and N-H stretching functional group bands. (Raman shift of 4000 to 500 cm^{-1} is shown). Relative peak heights are shown.

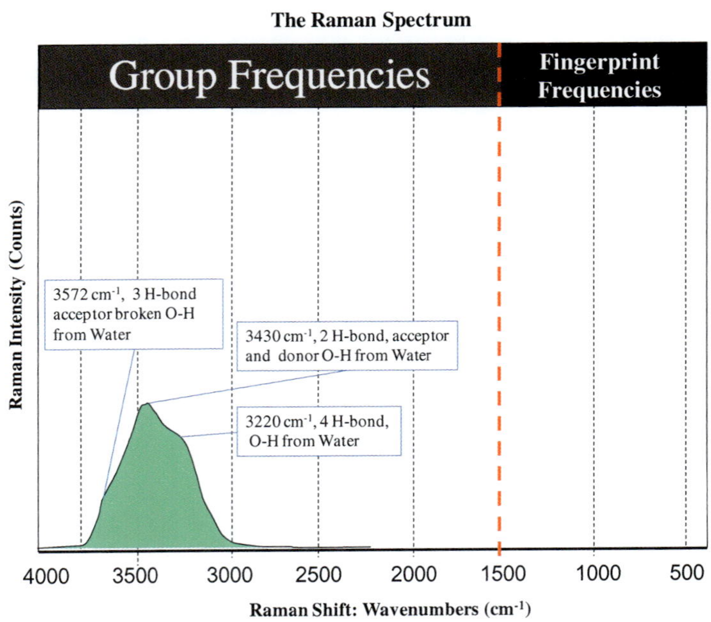

Fig. 7.14 Raman spectral chart: Raman fundamental group frequency and fingerprint frequency locations, showing a Raman intensity (in counts) spectrum for O-H stretching functional group band. (Raman shift of 4000 to 500 cm^{-1} is shown). Relative overlapping peak heights are shown.

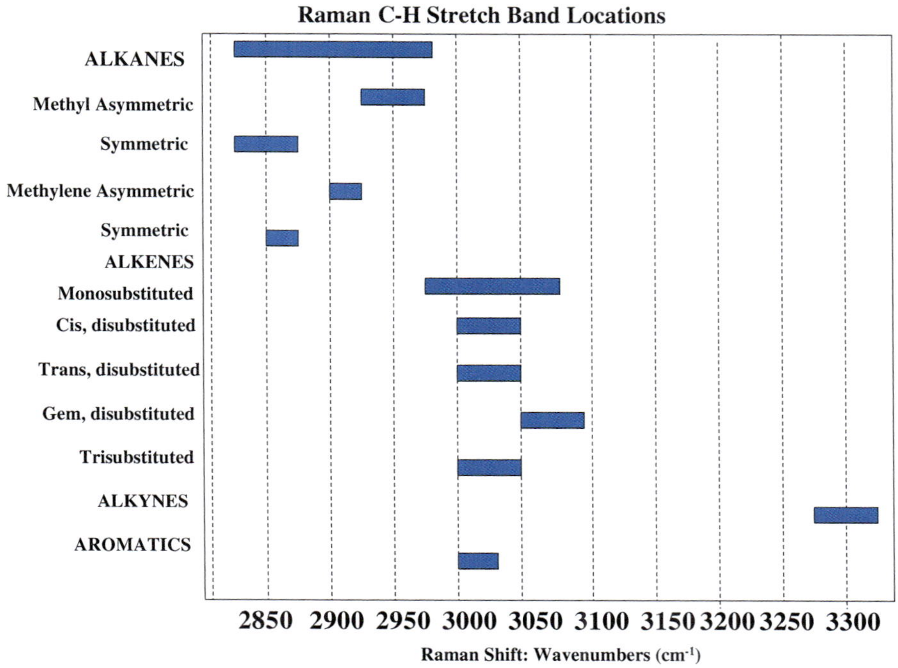

Fig. 7.15 Raman spectra-structure band positions chart — Raman fundamental C-H absorptions indicated relative to wavenumbers (as 3300 to 2800 cm^{-1}).

- No sample preparation is required
- No specific pathlength vessel or cuvette is required
- Specific identification of materials is an excellent application for sharp Raman bands
- The Raman technique is complementary to infrared spectroscopy
- Raman provides excellent micro sampling capabilities (i.e., low mg samples)
- There is access to low frequency molecular backbone vibrations to near 50 cm^{-1}
- Waveguides consisting of water guides or aqueous samples in Teflon tubing are possible
- High quality measurements may be made with low cost optical fibers
- Raman probes are useful for in situ measurements, which are difficult to sample for other methods

7.1.3 *Agricultural Sciences*

This review describes the Raman techniques of dispersive Raman, Fourier transform (FT-Raman), surface-enhanced Raman spectroscopy (SERS), and spatially offset

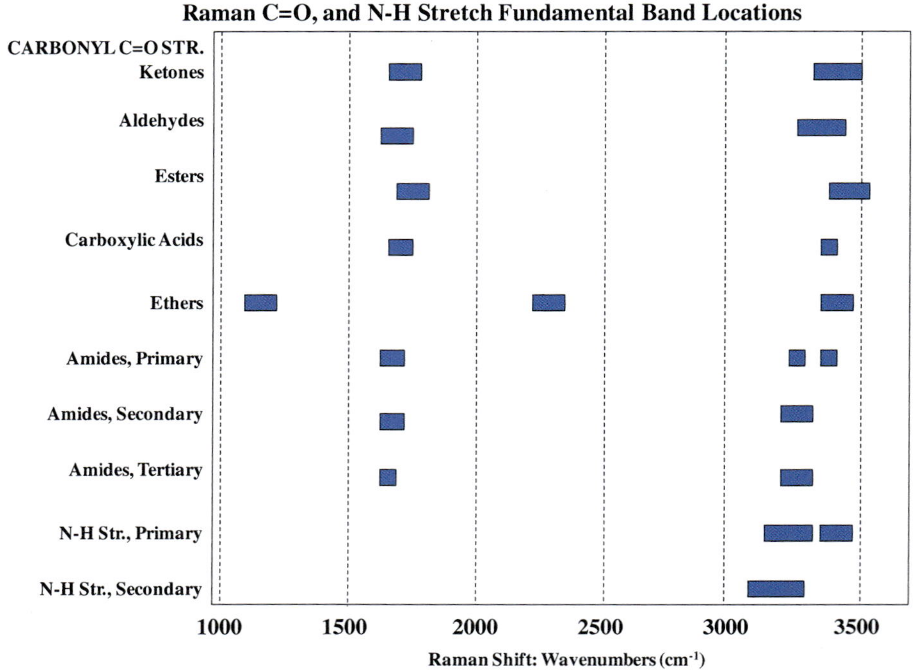

Fig. 7.16 Raman spectra-structure band positions chart: fundamental C=O, and N-H absorptions indicated relative to wavenumbers (as 3500 to 1000 cm^{-1}).

Raman spectroscopy (SORS) for agricultural and food products. Applications of Raman spectroscopy are described for fruits and vegetables, crop analysis, meat and dairy products, oils, and beverages. This chapter also discusses the importance of Raman spectroscopy for fundamental and applied research of agricultural and food products.[5]

To begin, Surface-Enhanced Raman Spectroscopy (SERS) has been described as a technique to detect pesticides extracted from the surface of fruit samples. Gold-coated SERS active nanosubstrates were used for SERS measurement and characterization of pesticide residues. Raman analysis was completed for three types of pesticides (i.e., carbaryl, phosmet, and azinphos-methyl) widely used for pesticide treatment of apples and tomatoes. The enhanced SERS Raman signals were subjected to multivariate analysis using partial least squares and principal components analysis for both quantitative and qualitative predictive models.

This technique was able to detect all three types of pesticides extracted from fruit samples at the parts per million level. The study of detection limits demonstrated that at 99.86% confidence interval, that SERS can detect carbaryl at 4.51 ppm, phosmet

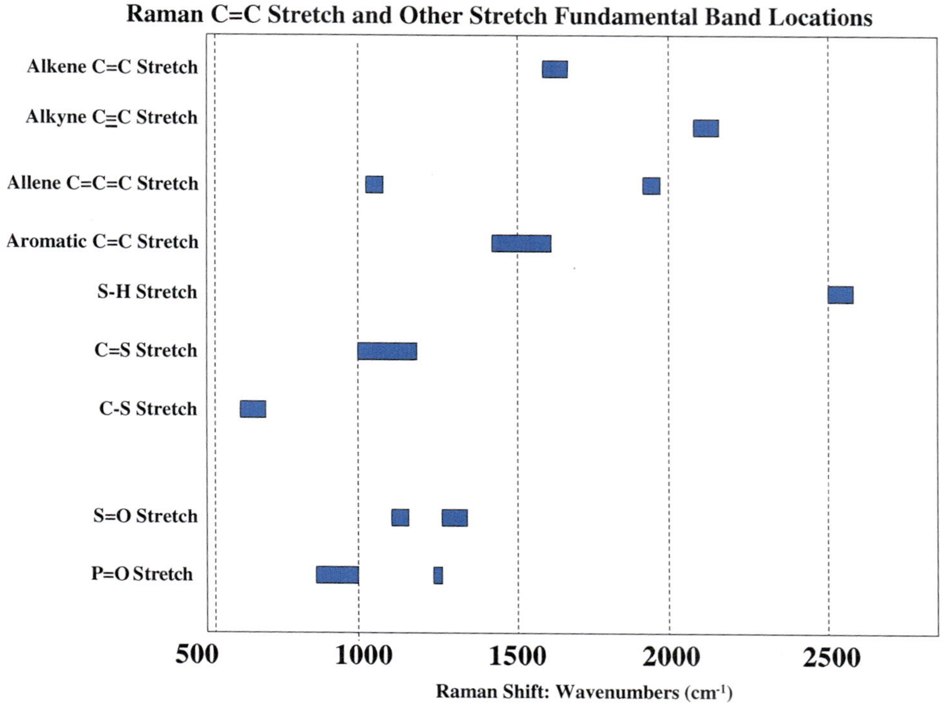

Fig. 7.17 Raman spectra-structure band positions chart — Raman fundamental absorptions indicated relative to wavenumbers (as 3000 to 500 cm^{-1}).

at 6.51 ppm, and azinphos-methyl at 6.66 ppm spiked on apples; and carbaryl at 5.35 ppm, phosmet at 2.91 ppm, and azinphos-methyl at 2.94 ppm on tomatoes. Satisfactory recoveries (78–124%) were achieved for samples with concentrations at or larger than the detection limit. These detection limits are compatible with the maximum residue limits established by Food and Agriculture Organization of the United Nations and World Health Organization.[6]

7.1.4 *Animal Sciences*

Raman and infrared spectroscopy are reviewed for the investigation of protein conformation in the study of silk proteins. The silk protein structures of the domestic silkworm (Bombyx mori), the wild silkworms (mainly tussah) and spider silk are studied. The characteristic Raman and infrared peaks of silk proteins as related to the various conformations are described.[7]

Carotenoids are considered antioxidants on skin. Resonance Raman spectroscopy (RRS) and reflection spectroscopy (RS) may be used non-destructively to measure carotenoids directly on living skin. For RS, a LED-based miniaturized

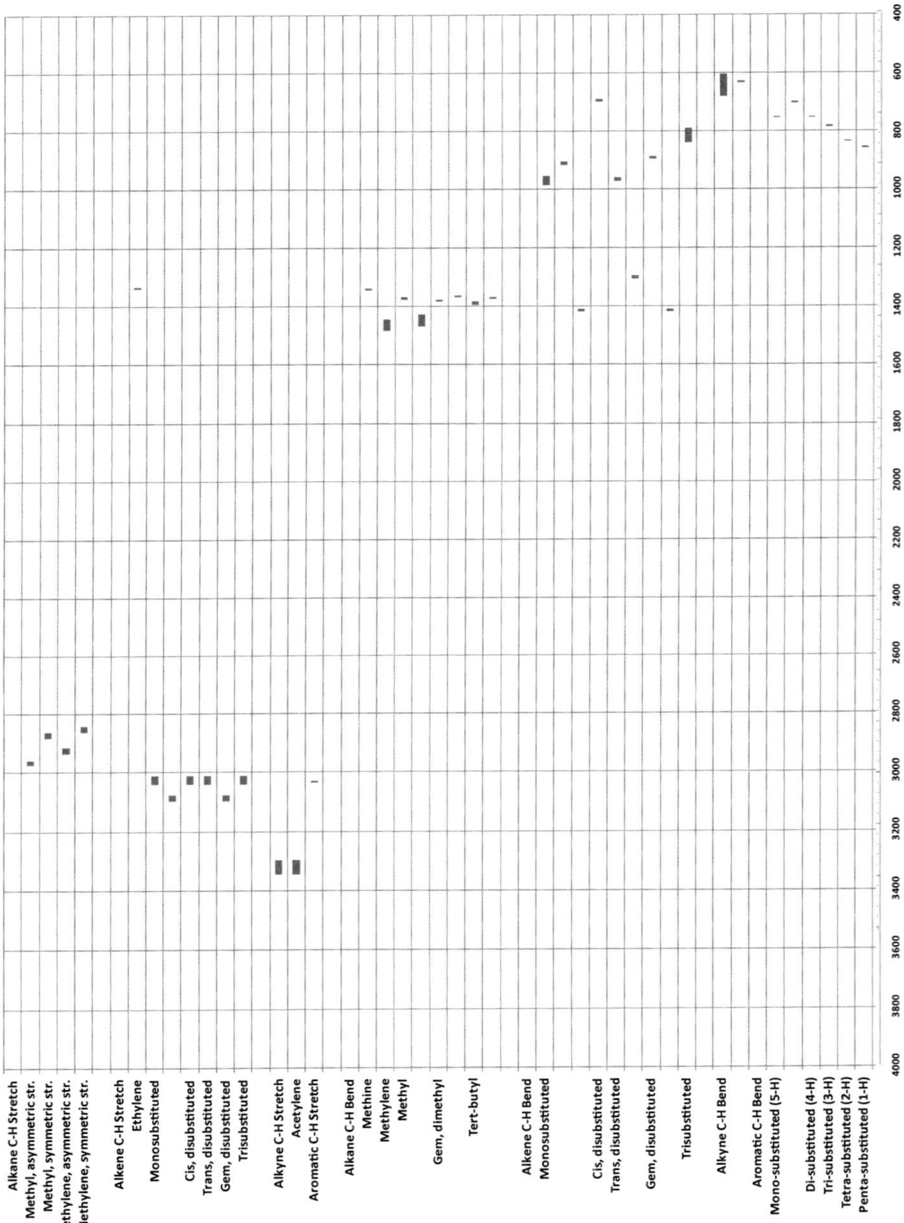

Fig. 7.18 Raman functional group band positions (in cm^{-1}), alkanes to aromatics.

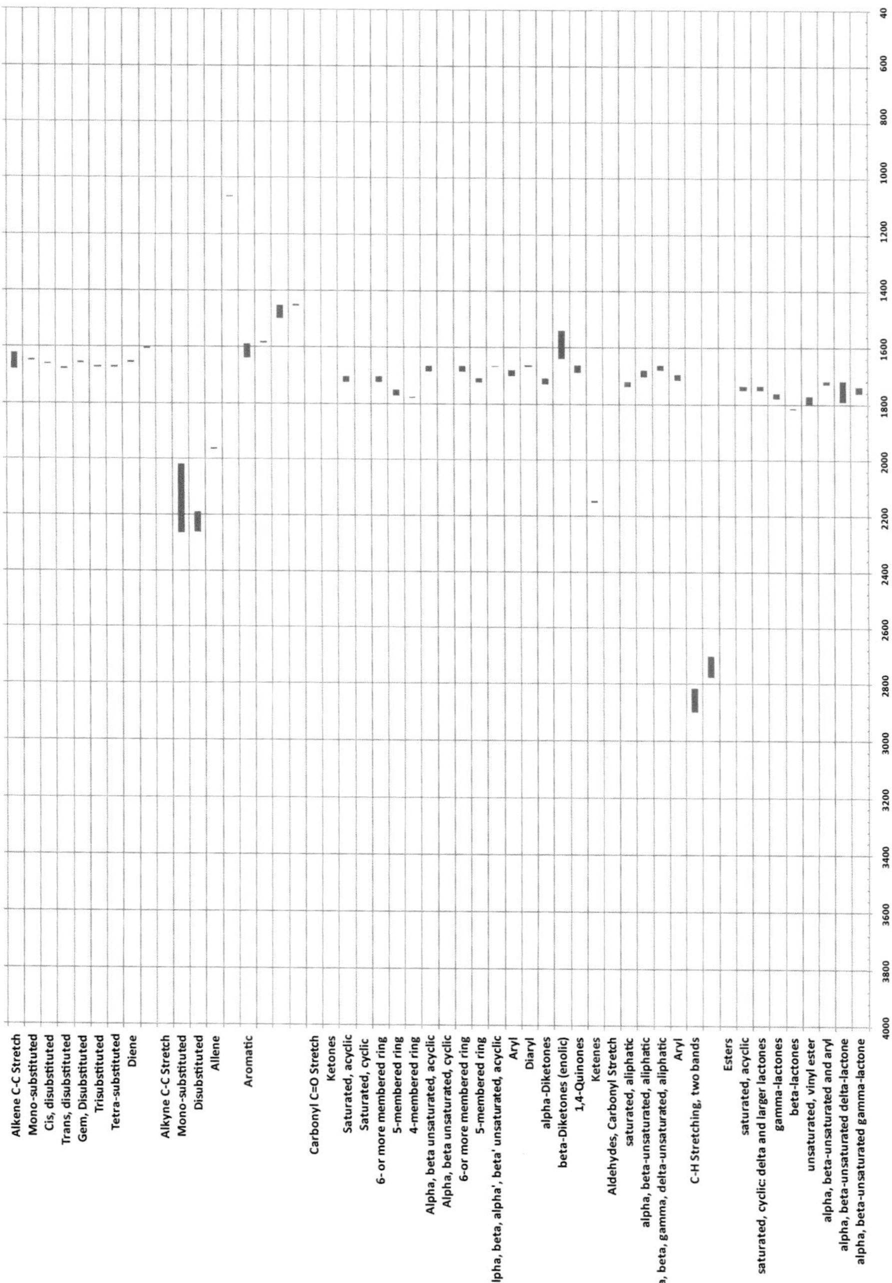

Fig. 7.19 Raman functional group band positions (in cm^{-1}), alkenes to esters.

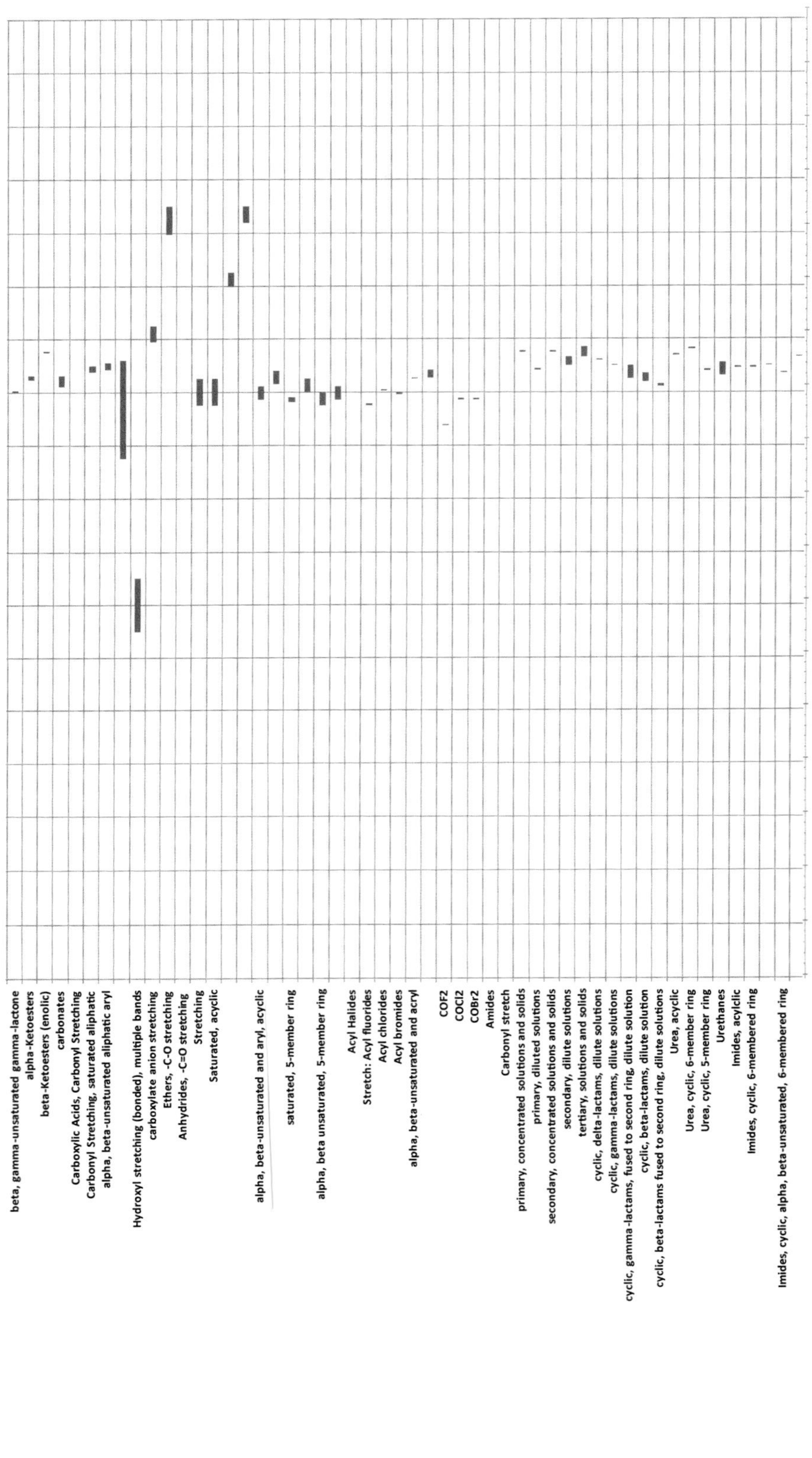

Fig. 7.20 Raman functional group band positions (in cm^{-1}), esters to urethanes.

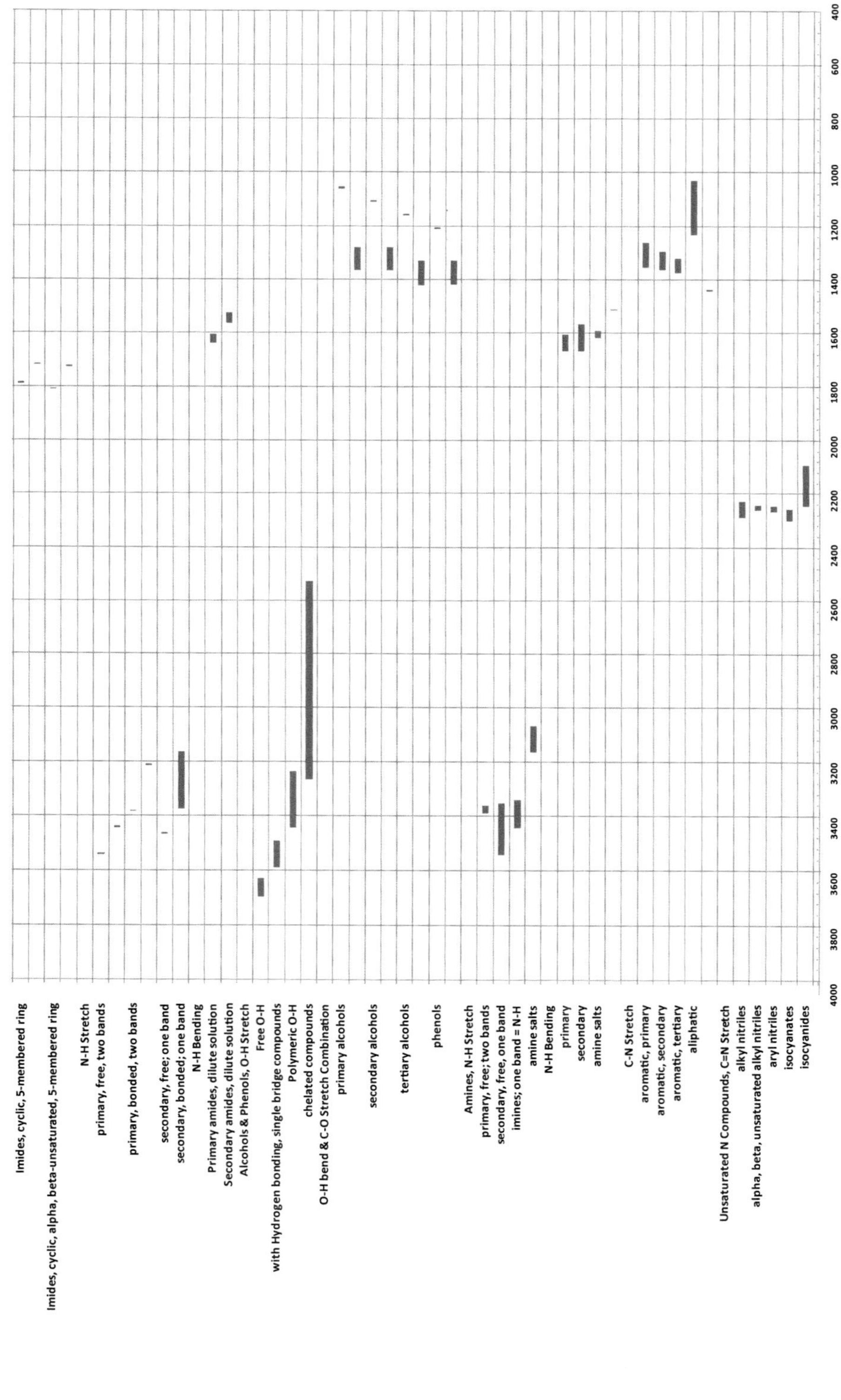

Fig. 7.21 Raman functional group band positions (in cm^{-1}), N-H Stretch to C=N stretch.

Fig. 7.22 Raman functional group band positions (in cm^{-1}), N=N stretch to S-H stretch.

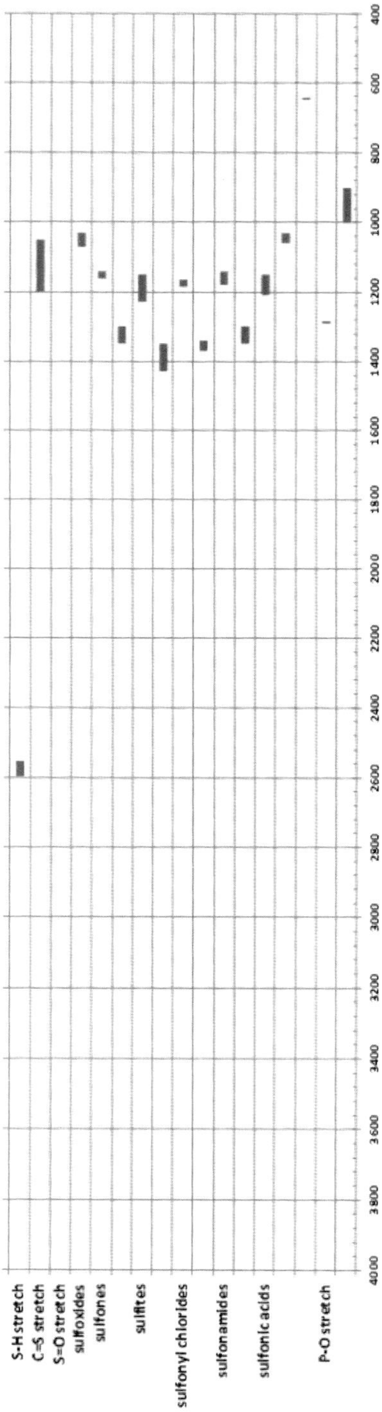

Fig. 7.23 Raman functional group band positions (in cm^{-1}), S-H stretch to P-O stretch.

Table 7.2 Individual H-O-H bands associated with Raman water spectrum.

Hydrogen bonding	Band position
None	3636 cm^{-1}
1 H-bond	3572 cm^{-1}
2 H-bond	3430 cm^{-1}
3 H-bond	3220 cm^{-1}
4 H-bond	3014 cm^{-1}

spectroscopic system (MSS) was developed for noninvasive measurement of carotenoids in human skin. The optimization and subsequent calibration of the MSS was performed with the use of RRS. A strong correlation between the carotenoid concentration determined by the RS and for the RRS system was achieved for human skin in vivo ($R = 0.88$), and for bovine udder skin in vitro ($R = 0.81$).[8]

7.1.5 *Biotechnology*

Lipid measurements in biological systems have often involved invasive techniques, population measurements, or nonspecific labeling. Coherent anti-Stokes Raman (CARS) microscopy is capable of accurate and nondestructive in vivo measurements of lipids within organisms and tissues. CARS is capable of high selectivity and sensitivity with real-time imaging of lipid-rich structures and demonstrating submicron three-dimensional spatial resolution. CARS microscopy is applied to studies of lipid biology in cell cultures, tissue biopsies, and model organisms. Technical advances and limitations of the technique are described and summarized.[9]

Limitations are discussed for the Raman techniques of Surface-enhanced Raman scattering (SERS) and tip-enhanced Raman spectroscopy (TERS). A novel Raman method referred to as shell-isolated nanoparticle-enhanced Raman spectroscopy has been described. In this technique, the Raman signal amplification is provided by gold nanoparticles with an ultrathin silica or alumina shell. A monolayer of such nanoparticles is spread as ‘smart dust’ over the surface that is to be probed. The ultrathin coating keeps the nanoparticles from agglomerating, separating them from direct contact with the probed material and allowing the nanoparticles to conform to different contours of substrates. High-quality Raman spectra were obtained on various molecules adsorbed at Pt and Au single-crystal surfaces and from Si surfaces with hydrogen monolayers. Substantial signal enhancement and improved sampling flexibility is reported using SERS for such applications as life sciences, food safety, drugs, explosives, and environment pollutants.[10]

Table 7.3 Functional groupings and calculated locations for Raman spectroscopy (adapted with permission from J. Workman, *Handbook of Organic Compounds: NIR, IR, Raman, and UV-Vis spectra featuring polymers and surfactants, Volume 1*, (Academic Press-Elsevier Publishers, 2001)).

	Molecular structure	Relative intensity	Fundamental position
1. C-H Stretch			
a. Alkane, C_nH_{2n+2}	CH_3-$(CH_2)_N$- CH_3, normal paraffins	m-s	
methyl	-C-H_3	m-s	
asymmetric str.			2973-2959 cm^{-1}
symmetric str.			2882-2862 cm^{-1}
methylene	-C-H_2	m-s	
asymmetric str.			2936-2915 cm^{-1}
symmetric str.			2864-2843 cm^{-1}
b. Alkene, C_nH_{2n}	>C=C<, olefinic group		
Ethylene	H2-C=C-H2	s	~1340 cm^{-1}
	H H		
monosubstituted	>C=C<	m	3040-3012 cm^{-1}
	R H	m	3096-3077 cm^{-1}
	H H		
cis, disubstituted	>C=C<	m	3040-3012 cm^{-1}
	R R		
	H R		
trans, disubstituted	>C=C<	m	3040-3012 cm^{-1}
	R H		
	R H		
gem, disubstituted	>C=C<	m	3096-3077 cm^{-1}
	R H		
	R R		
trisubstituted	>C=C<	m	3040-3012 cm^{-1}
	R H		

(Continued)

Table 7.3 (*Continued*).

	Molecular structure	Relative intensity	Fundamental position
c. alkyne, C_nH_{2n-2}	$-C \equiv C - H$	s	3345-3295 cm^{-1}
Acetylene	H-C=C-H	w	3345-3295 cm^{-1}
d. aromatic, C_nH_n	Ar(C)-H	v	3030 cm^{-1}
2. C-H Bend			
a. Alkane, C_nH_{2n+2}			
methine	R-CHR′R″	w	1340 cm^{-1}
methylene	R-CH_2R′	m	1483-1445 cm^{-1}
methyl	R-CH_3	m	1379-1370 cm^{-1}
		s	1470-1429 cm^{-1}
gem, dimethyl	CH_3-CHR-CH_3	s	1385-1379 cm^{-1}
		s	1370-1364 cm^{-1}
tert-butyl	$(CH_3)_3$-CH	m	1395-1385 cm^{-1}
		s	~ 1370 cm^{-1}
b. Alkene, C_nH_{2n}	>C=C<		
monosubstituted	H H >C=C< R H	s	952-985 cm^{-1}
		s	915-905 cm^{-1}
		s	1420-1410 cm^{-1}
cis, disubstituted	H H >C=C< R R	s	~690 cm^{-1}
trans, disubstituted	H R >C=C< R H	s	970-960 cm^{-1}
		m	1309-1295 cm^{-1}
gem, disubstituted	R H >C=C< R H	s	895-885 cm^{-1}
		s	1420-1410 cm^{-1}

(*Continued*)

Table 7.3 (*Continued*).

trisubstituted	R R >C=C< R H	s	840-790 cm^{-1}
c. Alkyne, C_nH_{2n-2}	$-C \equiv C - H$	s	~680-605 cm^{-1} ~629 cm^{-1}
d. Aromatic, C_nH_n	**Ar(C)-H**		
mono-substituted (5 hydrogens)	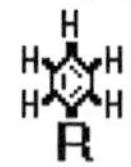	v, s	~752 cm^{-1} ~699 cm^{-1}
di-substituted (4 hydrogens)	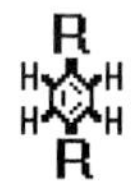	v, s	~752 cm^{-1}
tri-substituted (3 hydrogens)		v, s	781 cm^{-1}
tetra-substituted (2 hydrogens)		v, m	~833 cm^{-1}
penta-substituted (1 hydrogen)		v,w	~855 cm^{-1}
3. C-C Stretch			
a. Alkene, C_nH_{2n}	>C=C<	v	1681-1621 cm^{-1}
mono-substituted	As shown above	m	~1645 cm^{-1}
cis, disubstituted	"	m	~1658 cm^{-1}
trans, disubstituted	"	m	~ 1675 cm^{-1}
gem, disubstituted	"	m	~1653 cm^{-1}
trisubstituted	"	m	~1669 cm^{-1}
tetra-substituted	"	w	~1669 cm^{-1}
diene	>C=C<	w	~1650 cm^{-1}
	. . .>C=C<	w	~1600 cm^{-1}

(*Continued*)

Table 7.3 (*Continued*).

	Molecular structure	Relative intensity	Fundamental position
b. Alkyne, C_nH_{2n-2}			
mono-substituted		m	2265-2020 cm^{-1}
disubstituted		v,w	2262-2188 cm^{-1}
Allene	>C=C=C<	m	~1961 cm^{-1}
		m	~1064 cm^{-1}
Aromatic	containing	v	1640-1590 cm^{-1}
		v	~1580 cm^{-1}
		m	1499-1450 cm^{-1}
		m	~1449 cm^{-1}
4. Carbonyl >C=O			
a. Ketones	R >C=O R’		
saturated, acyclic	alkane (R) >C=O alkane(R’)	s	1724-1704 cm^{-1}
saturated, cyclic			
6- or more membered ring	 (or) AR >C=O R’	s	1724-1704 cm^{-1}
5-membered ring	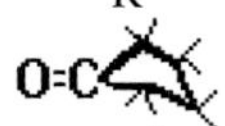	s	1751-1770 cm^{-1}
4-membered ring	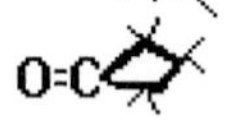	s	~1776 cm^{-1}

(*Continued*)

Table 7.3 (*Continued*).

alpha, beta unsaturated, acyclic		s	1684-1664 cm^{-1}
alpha, beta unsaturated, cyclic			
6- or more membered ring		s	1684-1664 cm^{-1}
5- member ring		s	1724-1709 cm^{-1}
alpha, beta, alpha′, beta′, unsaturated, acyclic		s	1669-1664 cm^{-1}
Aryl		s	1701-1681 cm^{-1}
Diaryl		s	1669-1661 cm^{-1}
alpha-Diketones		s	1730-1709 cm^{-1}
beta-Diketones (enolic)		s	1639-1538 cm^{-1}
1,4-Quinones			1689-1661 cm^{-1}
Ketenes			~2151 cm^{-1}
b. Aldehydes, Carbonyl stretch	>C=O		
saturated, aliphatic		s	1739-1721 cm^{-1}
alpha, beta-unsaturated, aliphatic		s	1704-1681 cm^{-1}
alpha, beta, gamma, delta-unstaturated, aliphatic		s	1681-1661 cm^{-1}
Aryl		s	1715-1695 cm^{-1}
C-H Stretching, two bands		w	2899-2817 cm^{-1}
		w	2778-2703 cm^{-1}
c. ESTERS	R-C=O-OR′		
saturated, acyclic		s	1751-1736 cm^{-1}
saturated, cyclic: delta and larger lactones		s	1751-1736 cm^{-1}
gamma-lactones		s	1779-1761 cm^{-1}

(Continued)

Table 7.3 (*Continued*).

Molecular structure	Relative intensity	Fundamental position
beta-lactones	s	~1818 cm^{-1}
unsaturated, vinyl ester	s	1799-1770 cm^{-1}
alpha, beta-unsaturated and aryl	s	1730-1718 cm^{-1}
alpha, beta unsaturated delta-lactone	s	1792-1718 cm^{-1}
alpha, beta unsaturated gamma-lactone	s	1761-1739 cm^{-1}
beta, gamma-unsaturated gamma-lactone	s	~1799 cm^{-1}
alpha-Ketoesters	s	1754-1739 cm^{-1}
beta-Ketoesters (enolic)	s	~1650 cm^{-1}
carbonates	s	1779-1739 cm^{-1}
5. CARBOXYLIC ACIDS -O-C=O –		
carbonyl stretching: saturated aliphatic	s	1724-1701 cm^{-1}
alpha, beta-unsaturated aliphatic aryl	s	1715-1689 cm^{-1} 2049-1681 cm^{-1}
hydroxyl stretching (bonded), several bands	w	2703-2500 cm^{-1}
carboxylate anion stretching	s	1610-1550 cm^{-1}

(*Continued*)

Table 7.3 (*Continued*).

6. ANHYDRIDES			
Stretching, -C=O		m	1850-1750 cm^{-1}
saturated, acyclic	-C=O stretching	m	1850-1750 cm^{-1}
	H3CC-O	m	1400-1350 cm^{-1}
	C-O-C stretching	m	1160-1100 cm^{-1}
alpha, beta-unsaturated and aryl, acyclic		m	1828-1779 cm^{-1}
			1770-1721 cm^{-1}
saturated, 5-member ring		m	1869-1821 cm^{-1}
			1767-1751 cm^{-1}
alpha, beta unsaturated, 5-member ring		m	1848-1799 cm^{-1}
			1828-1779 cm^{-1}
7. ACYL HALIDES			
Stretch: Acyl fluorides		s	~1848 cm^{-1}
Acyl chlorides		s	~1795 cm^{-1}
Acyl bromides		s	~1808 cm^{-1}
alpha, beta-unsaturated		s	1751-1748 cm^{-1}
and aryl		m	1748-1718 cm^{-1}
COF_2		s	~1927 cm^{-1}
$COCl_2$		s	~1828 cm^{-1}
$COBr_2$		s	~1828 cm^{-1}
8. AMIDES			
Carbonyl stretch			
primary, concentrated solutions and solids		s	~1650 cm^{-1}
primary, dilute solutions		s	~1718 cm^{-1}
secondary, concentrated solutions and solids		s	~1650 cm^{-1}
secondary, dilute solutions		s	1701-1669 cm^{-1}
tertiary, solutions and solids		s	1669-1629 cm^{-1}

(*Continued*)

Table 7.3 (*Continued*).

Molecular structure	Relative intensity	Fundamental position
cyclic, delta-lactams, dilute solutions	s	~1681 cm^{-1}
cyclic, gamma-lactams, dilute solutions	s	~1701 cm^{-1}
cyclic, gamma-lactams fused to second ring, dilute solution	s	1751-1701 cm^{-1}
cyclic, beta-lactams, dilute solution	s	1761-1730 cm^{-1}
Cyclic, beta-lactams fused to second ring, dilute solutions	s	1779-1770 cm^{-1}
Urea, acyclic	s	~1661 cm^{-1}
Urea, cyclic, 6-membered ring	s	~1639 cm^{-1}
Ureas, cyclic 5-member ring	s	~1721 cm^{-1}
Urethanes	s	1739-1689 cm^{-1}
Imides, acyclic	s	~1709 cm^{-1}
Imides, cyclic,	s	~1709 cm^{-1}
6-membered ring	s	~1701 cm^{-1}
Imides, cyclic, alpha,	s	~1730
beta-unsaturated, 6-membered ring	s	~1669 cm^{-1}

(*Continued*)

Table 7.3 (*Continued*).

Imides, cyclic,		s	~1770 cm^{-1}
5-membered ring		s	
Imides, cyclic, alpha,		s	~1792 cm^{-1}
beta-unsaturated,		s	~1709 cm^{-1}
5-membered ring			
N-H stretch			
primary, free, two bands		m	~3497 cm^{-1}
		m	~3401 cm^{-1}
primary, bonded, two		m	~3344 cm^{-1}
bands		m	~3175 cm^{-1}
secondary, free; one band		m	~3425 cm^{-1}
secondary, bonded; one band		m	3333-3125 cm^{-1}
N-H Bending			
Primary amides, dilute solution		s	1621-1590 cm^{-1}
secondary amides, dilute solution		s	1548-1511 cm^{-1}
9. ALCOHOLS & PHENOLICS, O-H stretch			
Free O-H		v, w, b	3650-3584 cm^{-1}
with hydrogen bonding, single bridge compounds	O-H....H-	v, w, b	3546-3448 cm^{-1}
Polymeric O-H		s,b	3401-3195 cm^{-1}
chelated compounds		w,b	3226-2500 cm^{-1}
O-H bend &			
C-O stretch Combination			
primary alcohols		s	~1053 cm^{-1}
		s	1351-1266 cm^{-1}
secondary alcohols		s	~1099 cm^{-1}
		s	1351-1266 cm^{-1}

(*Continued*)

Table 7.3 (*Continued*).

Molecular structure		Relative intensity	Fundamental position
tertiary alcohols		s	~1149 cm^{-1}
		s	1408-1316 cm^{-1}
phenols		s	~1198 cm^{-1}
		s	1405-1316 cm^{-1}
10. AMINES, N-H stretch			
primary, free; two bands		m	3350-3320 cm^{-1}
		m	
secondary, free; one band		m	3497-3311 cm^{-1}
Imines; one band	=N-H	m	3401-3300 cm^{-1}
amine salts		m	3125-3030 cm^{-1}
N-H bend			
primary		s-m	1650-1590 cm^{-1}
secondary		w	1650-1550 cm^{-1}
amine salts		s	1600-1575 cm^{-1}
		s	~1499 cm^{-1}
C-N stretch			
aromatic, primary		s	1340-1250 cm^{-1}
aromatic, secondary		s	1350-1280 cm^{-1}
aromatic, tertiary		s	1359-1309 cm^{-1}
aliphatic		w	1220-1020 cm^{-1}
		w	~1427 cm^{-1}
11. UNSATURATED N COMPOUNDS, C= N stretch			
alkyl nitriles	-C≡N	m	2262-2205 cm^{-1}
alpha, beta, unsaturated alkyl nitriles		m	2237-2217 cm^{-1}

(*Continued*)

Table 7.3 (*Continued*).

aryl nitriles	m	2242-2222 cm^{-1}
isocyanates	m	2273-2232 cm^{-1}
isocyanides	m	2222-2070 cm^{-1}
imines & oximes		
alkyl compounds	v	1689-1649 cm^{-1}
alpha, beta unsaturated	v	1661-1629 cm^{-1}
-N=N- stretch, azo-compounds	v	1629-1575 cm^{-1}
-N=C=N-, diimides	s	2155-2128 cm^{-1}
$-N_3$ stretch, azides	s	2160-2137 cm^{-1}
	w	1340-1179 cm^{-1}
$C-NO_2$, nitro- compounds		
aromatic	s	1570-1499 cm^{-1}
	s	1370-1299 cm^{-1}
aliphatic	s	1570-1550 cm^{-1}
	s	1379-1370 cm^{-1}
$O-NO_2$, nitrates	s	1650-1600 cm^{-1}
	s	1299-1250 cm^{-1}
C-NO, nitroso	s	1600-1499 cm^{-1}
O-NO, nitrites	s	1681-1600 cm^{-1}
	s	1626-1610 cm^{-1}
12. HALOGEN, C-X stretch and X-C= O stretch		
C-F	m	1408-1000 cm^{-1}
C-Cl	m	800-602 cm^{-1}
C-Br	m	602-500 cm^{-1}
C-I	m	~500 cm^{-1}
Aromatic, 1,3,5 meta substitution	v	1000-990 cm^{-1}

(*Continued*)

Table 7.3 (*Continued*).

Molecular structure		Relative intensity	Fundamental position
X-C=O stretch		w	1820-1790 cm^{-1}
13. SULFUR COMPOUNDS			
S-H stretch		w	2597-2551 cm^{-1}
C=S stretch		s	1200-1050 cm^{-1}
S=O stretch			
sulfoxides		s	1070-1030 cm^{-1}
sulfones		s	1160-1140 cm^{-1}
		s	1350-1300 cm^{-1}
sulfites		s	1230-1149 cm^{-1}
		s	1429-1350 cm^{-1}
sulfonyl chlorides		s	1185-1164 cm^{-1}
		s	1370-1340 cm^{-1}
sulfonamides		s	1179-1140 cm^{-1}
		s	1350-1300 cm^{-1}
sulfonic acids		s	1209-1149 cm^{-1}
		s	1060-1030 cm^{-1}
		s	~649
14. Other			
P-O stretch	P=O	s	1290 cm^{-1}
		s,b	1000-900 cm^{-1}

Note: The following abbreviations for band intensities: w = *weak*, m = medium, m-s = medium-strong, s = *strong*, υ = variable.

7.1.6 *Chemical Production*

Raman has been applied for the characterization of aligned carbon nanotubes of average diameter of 10–15 nm. These nanotubes were produced using chemical vapor deposition on a mesoporous substrate. The resonance behavior and higher-order Raman bands up to fourth order were compared with those of carbon nanotubes produced using an arc discharge with highly oriented pyrolytic graphite.[11]

Tip-enhanced Raman spectroscopy (TERS) exhibits enhanced sensitivity with high spatial resolution; extremely powerful as a technique for chemical analysis. A high vacuum TERS (HV-TERS) has been constructed to investigate the plasmon-driven in-situ chemical reaction of 4-nitrobenzenethiol dimerizing to dimercaptoazobenzene. The chemical reactions can be controlled by the plasmon intensity, which in turn can be controlled by the incident laser intensity, tunneling current and bias voltage. The temperature of such a chemical reaction can also be obtained by the clearly observed Stokes and Anti-Stokes HV-TERS peaks. The HV-TERS system may significantly extend the studies of certain chemical reactions.[12]

7.1.7 *Cosmetics*

A study using confocal Raman microscopy has been demonstrated for comparison of human and porcine follicle composition for imaging of drug delivery component distribution within skin follicle cross-sections. The Raman spectra were used to confirm the optical/spectral similarity of drug delivery for both species. Cyanoacrylate skin surface biopsies were also imaged and profiled using confocal Raman microscopy combined with optical profilometry for chemical and topographical surface mapping. This combined technique provided three dimensional information on the intactness and component distribution of the excised hair bulb. The paper concludes that confocal Raman microscopy demonstrates high potential for noninvasive analysis of trans-follicular drug delivery.[13]

A feasibility test was published using surface-enhanced Raman scattering (SERS) to study the cosmetic ingredient kinetin (KT) as (KT)-gold nanoparticle (AuNP) conjugates. This paper includes the use of quantum mechanical calculations for describing the basic analysis. The adsorptions of KT on AuNP surfaces were examined using absorption spectra SERS. The concentration-dependent SERS spectra of KT indicated that 5×10^{-5} M exhibited the highest SERS signals.[14]

7.1.8 *Earth Sciences and Mineralogy*

Raman confocal spectroscopy was reported to yield three-dimensional molecular-structural images of apatite grains and associated minerals embedded in a banded quartz-pyroxene-magnetite supracrustal rock. These specimens were obtained from Akilia, southern west Greenland. The paper reported the measurement of inclusions of graphite wholly contained within apatite for these rock specimens. The carbon isotopic composition of one such inclusion, its graphitic composition established by Raman spectroscopy, was measured by secondary ion mass spectrometry. Reported findings indicate that graphite-containing apatite grains may represent chemical fossils of early life forms.[15]

Micro Raman spectroscopy has been applied for the potential identification of biosignatures in Earth's ancient rock record. The work is intended to find a basis for the detection and understanding of extraplanetary life as one of the primary purposes of the field of astrobiology. To this end micro Raman spectroscopy has been evaluated as a nondestructive method for in situ identification of a variety of minerals and inorganic compounds. The use of D ($\sim$1350 cm^{-1}) and G ($\sim$1580 cm^{-1}) bands to infer the biogenicity of carbonaceous materials in fossils is widely accepted. However, carbonaceous compounds from a variety of different sources exhibit identical spectroscopic characteristics. For this research, microfossils from a selected set of sedimentary rock samples were analyzed using micro Raman spectroscopy to study the carbonaceous material. Micro Raman analyses was also performed on well-characterized meteorite samples as abiological controls. A summary of the work indicates a correlation between precursor carbonaceous material with D-band parameters, and of thermal specimen history with G-band parameters.[16]

A review describes the use of a cone penetrometer for in situ analysis of contaminants in soil and groundwater. The review, containing 64 references, details soil gas, groundwater and soil sampling and analysis for the determination of organic contaminants. The analytical methods reported for organic analyses include laser-induced fluorescence, Raman and infrared spectrometry and specialized sensors. Inorganic contaminants are analyzed using laser-induced breakdown spectrometry, XRF and passive gamma-ray spectrometry.[17]

7.1.9 *Environmental Sciences*

Raman spectroscopy studies are used to demonstrate chromium on iron oxide at 826 cm^{-1}, possibly attributed to interaction between chromium and iron oxide. Raman spectra combined with XPS studies indicates that electrostatic attraction and oxidation–reduction between chromium and mixed maghemite–magnetite might be a mechanism for the removal of Cr(VI) from aqueous solutions.[18]

Surface-enhanced Raman scattering (SERS) and surface-enhanced fluorescence (SEF) were used for the analysis of soil humic acids (HA). HA structure was observed at different pH and HA concentrations and the structural differences were studied. Raman and fluorescence were successfully employed for structural and dynamic analysis of the soil HA. The surface-enhanced emission (SEE) spectra, that is the sum of the Raman and the fluorescence emissions, were acquired by using both *macro*- and *micro*-experimental configurations in order to apply imaging and confocal Raman and fluorescence spectroscopy techniques on the analysis of HA.[19]

7.1.10 *Food and Beverages*

Raman spectroscopy is a nondestructive analytical technique for determination of chemical and physical information in food and beverages for rapid on-line analysis. The technique does not require sample preparation. One review paper describes the use of dispersive Raman spectroscopy, Fourier transform Raman (FT-Raman) spectroscopy, surface-enhanced Raman spectroscopy (SERS), and spatially offset Raman spectroscopy (SORS) for food product analysis. Raman applications discussed include agricultural and food products such as fruits and vegetables, crops, meat and dairy products, oils, and beverages. The topic of Raman spectroscopy for applied research of foods and agricultural products is discussed.[20]

Surface enhanced Raman spectroscopy (SERS) is reported for use in trace chemical detection. In one reported technique, filter membranes are used to trap silver nanoparticles forming a SERS-active substrate. Then samples are concentrated from a mL-scale into a μL-scale volume for measurement. A significant improvement in detection limit was reported using this method as compared to the typical colloidal SERS. Enhanced signal was found for the pesticide malathion and the food contaminant melamine. In addition to increased sensitivity, the technique exhibited higher precision as compared to traditional SERS techniques. The authors propose this technique as a practical method for a broad range of analytical applications, including field-based detection of toxins.[21]

Raman spectroscopy was reported for direct quantitative measurements of ethanol and methanol in distilled alcoholic beverages. Raman spectra of varying ethanol–methanol mixtures were measured, baseline corrected, and normalized using the intensity of acetonitrile as an internal standard (at 921 cm^{-1}). Calibration curves were constructed for ethanol and methanol concentrations in the ranges of 0–7 M and 0–10 M, respectively. Calibration results for the methods include R^2 values of 0.998 for ethanol and 0.998 for methanol. The method was validated based on linearity, sensitivity, intraday and interday repeatability, and recovery tests. The limit of detection and limit of quantification were 1.2 and 3.7 mM for ethanol concentration, and 3.4 and 10.3 mM for methanol concentration. For actual product samples the method correlated with the traditional reference method with an $R^2 = 0.926$.[22]

7.1.11 *Forensic Science*

Raman spectroscopy was studied to optimize the selection of excitation lasers for the analysis of black-grey and blue cotton fibers following a reactive dye process. A single blue cotton fiber was analyzed for three dye components, and compared to an undyed cotton fiber as a control, using five different laser wavelengths and two different Raman microprobe spectrometers. It was determined that the 785 nm and 830 nm lasers gave the best results. It was reported that the major dye components could be identified using the Raman technique.[23]

Raman and Infrared spectroscopic methods were combined with statistical techniques for characterization and discrimination of paint samples. A set of 34 red paint samples were measured using both infrared and Raman spectroscopy. Data pretreatments involved the use of Standard Normal Variate (SNV), as well as spectral windowing for infrared spectral data from 650 to 1830 cm^{-1} and 2730–3600 cm^{-1}. Discrimination was performed using principal component analysis (PCA) and hierarchical clusters analysis (HCA). The authors conclude that Raman and infrared spectroscopic techniques are useful for the forensic analysis of paints to assist in objective decision-making, a reduction of the possible classification errors, and improved efficiency.[24]

7.1.12 *Gas Phase Analysis*

In the early stages of development, a special Raman gas cell was designed for use to measure the concentration of multiple polyatomic gases. Details of the design and methods of measurement are given for an early version of a gas phase monitoring Raman spectrometer.[25]

Raman spectroscopy is demonstrated for a high sensitivity method for gas concentration analysis is demonstrated at the early stages of commercial Raman development. Intense light scattering is obtained through a stimulated interaction. The method appears more practical than spontaneous Raman scattering for gas flow analysis with fine spatial resolution. Experimental results with the detection of H_2 are given.[26]

For recent advances in gas-phase measurements using Raman, a paper describing the application of Coherent anti-Stokes Raman scattering (CARS) spectroscopy using ultrashort pulses has been demonstrated for the measurement of gas-phase molecules in reacting flows. The method reportedly eliminates the nonresonant interference contribution that generally is found in CARS measurements. The use of ultra-short excitation pulses provides access to high data-acquisition rates. The report demonstrated that spectral focusing, coupled with time-resolved measurements based upon probe delay, allows the observation of interference-free coherence dynamics of multiple molecules and gas-phase temperature making this technique ideal for gas-phase measurements of reacting flows and combustion processes.[27]

7.1.13 *Instrument Physics*

An early Raman backscattering instrument is described using a modern holographic transmission grating spectrograph. The initial design of a, then, new generation of Raman spectrometers is described. The novel dual-channel design allowed simultaneous measurements of right and left circular scattered light, which had the effect of reducing particle and Schlieren sensitivities in the sampling matrix.[28]

Raman–*laser-induced-breakdown-spectroscopy* (LIBS) was selected as a Pasteur payload module for the ExoMars mission. The instrument incorporated both Raman spectroscopy and LIBS in the same spectrometer. The spectrometer was designed to exhibit high resolution over a relatively large spectral range. The specification requirements included low mass, miniature size, and low power consumption. A summary of the specifications, the optical design, and performance criteria are included.[29]

7.1.14 *Medicine and Clinical Chemistry*

Raman spectroscopy has been studied to measure the concentration of multiple analytes in human serum and whole blood since the late 1990s. It has been tested for physiological concentrations in vitro across multipatient data sets. A near infrared excitation diode laser at 830 nm has been demonstrated, with calibrations developed for glucose, cholesterol, urea, and other analytes using partial least-squares (PLS) cross validation. Six analytes were predicted in serum, including hematocrit in one early paper.[30]

Near infrared (laser) Fourier-transform Raman spectroscopy has been used with partial least squares (PLS) regression to analyze cholesterol/cholesterol, linoleate/cholesterol oleate and cholesterol palmitate/cholesterol stearate mixtures, and to determine single components within the ternary mixtures. The method was being developed in the mid-1990s for eventual application to in situ measurements of arterial walls to assess plaque formation.[31]

Near infrared Raman spectrometer systems have been proposed for in vivo studies of diseased human tissue. Safe and rapid data acquisition are benefits of such technology. Two designs of such systems were described starting in the mid-1990s, for example, (1) a high-quality-spectra instrument for laboratory use, and (2) a trolley-mounted optical-fiber instrument for clinical use.[32]

Raman, MIR DRIFTS and NIR spectrometry were used in the mid-1990s to analyze fibrous (asbestos) and non-fibrous forms of serpentine and amphibale minerals. Raman spectra were obtained with excitation at 785 nm, 632.8 nm and 1064 nm. Results indicated that NIR spectrometry was the most useful method due to: (1) relatively easy to interpret spectra in the 7400–6900 cm^{-1} range; (2) a higher single-to-noise ratio; (3) the use of low cost silica optical-fiber cables; and (4) the short analysis time in comparison to Raman and FT-IR. The paper reports that the method compares well with other conventional methods of asbestos analysis, such as optical microscopy and transmission electron microscopy.[33]

Surface-enhanced Raman scattering (SERS) has been used in biomedical and genomics analysis. The technology uses DNA gene probes based on SERS labels for gene detection and DNA mapping. Surface-enhanced Raman gene (SERGen) probes can be used to detect DNA targets via hybridization to DNA sequences complementary to these probes. The SERGen approach has been evaluated for biomedical diagnostics, high-throughput analysis, and DNA mapping and sequencing.[34]

7.1.15 *Military, Explosives, and Propellants*

Confocal Raman microscopy has been applied using 90 second measurements for the identification of explosives and their precursors in situ on undyed natural and synthetic fibers and colored textile specimens. Raman spectra were obtained from explosives particles of 5–10 μm diameter that were trapped within the specimens. Pentaerythritol tetranitrate (PETN), trinitrotoluene (TNT), and ammonium nitrate were discriminated, as were hexamethylenetetraamine (HMTA) and pentaerythritol precursors. The explosive substances could be identified within a variety of fiber types using their characteristic Raman bands.[35]

Raman microspectroscopy was reported as an analysis method to discriminate explosive particulates from other particulate residue in a microscopic image. The technique filters out other spectral signatures, except the region of the strong bands exhibited by PETN and RDX. The laser used for this purpose was a helium-neon (He:Ne) excitation at 633 nm. In this work, line images of the sample are combined with high spectral resolution Raman spectra to distinguish chemical components within the image scene. Images are captured using a CCD camera.

Confocal scanning combined with imaging, can be used to derive a quantitative measure of the concentration over the image plane.[36]

Surface-enhanced Raman scattering (SERS) spectroscopy has been used to detect traces of RDX with high sensitivity and reproducibility. Using gold (Au) nanoparticles of 90–100 nm in diameter, RDX was detected at concentrations as low as 0.15 mg/L within contaminated groundwater. A signal enhancement factor of $\sim 6 \times 10^4$ was obtained, yielding a detection limit of approximately two orders of magnitude lower than previous SERS techniques.[37]

7.1.16 *Petroleum, Natural Gas, and Fuels Research*

In the mid-1990s, remote fiber optic Raman spectroscopy with CCD detection and near infrared diode laser excitation was reported for the analysis of synthetic petroleum fuel samples for the analytes of benzene, toluene, and ethylbenzene (BTE), ranging from 0 to 16% in concentration. Partial Least Squares (PLS) regression analysis was used for calibration modeling. Data windowing was found to improve calibration data for this application, rather than using the entire spectrum for calibration model computations.[38]

Early work using Raman spectrometers proved valuable for measuring water in hydrocarbons. Water dissolved in a hydrocarbon phase appears at 3630 cm^{-1} on the Raman spectrum and its Raman fingerprint is totally different from the Raman signature of liquid water with a band located in the 3250–3450 cm^{-1} range. The use of Raman microscopy has been used to measure an increase in water solubility within hydrocarbons as a function of an increase in temperature. However, Raman analysis cannot be applied to petroleum inclusions; their fluorescence masks the Raman signal. Similar experiments have been made using FT-IR microspectroscopy on synthetic and natural inclusions, noting that intensity of the liquid water infrared band decreases with a temperature increase due to solubility of water in the oil phase at high enough temperatures. It is now recognized that homogenization temperature of a petroleum inclusion may be observed when the liquid water band, centered at 3400 cm^{-1}, disappears from an FT-IR spectrum of the water-oil phase mixture.[39]

7.1.17 *Pharmaceutical Industry*

Early reports for Raman spectroscopy have described it as a non-destructive and promising tool for on-line process monitoring and analysis in the pharmaceutical industry. One paper gives a short introduction to Raman spectroscopy and presents several applications in the pharmaceutical field.[40]

FT-Raman has been reported to be capable of measuring a pharmaceutical active compound in its polymorph forms. In one report, both the polymorphs, as A- and B- forms were found to have distinct FT-Raman spectra. A band at 1716 cm^{-1} (A-form) or 1724 cm^{-1} (B-form) were reported for distinguishing the forms. Mixtures of A- in B-form were prepared, and the relative intensity of the characteristic bands at 1716 and 1724 cm^{-1} were found to be proportional to the relative amounts of A- and B-form in the mixtures. A calibration was made which was linear in the range from 1.8 to 15.4% (w/w) of A- in B-form. Multiple analytical techniques were compared for analysis of polymorphs, including FT-IR spectrometry (FT-IR), differential scanning calorimetry (DSC), and powder X-ray diffractometry (XRD).[41]

An article reviews recent advances in transmission Raman spectroscopy and its applications, from the perspective of pharmaceutical analysis. Rapid non-invasive volumetric analysis of pharmaceutical formulations are possible for quantitative bulk analysis pharmaceutical tablets and capsules. This technique is useful for quality and process control of mixtures and final dosage forms.[42]

7.1.18 *Plant Sciences, Paper and Pulp*

Rapid Lignin Measurement in Hardwood Pulp Samples is reported using near infrared Fourier Transform Raman (FT-Raman) Spectroscopy and is presented from a late 1990s perspective in reference.[43]

As a complementary technique to infrared spectroscopy, a method is described for the rapid prediction of native wood pulp properties by Fourier transform (FT) Raman spectroscopy using two Eucalyptus species. The pulp samples included various ages and colors. Pulp properties such as pulp yield, sheet density, burst factor, breaking length, tear factor, unbleached brightness and Kappa number, were measured. The Raman spectroscopic data revealed highly significant correlations between measured and Raman-predicted values for all traits, with correlation coefficients (r) greater than 0.998 for calibration samples, and greater than 0.882 in the prediction for unknown samples, respectively. This rapid method has suggested its validity for analyzing Eucalyptus native wood meal samples, regardless of their age and color, for the assessment of pulp properties.[44]

Fourier transform (FT) Raman and FT infrared (IR) spectroscopic methods have been used to measure a band at 1675 cm^{-1}; most likely related to the yellow chromophores in photoexposed thermomechanical pulps (TMPs). On the basis of spectroscopic studies that involved both o- and p-quinone models and Fremy's salt-oxidized TMP, the 1675 cm^{-1} band is assigned to the p-quinone functional group. Moreover, in the presence of known photoyellowing inhibitors, the photoyellowing behavior of methyl hydroquinone was similar to that of TMPs. Another important finding was that the molecular oxygen sensitivity of the laser-induced fluorescence (excited at 514.5 nm) of p-quinone and hydroquinone models was similar to what had been previously observed for yellowed and unyellowed TMPs. These results provide strong support for a yellowing hypothesis in which a hydroquinone/p-quinone couple was seen as an important yellowing color system in mechanical pulps.[45]

Near IR Fourier transform (FT) Raman spectroscopy was used as a technique complementary to infrared spectroscopy to quantify lignin concentration in unbleached and partially bleached pine (*Pinus*) kraft pulps. The bleached pulps were treated in CEDED and polyoxometalate sequences. The intensity of the Raman band at 1600 cm^{-1}, representing a phenyl ring vibration, was used to measure total lignin content in the pulp samples. A linear relationship was observed between the band intensities at 1600 cm^{-1} and pulp kappa number. This relationship existed for pulps bleached in both sequences. The results indicate that Raman intensity (1600 cm^{-1}) is dependent upon pulp phenyl-group content and generally independent of bleaching chemistry. These findings suggest that FTIR spectroscopy might also offer a rapid, nondestructive method for determining lignin concentration in unbleached and partially bleached kraft pulps.[46]

In order to expand the use of vibrational spectroscopy as a meaningful tool for the investigation of polysaccharide conformations, normal coordinate calculations have been performed for isolated chains of cellulose I and V-amylose. The calculated frequencies for both polysaccharides show good agreement with the observed IR and Raman data, while the computed potential energy distribution is compatible with previous assignments, based on deuterium exchange. Calculation of the atomic displacement coordinates indicate that all the predicted vibrational modes in the region below 1500 cm^{-1} result from complex motions of a number of atoms or groups, rather from single group vibrations. Most of the modes above 1200 cm^{-1} localized within the residue and are similar to those predicted for the respective monomers. Below 1200 cm^{-1} inter-residue coupling of vibrations become more appreciable. Polarized IR spectra of oriented crystalline glycosaminoglycans have permitted the determination of the orientation of the acetamido and carboxyl side chains. In addition, the dichroism of the symmetric stretching of the carboxyl side chain at 1410 cm^{-1} has been correlated with the degrees of extension of the glycosaminoglycan helix. IR and Raman studies of chondroitin 6-sulfate in aqueous solution suggest the presence of an ordered conformation. Lowering the pH of such solutions results in a cooperative transition at pH 4.1 which has been interpreted as further ordering of the macromolecule due to the reduction in electrostatic repulsion of the charges carboxyl groups.[47]

Another paper has assigned the photoyellowing-related 1675 cm^{-1} Raman/IR band to p-quinones and describes its implications to the mechanism of color reversion in mechanical pulps. The author uses FT (Fourier-transform) Raman and FT infrared (IR) spectroscopies, to detect the new band at 1675 cm^{-1}; this was most likely to have come from the yellow chromophores in photoexposed thermomechanical pulps (TMPs). On the basis of spectroscopic studies that involved both o- and p-quinone models and Fremy's salt-oxidized TMP, the 1675 cm^{-1} band is assigned to the p-quinone functional group. Moreover, in the presence of known photoyellowing inhibitors, the photoyellowing behavior of methyl hydroquinone was similar to that of TMPs. Another important finding was that the molecular oxygen sensitivity of the laser-induced fluorescence (excited at 514.5 nm) of p-quinone and hydroquinone models was similar to what had been previously observed for yellowed and unyellowed TMPs. Taken together, these results provided strong support for a previously suggested yellowing hypothesis in which a hydroquinone/p-quinone couple was seen as an important leucochromophore/chromophore system in mechanical pulps.[48]

The first attempt to investigate polymer-surfactant interactions in gelling and nongelling aqueous mixtures of a nonionic cellulose ether and a surfactant by means of vibrational spectroscopy has been presented. A series of aqueous solutions of ethyl (hydroxyethyl) cellulose (EHEC) with addition of anionic surfactant sodium dodecyl sulfate (SDS) of different concentrations was thoroughly investigated by means of Raman scattering and FTIR absorption techniques. Data was reported that even in the sol state (i.e., below the gel point) there are interactions between the polymer and the surfactant, and both bound and free surfactant molecules have been detected. This interaction, which cannot be characterized as a chemical one, occurs mainly between the side chains of the polymer and the sulfonic acid groups of SDS. Above the gel point, a new type of interaction appears, which mainly involves the SO_3^- groups and water molecules. The intermolecular interactions are studied vs changes of both temperature and polymer-surfactant compositions, and a possible model for the gelation process is discussed.[49]

A nondestructive test for the characterization of lignins in wood was developed based upon Raman spectroscopy using a near IR (NIR) excitation laser and Fourier transform (FT) Raman. The technique was used to characterize a series of lignin model compounds and dehydrogenated polymerizates (DHPs) (i.e., synthetic lignins) and samples of untreated wood meal; extract-free wood meal, holocellulose, and alpha-cellulose; and wood blocks prepared from a variety of hardwoods and softwoods. Only 3–5 mg of each sample was required for spectral determination. Results indicated that the vibration band for the aromatic carbon-carbon double bond was shifted when the aromatic ring was substituted by a methoxyl group. The Raman spectra of guaiacyl and syringyl DHPs clearly showed that the carbon double-bond vibrations for guaiacyl and syringyl nuclei exhibited different Raman shifts at 1599 and 1594 cm^{-1}, respectively. These Raman shifts were designated as guaiacyl and syringyl marker bands, respectively. The marker bands existed clearly in the spectra of wood meals and wood blocks. The syringyl:guaiacyl ratios of the samples were calculated from the area intensity of these markers. Results were in good agreement with those obtained from conventional chemical analysis based on nitrobenzene oxidation.[50]

7.1.19 *Polymer Science*

Near IR FT Raman spectroscopy has been demonstrated for on-line monitoring the polymerization of vinyl acetate at 54°C in H_2O containing sodium lauryl sulfate, NaH_2PO_4, and $K_2S_2O_8$ in a 150 ml reaction vessel. The 1064 nm line from a Nd:YAG laser passed through the side arm of the reaction vessel using backscattering geometry. Spectra were recorded at 8 min intervals with scanning from 20-3500 cm^{-1} and a resolution of 4 cm^{-1}. The method is applicable to the monitoring of industrial production of the polymer.[51]

Infrared and resonance Raman spectroscopic studies on the photopolymerization process of the Langmuir-Blodgett films of a diacetylene monocarboxylic acid are described in reference 52.

7.1.20 *Process Control Applications*

A selective Review of the application of Raman spectroscopy for process control has been described in the early literature. This review includes monitoring quality of hard carbon coatings on computer hard discs; chemical composition (as mixtures, solvent separations, reactions such as polymerization, hydrogenation and curing, gas composition, fermentation, polymorphy screening in pharmaceuticals, and basic polymer morphology.[53] A variety of molecular spectroscopic techniques such as UV, IR (MIR), and Raman spectroscopy are compared as measurement techniques for studies of the double bond positional isomerization process in linseed oil.[54]

Different aspects of the optimization of NIR FT-Raman spectroscopy (excitation at 1064 nm) are presented. Methods for overcoming the fluorescence of impurities, increasing Raman intensity, and the efficiency of Raman scattering are summarized. Applications of the technique for routine analysis, process control, and quality control are discussed and possible future developments of process Raman are considered.[55]

7.1.21 *Semiconductors, Electronics*

The application of Raman spectroscopy useful for analysis of electronic band bending at surfaces and interfaces of III–V semiconductors has been described. Raman theory and measurements have been described for gas-covered surfaces, semiconductor and metal interfaces, as well as heterostructures of semiconductors. Raman spectroscopy is compared with photoelectron spectroscopy for the determination of band bending. Special attention is given to buried interfaces, and to simultaneous analysis of band bending and structural properties.[56]

First-order Raman spectra from nanocrystalline semiconductors are described. A micro-Raman probe to analyze porous-silicon samples was prepared using electrochemical anodization. Both Raman and photoluminescence measurements were carried out on the *same* sampling spots. The influence of crystallite sizes on the Raman shifts and line shapes are discussed. A Gaussian distribution in crystallite sizes is explicitly included to calculate the Raman spectra of porous silicon. The size distribution obtained from fitting the Raman data is able to predict the photoluminescence accurately.[57]

7.1.22 *Surface Analysis*

An optical multichannel apparatus has been used to obtain Raman spectra from electrochemically oxidized pyrite in a neutral solution. The pyrite Raman spectra are compared with the spectra from electrochemically oxidized galena and sulfur electrodeposited on platinum. Elemental sulfur and polysulfides are found in the range 0.42–1.0 V vs. the saturated calomel electrode (SCE) for pyrite and 0.13–1.0 V (SCE) for galena.[58]

Raman spectroscopy is shown to provide direct information on the composition and structure of surface oxides formed on alloys. Previously, identification of specific compounds present in these oxides only could be inferred from conventional surface analysis methods. Backscattering spectra were obtained from thick oxide films ($\gtrsim 10$ μm) present on high purity and commercial stainless steels after exposure to air and coal gasifier environments at 980 °C. With as-grown oxides, data were obtained from the outer oxide regions. Spectra from inner regions also were measured using grazing-angle sectioning techniques. The presence of impurities and minor alloying constituents in the substrates are shown to have profound effect on results; for example, the addition of a few percent of titanium and manganese resulted in the formation of a duplex scale with a thin, complex, outer layer and a relatively thick α–Cr_2O_3 inner scale.[59]

7.1.23 *Textiles*

A preliminary investigation of the novel application of Fourier transform Raman spectroscopy to the non-destructive analysis of archaeological fabrics is illustrated with linen samples from two sites, one a mummy-wrapping from a XIIth Dynastic Egyptian rock tomb burial (*ca.* 1900 BC) and the other some cloth from a mass grave at Kasr-el-Yahud, near Jericho in the Dead Sea region (AD 614). Comparison of the spectra with that of pure cotton, another cellulose-based natural fiber, was effected and a distinction between the two was demonstrated. The samples were not pretreated chemically before analysis, and were used directly from their respective storage collections.[60]

An in situ surface enhanced resonance Raman scattering (SERRS) procedure is described for the analysis of a reactive dye covalently bound to a single strand of a cotton fiber. This procedure can be completed in 5 h, whereas an alternative enzyme digestion method takes approximately 21 h. These two fiber preparation methods give similar spectra from picogram quantities of dye present on a 2–5 mm length of fiber. The in situ nature of the analysis and the small sample size make this method particularly suitable for forensic applications.[61]

7.2 Raman Spectra-Structure Correlation Charts

The following figures demonstrate the spectra-structure correlation features of Raman spectra for model hydrocarbon compounds. The figures are described as follows.

Fig. 7.1. Raman Spectral Chart: Raman Fundamental Group Frequency and Fingerprint Frequency locations, showing a Raman Intensity (in Counts) Spectrum with Typical Functional Groups and Band Locations Labeled (Raman Shift of 4000 to 100 cm^{-1}).

Fig. 7.2. Raman Spectral Chart: Raman Fundamental Group Frequency locations, showing a Raman Intensity (in Counts) Spectrum for n-Alkanes with Typical Functional Groups and Band Locations Labeled (Raman Shift of 3000 to 2800 cm^{-1} is shown).

Fig. 7.3. Raman Spectral Chart: Raman Fundamental Group Frequency locations, showing a Raman Intensity (in Counts) Spectrum for n-Alkanes C-H Stretching for Methyl (CH_3) and Methylene (CH_2) Functional Group Bands (Raman Shift of 3000 to 2800 cm^{-1} is shown).

Fig. 7.4. Raman Spectral Chart: Raman Fundamental Group Frequency locations, showing a Raman Intensity (in Counts) Spectrum for n-Alkanes Asymmetric and Symmetric C-H Stretching for Methyl (CH_3) Functional Group Bands (Raman Shift of 3000 to 2800 cm^{-1} is shown).

Fig. 7.5. Raman Spectral Chart: Raman Fundamental Group Frequency locations, showing a Raman Intensity (in Counts) Spectrum for n-Alkanes Asymmetric and Symmetric C-H Stretching for Methylene (CH_2) Functional Group Bands (Raman Shift of 3000 to 2800 cm^{-1} is shown).

Fig. 7.6. Raman Spectral Chart: Raman Fundamental Group Frequency and Fingerprint Frequency locations, showing a Raman Intensity (in Counts) Spectrum for C-H and C=C Functional Group Bands. These include Aromatic =C-H, Aliphatic C-H, and C=C Groups (Raman Shift of 4000 to 500 cm^{-1} is shown).

Fig. 7.7. Raman Spectral Chart: Raman Fundamental Group Frequency and Fingerprint Frequency locations, showing a Raman Intensity (in Counts) Spectrum for O-H and C-H Stretching, C-H Bending, and C-O Stretching Functional Group Bands. (Raman Shift of 4000 to 500 cm^{-1} is shown). Relative Peak Heights are shown.

Fig. 7.8. Raman Spectral Chart: Raman Fundamental Group Frequency and Fingerprint Frequency locations, showing a Raman Intensity (in Counts) Spectrum for >C=O carbonyl Stretching Functional Group Band. (Raman Shift of 4000 to 500 cm^{-1} is shown).

Fig. 7.9. Raman Spectral Chart: Raman Fundamental Group Frequency and Fingerprint Frequency locations, showing a Raman Intensity (in Counts) Spectrum for O-H and C-H Stretching, and C-O Stretching Functional Group Bands. (Raman Shift of 4000 to 500 cm^{-1} is shown). Relative Peak Heights are shown.

Fig. 7.10. Raman Spectral Chart: Raman Fundamental Group Frequency and Fingerprint Frequency locations, showing a Raman Intensity (in Counts) Spectrum for C-O Stretching Functional Group Bands. (Raman Shift of 4000 to 500 cm^{-1} is shown).

Fig. 7.11. Raman Spectral Chart: Raman Fundamental Group Frequency and Fingerprint Frequency locations, showing a Raman Intensity (in Counts) Spectrum for >C=O Carbonyl and C-O Stretching Functional Group Bands. (Raman Shift of 4000 to 500 cm^{-1} is shown). Relative Peak Heights are shown.

Fig. 7.12. Raman Spectral Chart: Raman Fundamental Group Frequency and Fingerprint Frequency locations, showing a Raman Intensity (in Counts) Spectrum for O-H and C-H Stretching, as well as >C=O Carbonyl and C-O Stretching Functional Group Bands. (Raman Shift of 4000 to 500 cm^{-1} is shown). Relative Peak Heights are shown.

Fig. 7.13. Raman Spectral Chart: Raman Fundamental Group Frequency and Fingerprint Frequency locations, showing a Raman Intensity (in Counts) Spectrum for C-H and N-H Stretching Functional Group Bands. (Raman Shift of 4000 to 500 cm^{-1} is shown). Relative Peak Heights are shown.

Fig. 7.14. Raman Spectral Chart: Raman Fundamental Group Frequency and Fingerprint Frequency locations, showing a Raman Intensity (in Counts) Spectrum for O-H Stretching Functional Group Band. (Raman Shift of 4000 to 500 cm^{-1} is shown). Relative Overlapping Peak Heights are shown.

Fig. 7.15. Raman Spectra-Structure Band Positions Chart - Raman Fundamental C-H absorptions indicated relative to wavenumbers (as 3300 to 2800 cm^{-1}).

Fig. 7.16. Raman Spectra-Structure Band Positions Chart: Fundamental C=O, N-H, and O-H absorptions indicated relative to wavenumbers (as 3500 to 1000 cm^{-1}).

Fig. 7.17. Raman Spectra-Structure Band Positions Chart - Raman Fundamental absorptions indicated relative to wavenumbers (as 3000 to 500 cm^{-1}).

Fig 7.18. Raman Functional Group Band Positions (in cm^{-1}), alkanes to aromatics.

Fig 7.19. Raman Functional Group Band Positions (in cm^{-1}), alkenes to esters.

Fig 7.20. Raman Functional Group Band Positions (in cm^{-1}), esters to urethanes.

Fig 7.21. Raman Functional Group Band Positions (in cm^{-1}), N-H Stretch to C=N Stretch.

Fig 7.22. Raman Functional Group Band Positions (in cm^{-1}), N=N Stretch to S-H Stretch.

Fig 7.23. Raman Functional Group Band Positions (in cm^{-1}), S-H Stretch to P-O Stretch.

The H-O-H of water yields a complex broad-banded spectral shape, which after deconvolution is composed of multiple Gaussian-shaped bands. A band at 3430 cm^{-1}, is due to symmetric and asymmetric H-O-H stretching. Local hydrogen bonding has a great deal of effect on the broadness (or narrowness) of the involved bands. Figure 7.14 and Table 7.2 are indicative of the individual bands contributing to the broad single water spectrum band, depending upon hydrogen bonding effects.[62–66]

7.3 Published Raman Standard Reference Methods and Practices

The following information is provided to indicate the standard reference methods, practices, and guidelines published which relate directly to Raman spectroscopy.

7.3.1 *European Pharmacopoeia Commission*

Non-destructive analysis allows process measurements to be made on dosage forms for quality and manufacturing assessment. Rapid non-destructive techniques include near infrared (NIR) spectroscopy and Raman spectroscopy; both referred to as Process Analytical Technologies (PAT) methods. These methods may be used to assess the uniformity of dosage units. A document published as a general chapter in the European Pharmacopoeia has been published in Pharmeuropa as, "2.9.47. Demonstration of uniformity of dosage units using large sample sizes".[67–69]

7.3.2 *United States Pharmacopeia (USP) Methods*[73–75]

The USP has been working to develop methods and procedures for the use of non-destructive Raman testing of Pharmaceutical materials. The technique is useful in the detection of counterfeit medicines and substandard products.[70] The USP has also been working to produce spectral libraries for identification of pharmaceutical raw materials using Raman spectroscopy.[71–73]

SP 29 (2005) General Chapter <1120>, Raman Spectrophotometry, discusses both qualitative and quantitative Raman spectrometry for use in pharmaceutical materials identification and analysis. The method stipulates that qualitative Raman measurements may be used for compendial ID testing, as well as for structural elucidation. Quantitative Raman measurements may be applied as appropriate for organic compound analysis, both for active and inactive (i.e., excipient) ingredients. The quantitative Raman method is described as using the following relationship:

$$I_V = KLCI_0 \tag{7.1}$$

where I_V is the Raman peak intensity at a given wavelength (wavenumber); K represents instrument and sample constants (such as the Raman Stokes shift); L is the active pathlength; C is the molar concentration of a particular component in the sample (in mass per unit volume); and I_0 is the excitation laser intensity.

The method discusses the use of several laser types, including both NIR and visible excitation types as indicated in Table 7.4.

Table 7.4 Typical lasers used in pharmaceutical applications.[74–75]

Region	Laser type	Power	Wavelength range, nm*
NIR	1064	Up to 3W	1075–1563
	785	Up to 500 mW	791–1027
Visible	488–632.8	Up to 1W	488–781

*(Stokes Region, 100 cm^{-1} to 3000 cm^{-1} shift).

7.3.3 *International Organization for Standardization (ISO) Standards*[76–77]

For the International Standard Organization (ISO) there were no specific Raman methods listed for 2015 as "Raman methods" in the ISO search site. There has been a paper published related to ISO 6579:2002 standards to identify Salmonella spp. in food. The paper reported that the variability between Raman spectral comparisons was related to the microbial species and the growth phase of bacteria in food.[78]

7.3.4 *Association of Official Analytical Chemists (AOAC) International Method*[79]

The Association of Official Analytical Chemists (AOAC) International has several methods for using Raman spectroscopy for analysis. Raman spectroscopy methods are being investigated for dairy and food analysis.

Non-destructive methods have been investigated for food analysis using Raman spectroscopy. These include measurements of total dietary fiber,[80] edible oils and fats discrimination,[81] diazinon pesticide formulations,[82] caffeine in energy drinks,[83] and ethanol in biofuels and beverages.[84] There are other constituents being evaluated for standard Raman method applications.

7.3.5 *ASTM International Standards*[85]

The American Society for Testing and Materials (ASTM) International has published several standards for calibrating Raman spectrometers. The most cited Raman Practices include the following, with designation numbers, titles, and brief descriptions.

ASTM E131-10 Standard Terminology Relating to Molecular Spectroscopy
The values stated in SI units are to be regarded as standard. No other units of measurement are included in this standard.[86]

ASTM E1654-94(2013) Standard Guide for Measuring Ionizing Radiation-Induced Spectral Changes in Optical Fibers and Cables for Use in Remote Raman FiberOptic Spectroscopy. This guide covers the method for measuring the real time, in situ radiation-induced alterations to the Raman spectral signal transmitted by a multimode, step index, silica optical fiber.[87]

ASTM E1683-02(2007) Standard Practice for Testing the Performance of Scanning Raman Spectrometers. This practice covers routine testing of scanning Raman spectrometer performance and to assist in locating problems when performance has degraded. It is also intended as a guide for obtaining and reporting Raman spectra.[88]

ASTM E1840-96(2007) Standard Guide for Raman Shift Standards for Spectrometer Calibration. This guide covers Raman shift values for common liquid and solid chemicals that can be used for wavenumber calibration of Raman spectrometers.[89]

ASTM E2056-04(2010) Standard Practice for Qualifying Spectrometers and Spectrophotometers for Use in Multivariate Analyses, Calibrated Using Surrogate Mixtures. This practice relates to the multivariate calibration of spectrometers and spectrophotometers used in determining the physical and chemical characteristics of materials. A detailed description of general multivariate analysis is given in Practice E1655. This standard refers only to those instances where surrogate mixtures can be used to establish a suitable calibration matrix.[90]

ASTM E2529-06 Standard Guide for Testing the Resolution of a Raman Spectrometer. This guide is designed for routine testing and assessment of the spectral resolution of Raman spectrometers using either a low-pressure arc lamp emission lines or a calibrated Raman band of calcite.[91]

ASTM E2642-09 Standard Terminology for Scientific Charge-Coupled Device (CCD) Detectors. This terminology brings together and clarifies the basic terms and definitions used with scientific grade cooled charge-coupled device (CCD) detectors, thus allowing end users and vendors to use common documented terminology when evaluating or discussing these instruments. CCD detectors are sensitive to light in the region from 200 to 1100 nm.[92]

ASTM E2911 - 13 Standard Guide for Relative Intensity Correction of Raman Spectrometers. This guide is designed to enable the user to correct a Raman spectrometer for its relative spectral-intensity response function using NIST Standard Reference Materials[2] in the 224X series (currently SRMs 2241, 2242, 2243, 2244, 2245, 2246), or a calibrated irradiance source. This relative intensity correction procedure will enable the intercomparison of Raman spectra acquired from differing instruments, excitation wavelengths, and laboratories.[93]

References

1. N.B. Colthup, L.H. Daly and S.E. Wiberley, *Introduction to Infrared and Raman Spectroscopy*, 3rd edn. (Academic Press, Boston, 1990).

2. D. Lin-Vien, N.B. Colthup, W.G. Fateley and J.G. Grasselli, *The handbook of infrared and Raman characteristic frequencies of organic molecules,* 1st edn. (Academic Press Elsevier, Boston, 1991).
3. J. Grasselli, R. Hannah, D.W. Mayo and F.A. Miller, *Infrared Spectroscopy* (Bowdoin College Press, Brunswick, Maine, 1994, p. 44).
4. G. Socrates, *Infrared Characteristic Group Frequencies*, 2nd edn. (John Wiley & Sons, Chichester, 1997, pp. 121–129).
5. D. Yang and Y. Ying, Applications of Raman spectroscopy in agricultural products and food analysis: a review, *Applied Spectroscopy Reviews* **46**(7) (2011) 539–560.
6. B. Liu, P. Zhou, X. Liu, X. Sun, H. Li and M. Lin, Detection of pesticides in fruits by surface-enhanced Raman spectroscopy coupled with gold nanostructures, *Food and Bioprocess Technology* **6**(3) (2013) 710–718.
7. Z. Xin, W. Chen and S. Zhengzhong, Conformation Studies of Silk Proteins with Infrared and Raman Spectroscopy, *Progress in Chemistry* **11** (2006) 013.
8. M.E. Darvin, C. Sandhagen, W. Koecher, W. Sterry, J. Lademann and M.C. Meinke, Comparison of two methods for noninvasive determination of carotenoids in human and animal skin: Raman spectroscopy versus reflection spectroscopy, *Journal of Biophotonics* **5**(7) (2012) 550–558.
9. T.T. Le, S.Yue and J.-X. Cheng, Shedding new light on lipid biology with coherent anti-Stokes Raman scattering microscopy, *Journal of Lipid Research* **51**(11) (2010) 3091–3102.
10. J.F. Li, Y.F. Huang, Y. Ding, Z.L.Yang, S.B. Li, X.S. Zhou and F.R. Fan *et al.*, Shell-isolated nanoparticle-enhanced Raman spectroscopy, *Nature* **464**(7287) (2010) 392–395.
11. W. Li, H. Zhang, C. Wang, Y. Zhang, L. Xu, K. Zhu and S. Xie, Raman characterization of aligned carbon nanotubes produced by thermal decomposition of hydrocarbon vapor, *Applied Physics Letters* **70**(20) (1997) 2684–2686.
12. M. Sun, Z. Zhang, H. Zheng and H. Xu, In-situ plasmon-driven chemical reactions revealed by high vacuum tip-enhanced Raman spectroscopy, *Scientific Reports* **2** (2012).
13. L. Franzen, C. Mathes, S. Hansen and M. Windbergs, Advanced chemical imaging and comparison of human and porcine hair follicles for drug delivery by confocal Raman microscopy, *Journal of Biomedical Optics* **18**(6) (2013) 061210–061210.
14. A. Lam, T. Ngoc, E.-O. Ganbold, K.H. Cho, D. Kang and S.-W. Joo, Raman spectroscopy of gold nanoparticle conjugates of cosmetic ingredient kinetin, *Vibrational Spectroscopy* **73** (2014) 15–18.
15. K.D. McKeegan, A.B. Kudryavtsev and J.W. Schopf, Raman and ion microscopic imagery of graphitic inclusions in apatite from older than 3830 Ma Akilia supracrustal rocks, west Greenland, *Geology* **35**(7) (2007) 591–594.
16. D.M. Bower, A. Steele, M. D. Fries and L. Kater, Micro Raman spectroscopy of carbonaceous material in microfossils and meteorites: improving a method for life detection, *Astrobiology* **13**(1) (2013) 103–113.
17. J.H. Aldstadt and A.F. Martin, Analytical chemistry and the cone penetrometer: In situ chemical characterization of the subsurface, *Microchimica Acta* **127**(1-2) (1997) 1–18.
18. S.R. Chowdhury, E.K. Yanful and A.R. Pratt, Chemical states in XPS and Raman analysis during removal of Cr (VI) from contaminated water by mixed maghemite–magnetite nanoparticles, *Journal of Hazardous Materials* **235** (2012) 246–256.
19. G. Corrado, S. Sanchez-Cortes, O. Francioso and J. V. Garcia-Ramos, Surface-enhanced Raman and fluorescence joint analysis of soil humic acids, *Analytica Chimica Acta* **616**(1) (2008) 69–77.

20. D. Yang and Y. Ying, Applications of Raman spectroscopy in agricultural products and food analysis: a review, *Applied Spectroscopy Reviews* **46**(7) (2011) 539–560.
21. W.Y. Wei and I.M. White, A simple filter-based approach to surface enhanced Raman spectroscopy for trace chemical detection, *Analyst* **137**(5) (2012) 1168–1173.
22. I.H. Boyaci, H.E. Genis, B. Guven, U. Tamer and N. Alper, A novel method for quantification of ethanol and methanol in distilled alcoholic beverages using Raman spectroscopy, *Journal of Raman Spectroscopy* **43**(8) (2012) 1171–1176.
23. J. Thomas, P. Buzzini, G. Massonnet, B. Reedy and C. Roux, Raman spectroscopy and the forensic analysis of black/grey and blue cotton fibres: Part 1. Investigation of the effects of varying laser wavelength, *Forensic Science International* **152**(2) (2005) 189–197.
24. C. Muehlethaler, G. Massonnet and P. Esseiva, The application of chemometrics on Infrared and Raman spectra as a tool for the forensic analysis of paints, *Forensic Science International* **209**(1) (2011) 173–182.
25. R.E. Benner, J.D. Andrade, R.A. Van Wagenen and D.R. Westenskow, Molecular gas analysis by Raman scattering in intracavity laser configuration, U.S. Patent 4,648,714, issued March 10, 1987.
26. P.R. Regnier and JP-E. Taran, On the possibility of measuring gas concentrations by stimulated anti-Stokes scattering, *Applied Physics Letters* **23**(5) (1973) 240–242.
27. P.J. Wrzesinski, S. Roy and J.R. Gord, Interference-free coherence dynamics of gas-phase molecules using spectral focusing, *Optics Express* **20**(21) (2012) 23390–23397.
28. W. Hug and G. Hangartner, A novel high-throughput Raman spectrometer for polarization difference measurements, *Journal of Raman Spectroscopy* **30**(9) (1999) 841–852.
29. I. Escudero-Sanz, B. Ahlers and G.B. Courreges-Lacoste, Optical design of a combined Raman–laser-induced-breakdown-spectroscopy instrument for the European Space Agency ExoMars Mission, *Optical Engineering* **47**(3) (2008) 033001–033001.
30. A.J. Berger, T.-W. Koo, I. Itzkan, G. Horowitz and M.S. Feld, Multicomponent blood analysis by near infrared Raman spectroscopy, *Applied Optics* **38**(13) (1999) 2916–2926.
31. P. Le Cacheux, G. Menard, H. Nguyen Quang, P. Weinmann, M. Jouan and Nguyen Quy Dao, Quantitative analysis of cholesterol and cholesterol ester mixtures using near infrared Fourier transform Raman spectroscopy, *Applied Spectroscopy* **50**(10) (1996) 1253–1257.
32. J.F. Brennan, Y. Wang, R.R. Dasari and Michael S. Feld, Near infrared Raman spectrometer systems for human tissue studies, *Applied Spectroscopy* **51**(2) (1997) 201–208.
33. I.R. Lewis, N.C. Chaffin, M.E. Gunter and P.R. Griffiths, Vibrational spectroscopic studies of asbestos and comparison of suitability for remote analysis, *Spectrochimica Acta Part A: Molecular and Biomolecular Spectroscopy* **52**(3) (1996) 315–328.
34. T. Vo-Dinh, D.L. Stokes, G. D. Griffin, M. Volkan, U. J. Kim and M. I. Simo, Surface-enhanced Raman Scattering (SERS) method and instrumentation for genomics and biomedical analysis, *Journal of Raman Spectroscopy* **30**(9) (1999) 785–793.
35. E. Ali, H.G.M. Edwards and I. J. Scowen, Raman spectroscopy and security applications: the detection of explosives and precursors on clothing, *Journal of Raman Spectroscopy* **40**(12) (2009) 2009–2014.
36. A.G. Mercado, J.A. Jani and B. Gilbert, Image analysis of explosives fingerprint contamination using a Raman imaging spectrometer, In European Symposium on Optics for Environmental and Public Safety, *International Society for Optics and Photonics* (1995, September) 142–152.
37. N.A. Hatab, G. Eres, P.B. Hatzinger and B. Gu, Detection and analysis of cyclotrimethylenetrinitramine (RDX) in environmental samples by surface-enhanced Raman spectroscopy, *Journal of Raman Spectroscopy* **41**(10) (2010) 1131–1136.

38. P.E. Flecher, J.B. Cooper, T.M. Vess and W.T. Welch, Remote fiber optic Raman analysis of benzene, toulene and ethylbenzene in mock petroleum fuels using partial least squares regression analysis, *Spectrochimica Acta Part A: Molecular and Biomolecular Spectroscopy* **52**(10) (1996) 1235–1244.
39. J. Pironon, R. Thiery, S. Teinturier and F. Walgenwitz, Water in petroleum inclusions: evidence from Raman and FT-IR measurements, PVT consequences, *Journal of Geochemical Exploration* **69** (2000) 663–668.
40. T. Vankeirsbilck, A. Vercauteren, W. Baeyens, G. Van der Weken, F.Verpoort, G.Vergote and J. P. Remon, Applications of Raman spectroscopy in pharmaceutical analysis, *TrAC trends in Analytical Chemistry* **21**(12) (2002) 869–877.
41. F.W. Langkilde, J. Sjöblom, L. Tekenbergs-Hjelte and J. Mrak, Quantitative FT-Raman analysis of two crystal forms of a pharmaceutical compound, *Journal of Pharmaceutical and Biomedical Analysis* **15**(6) (1997) 687–696.
42. K. Buckley and P. Matousek, Recent advances in the application of transmission Raman spectroscopy to pharmaceutical analysis, *Journal of Pharmaceutical and Biomedical Analysis* **55** (4) (2011) 645–652.
43. Z. Sun, A. Ibrahim, P.B. Oldham, T.P. Schultz and T.E. Conners, Rapid Lignin Measurement in Hardwood Pulp Samples by Near Infrared Fourier Transform Raman Spectroscopy, *Journal of Agricultural and Food Chemistry* **45**(8) (1997) 3088–3091.
44. K. Ito, T Kato, Y. Kojima, T. Ona, Y. Ootake, M. Shibata, T. Sonoda, Y. Tamai, Rapid prediction of native wood pulp properties by Fourier transform Raman spectroscopy, *Journal of Pulp and Paper Science* **26**(2) (2000) 43–47.
45. U. P. Agarwal, Assignment of the photoyellowing-related 1675 cm^{-1} Raman/IR band to p-quinones and its implications to the mechanism of color reversion in mechanical pulps, *Journal of Wood Chemistry and Technology* **18**(4) (1998) 381–402.
46. U. P. Agarwal, R. H. Atalla, I. A. Weinstock, FT Raman Spectroscopy; Rapid, Noninvasive Technique for Direct Measurement of Lignin in Kraft Pulp, *Proceedings 1996 International Pulp Bleaching Conference*, Washington, DC US, April 14–18 (1996) 531–536.
47. J. J. Cael, Jr., Vibrational analysis of polysaccharides, Case Western Reserve Univ., *Ph.D. Thesis*, 1976: 193 pp. [Avail. from Univ. Microfilms, Ann Arbor, Mich. 48106]
48. U. P. Agarwal, Assignment of the photoyellowing-related 1675 cm^{-1} Raman/IR band to p-quinones and its implications to the mechanism of color reversion in mechanical pulps, *Journal of Wood Chemistry and Technology* **18**(4) (1998) 381–402.
49. A.-L. Kjoniksen, B. Nystrom, D. Ostrovskii and L. M. Torrell, Association and thermal gelation in aqueous mixtures of ethyl (hydroxyethyl) cellulose and ionic surfactant: FTIR and Raman study, *Macromolecules* **32**(5) (1999) 1534–1540.
50. T. Johjima, M. Takayama, H. Tanaka, H. Wariishi and T. Yamanaka, Fourier Transform Raman Assignment of Guaiacyl and Syringyl Marker Bands for Lignin Determination, *Spectrochimica Acta, Part A: Molecular Spectroscopy* **53A**(10) (1997) 1621–1628.
51. T. Özpozan, B. Schrader and S. Keller, Monitoring of the polymerization of vinylacetate by near IR FT Raman spectroscopy." *Spectrochimica Acta Part A: Molecular and Biomolecular Spectroscopy* **53**(1) (1997) 1–7.
52. A. Saito, Y. Urai and K. Itoh, Infrared and resonance Raman spectroscopic study on the photopolymerization process of the Langmuir-Blodgett films of a diacetylene monocarboxylic acid, 10, 12-pentacosadiynoic acid, *Langmuir* **12**(16) (1996) 3938–3944.
53. F. Adar, R. Geiger and J. Noonan, Raman spectroscopy for process/quality control, *Applied Spectroscopy Reviews* **32**(1-2) (1997) 45–101.

54. B. Chmielarz, K. Bajdor, A. Labudzinska and Z. Klukowska-Majewska, Studies on the double bond positional isomerization process in linseed oil by UV, IR and Raman spectroscopy, *J. Mol. Struct.* **348** (1995) 313–16.
55. B. Schrader, Raman spectroscopy in the near- infrared: a most capable method of vibrational spectroscopy, *Fresenius' J. Anal. Chem.* **355**(3–4) (1996) 233–239.
56. J. Geurts, Analysis of band bending at III–V semiconductor interfaces by Raman spectroscopy, *Surface Science Reports* **18**(1) (1993) 1–89.
57. Md.N. Islam and S. Kumar, Influence of crystallite size distribution on the micro-Raman analysis of porous Si, *Applied Physics Letters* **78**(6) (2001) 715–717.
58. S.B. Turcotte, R. E. Benner, A. M. Riley, J. Li, M. E. Wadsworth and D. M. Bodily, Surface analysis of electrochemically oxidized metal sulfides using Raman spectroscopy, *Journal of Electroanalytical Chemistry* **347**(1) (1993) 195–205.
59. R.L. Farrow, P. L. Mattern and A. S. Nagelberg, Characterization of surface oxides by Raman spectroscopy, *Applied Physics Letters* **36**(3) (1980) 212–214.
60. H. G. M. Edwards, E. Ellis, D. W. Farwell and R. C. Janaway, Preliminary study of the application of Fourier transform Raman spectroscopy to the analysis of degraded archaeological linen textiles, *Journal of Raman Spectroscopy* **27**(9) (1996) 663–669.
61. P.C. White, C. H. Munro and W. E. Smith, In situ surface enhanced resonance Raman scattering analysis of a reactive dye covalently bound to cotton, *Analyst* **121**(6) (1996) 835–838.
62. Raman Water Spectrum http://www1.lsbu.ac.uk/water/water_methods.html
63. Y. Raichlin, A. Millo and A. Katzir, Investigation of the structure of water using mid-IR fiberoptic evanescent wave spectroscopy, *Phys. Rev. Lett.* **93** (2004) 185703.
64. N. A. Chumaevskii and M. N. Rodnikova, Some peculiarities of liquid water structure, *J. Mol. Liq.* **106** (2003) 167–177.
65. A. Millo, Y. Raichlin and A. Katzir, Mid-IR fiberoptic ATR spectroscopy of the solid-liquid phase transition of water, *Appl. Spectroscopy* **59** (2005) 460–466.
66. Q. Sun, Raman spectroscopic study of the effects of dissolved NaCl on water structure, *Vibrational Spectr.* **62** (2012) 110–114.
67. European Pharmacopoeia Commission, Group PAT, Demonstration of uniformity of dosage units using large sample sizes (2.9.47), *Pharmeuropa* **23** (2011) 285–293.
68. E. Ziemons, J. Mantanus, P. Lebrun, E. Rozet, B. Evrard and Ph. Hubert, *J. Pharm. Biomed. Anal.* **53** (2010) 510–516.
69. J. Mantanus, E. Ziémons, P. Lebrun, E. Rozet, R. Klinkenberg, B. Streel, B. Evrard and Ph. Hubert, *Talanta* **80** (2010) 1750–1757.
70. M. Hajjou and P. Lukulay, Editorial: Potential use of handheld Raman devices as tools for screening medicines for quality, *BioPharma Asia* January/February 2014, 14–21. Found at http://www.usp.org
71. http://www.pharmamanufacturing.com/assets/Media/1008/Thermo_fromlabtodock.pdf
72. http://www.europeanpharmaceuticalreview.com/wp-content/uploads/Raman-Supplement-2012.pdf
73. B. Ma, USP Spectral Library Updates, The USP Excipients Stakeholder Forum, June 7, 2013, http://www.usp.org
74. USP Raman Spectrometry Reference: http://www.pharmacopeia.cn/v29240/usp29nf24s0_c1120.html

75. USP 29 (2005) General Chapter <1120>, Raman Spectrophotometry, "Spectrophotometric Identification Tests". United States Pharmacopeia and National Formulary (USP 29-NF 24). Rockville, MD: United States Pharmacopeia Convention; 2979.
76. International Organization for Standardization, ISO Central Secretariat, 1, ch. de la Voie-Creuse, CP 56 - CH-1211 Geneva 20, Switzerland.
77. ISO Link for search using "Infrared Methods", 34 Methods (2014). http://www.iso.org/iso/home/search.htm?qt=infrared+method&sort=rel&type=simple&published=on
78. A. Assaf, C. B. Y. Cordella and G. Thouand, Raman spectroscopy applied to the horizontal methods ISO 6579: 2002 to identify Salmonella spp. in the food industry, *Analytical and Bioanalytical Chemistry* **406**(20) (2014) 4899–4910.
79. Association of Official Analytical Chemists (AOAC), Official Methods for Analysis. 18th edn. (AOAC International, 481 N. Frederick Avenue, Suite 500, Gaithersburg, MD, USA 2005).
80. D.D. Archibald, S. E. Kays, D. S. Himmelsbach and F. E. Barton, Raman and NIR spectroscopic methods for determination of total dietary fiber in cereal foods: A comparative study, *Applied Spectroscopy* **52**(1) (1998) 22–31.
81. H. Yang, J. Irudayaraj and M. M. Paradkar, Discriminant analysis of edible oils and fats by FTIR, FT-NIR and FT-Raman spectroscopy, *Food Chemistry* **93**(1) (2005) 25-32.
82. S.G. Skoulika, A.G. Constantinos and M.G. Polissiou, FT-Raman spectroscopy—Analytical tool for routine analysis of diazinon pesticide formulations, *Talanta* **51**(3) (2000) 599–604.
83. S. Armenta, S. Garrigues and M. de la Guardia, Solid-phase FT-Raman determination of caffeine in energy drinks, *Analytica Chimica Acta* **547**(2) (2005) 197–203.
84. L.S. Mendes, F.C.C. Oliveira, P.A.Z. Suarez and J.C. Rubim, Determination of ethanol in fuel ethanol and beverages by Fourier transform (FT)-near infrared and FT-Raman spectrometries, *Analytica Chimica Acta* **493**(2) (2003) 219–231.
85. ASTM Practices for Raman Spectroscopy, *Annual Book* of *ASTM Standards*, Volume 03.06, 2014. ASTM International, 100 Barr Harbor Drive, PO Box C700, West Conshohocken, PA, 19428–2959 USA.
86. ASTM E131-10 Standard Terminology Relating to Molecular Spectroscopy, ASTM International, 100 Barr Harbor Drive, PO Box C700, West Conshohocken, PA, 19428–2959 USA (2010).
87. ASTM E1654-94(2013) Standard Guide for Measuring Ionizing Radiation-Induced Spectral Changes in Optical Fibers and Cables for Use in Remote Raman FiberOptic Spectroscopy, ASTM International, 100 Barr Harbor Drive, PO Box C700, West Conshohocken, PA, 19428–2959 USA (2013).
88. ASTM E1683-02(2007) Standard Practice for Testing the Performance of Scanning Raman Spectrometers, ASTM International, 100 Barr Harbor Drive, PO Box C700, West Conshohocken, PA, 19428–2959 USA (2007).
89. ASTM E1840-96(2007) Standard Guide for Raman Shift Standards for Spectrometer Calibration, ASTM International, 100 Barr Harbor Drive, PO Box C700, West Conshohocken, PA, 19428–2959 USA (2007).
90. ASTM E2056-04(2010) Standard Practice for Qualifying Spectrometers and Spectrophotometers for Use in Multivariate Analyses, Calibrated Using Surrogate Mixtures, ASTM International, 100 Barr Harbor Drive, PO Box C700, West Conshohocken, PA, 19428-2959 USA (2010).
91. ASTM E2529-06 Standard Guide for Testing the Resolution of a Raman Spectrometer, ASTM International, 100 Barr Harbor Drive, PO Box C700, West Conshohocken, PA, 19428–2959 USA (2006).

92. ASTM E2642-09 Standard Terminology for Scientific Charge-Coupled Device (CCD) Detectors, ASTM International, 100 Barr Harbor Drive, PO Box C700, West Conshohocken, PA, 19428–2959 USA (2009).
93. ASTM E2911-13 Standard Guide for Relative Intensity Correction of Raman Spectrometers, ASTM International, 100 Barr Harbor Drive, PO Box C700, West Conshohocken, PA, 19428–2959 USA (2013).

Chapter 8

Raman Reference Material Spectra and Tables

Contents

8.1 Standard Reference Materials (SRMs) for Raman Intensity Correction 229
8.1.1 The Meaning of Raman Intensity Calibration 230
8.2 Wavenumber Standard Reference Materials for Raman 233
8.2.1 The Meaning of the NIST Uncertainty Number for SRMs . . . 237
8.3 Raman Materials Data 240
8.3.1 Raman Detector Response Curves 241
8.3.2 Raman Laser Excitation Ranges 243
8.3.3 Raman Optical Material Transmittance Ranges 245
References 248

8.1 Standard Reference Materials (SRMs) for Raman Intensity Correction

The National Institute for Standards and Technology (NIST) previously known as the National Bureau of Standards (NBS) provides materials designed as reference sources to verify the performance characteristics of Raman spectrometers of any design. The main standards are currently used for intensity standardization. Table 8.1 lists the current materials available for Raman intensity calibration. The materials are termed SRMs or Standard Reference Materials. In the case of Raman, there are five active NIST SRMs for measuring Raman intensity, depending upon the excitation laser wavelength used.[1–6]

Table 8.1 NIST SRMs for Raman spectrometer calibration.[1]

SRM No.	Laser excitation (nm)	Material	SRM dimensions
2241	785 nm Excitation	Chromium-doped (mole fraction of 0.02% Cr_2O_3) sodium borosilicate matrix glass	10 mm in width × 10 mm in length × 1.65 mm in thickness
2242	532 nm Excitation	Manganese-doped (0.15 wt % MnO2) borate matrix glass	10.7 mm in width × 30.4 mm in length × 2.0 mm in thickness
2243	488 nm and 514.5 nm Excitation	Discontinued	NA
2244	1064 nm Excitation	Doped (mass fraction 0.7% Cr_2O_3) oxide in a borosilicate matrix glass	0.7 mm in width × 30.4 mm in length × 2.0 mm in thickness
2245	633 nm Excitation	Bismuth-doped (0.11% mole fraction) oxide in a phosphate matrix glass	10 mm in width × 10 mm in length × 1.65 mm in thickness
2246	830 nm Excitation	Chromium-doped (0.30% mole fraction) oxide in a borosilicate-matrix glass	10 mm in width × 10 mm in length × 1.65 mm in thickness

8.1.1 *The Meaning of Raman Intensity Calibration*

The Raman intensity standards are certified SRMs useful for the correction of the relative intensity of Raman spectra obtained with instruments employing specific laser excitation sources. Individual intensity standards have been provided for excitation wavelength of 488, 532, 633, 785, 830, and 1064 nm. For these standards, the relative intensity of the glass luminescence has been calibrated at NIST using a uniform- white-light -source, with an integrating sphere collection design. The shape of the mean luminescence spectrum of the SRM glass is described using a polynomial expression for relative spectral intensity versus Raman shift wavenumber (cm^{-1}) based on the specific excitation laser wavelength used (in nanometers). The spectral correction for any Raman spectrometer is determined by measuring the luminescence spectrum of the SRM, applying the polynomial model, and computing the spectral intensity-response correction for any Raman instrument. The application of the spectral intensity correction eliminates the instrument induced spectral artifacts to provide a more uniform calibrated Raman spectrum.

In order for a Raman spectrometer to be calibrated for intensity, the Raman wavenumber axis is corrected using ASTM E1840-96.[7] The laser excitation is aimed at the frosted surface of the glass, to minimize variation in the scattering response.

The intensity correction must be completed over the same Raman shift range as that intended for sample measurement.

The relative intensity of the measured Raman spectrum of the sample can be corrected for the instrument-specific response by a computational procedure that uses a correction curve. This curve is generated using the certified model and the measured luminescence spectrum of the SRM glass. For the spectral range of certification, $\Delta\upsilon = 200\,\text{cm}^{-1}$ to $3500\,\text{cm}^{-1}$, compute the elements of the certified relative mean spectral intensity of SRM 2241, ISRM($\Delta\upsilon$), according to Equation (8.1):

$$\text{ISRM}(\Delta\upsilon) = A_0 + A_1 \times (\Delta\upsilon)^1 + A_2 \times (\Delta\upsilon)^2 + A_3 \times (\Delta\upsilon)^3 + A_4 \times (\Delta\upsilon)^4 + A_5 \times (\Delta\upsilon)^5 \tag{8.1}$$

where ($\Delta\upsilon$) is the wavenumber in units of Raman shift (cm^{-1}) and the An's are the coefficients listed in Table 8.2. The elements of ISRM($\Delta\upsilon$) are obtained by evaluating Equation 8.1 at the same data point spacing used for the acquisition of the luminescence spectrum of the SRM and of the Raman spectrum of

Table 8.2 Coefficients of the Certified Polynomial (a) and of the 95% Confidence Curves (b) Describing the Mean Luminescence Spectrum of SRM 2241 for 785 nm excitation. (Valid for Temperatures of 20 °C to 25 °C).

Polynomial coefficient	Certified value polynomial coefficient (c)	Polynomial coefficient (c) of the ±95% confidence curves (b)	
		+95% CC	−99% CC
A0	9.71 937 E–02	1.04 276 E–01	9.01 111 E–02
A1	2.28 325 E–04	2.39 131 E–04	2.17 519 E–04
A2	-5.86 762 E–08	-7.81 489 E–08	-3.92 035 E–08
A3	2.16 023 E–10	2.32 243 E–10	1.99 803 E–10
A4	-9.77 171 E–14	-1.03 769 E–13	-9.16 653 E–14
A5	1.15 596 E–17	1.23 774 E–17	1.07 417 E–17

(a) A NIST certified value represents data, reported in an SRM Certificate, for which NIST has the highest confidence in its accuracy in that all known or suspected sources of bias have been fully investigated or taken into account.[9]
(b) The consensus curve is a point-by-point weighted mean of the average responses of three instruments,[10–11] fitted by the polynomial model. The uncertainty curves are polynomial models of point-by-point expanded uncertainties, with coverage factor $k = t0.975, 2 = 4.303$ (95% confidence), calculated by combining a between-instrument variance incorporating inter instrumental Type B uncertainties with a pooled within-method variance,[12] following the ISO/JCGM Guide.[13–14]
(c) Where $\text{ISRM}(\Delta\upsilon) = A_0 + A_1 \times (\Delta\upsilon)^1 + A_2 \times (\Delta\upsilon)^2 + A_3 \times (\Delta\upsilon)^3 + A_4 \times (\Delta\upsilon)^4 + A_5 \times (\Delta\upsilon)^5$, for $\Delta\upsilon =$ **200 cm-1 to 3500 cm-1** Raman Shift relative to 785 nm excitation. ISRM($\Delta\upsilon$) has units of photons s^{-1} (cm^{-2}) $(\text{cm}^{-1})^{-1}$ (Data used with permission from NIST).

the sample. ISRM($\Delta\upsilon$) has been normalized to unity and is a relative unit expressed in terms of photons s^{-1} (cm^{-2}) $(cm^{-1})^{-1}$. The data sets that are the measured glass luminescence spectrum, SSRM, and the measured Raman spectrum of the sample, SMEAS, must have the units of Raman shift (cm^{-1}). The elements of the correction curve ICORR($\Delta\upsilon$), defined by Equation 8.2, are obtained from ISRM($\Delta\upsilon$) and the elements of the glass luminescence spectrum, SSRM($\Delta\upsilon$), by:

$$\mathrm{ICORR}(\Delta\upsilon) = \mathrm{ISRM}(\Delta\upsilon)/\mathrm{SSRM}(\Delta\upsilon) \tag{8.2}$$

The elements of the intensity-corrected Raman spectrum, SCORR($\Delta\upsilon$), are derived by multiplication of the elements of the measured Raman spectrum of the sample, SMEAS($\Delta\upsilon$), by the elements of the correction curve as Equation 8.3.[8]

$$\mathrm{SCORR}(\Delta\upsilon) = \mathrm{SMEAS}(\Delta\upsilon) \times \mathrm{ICORR}(\Delta\upsilon) \tag{8.3}$$

The certified model, Equation 8.1, is certified for use between 200 cm^{-1} and 3500 cm^{-1}. The certified model is intended as a simple numerical descriptor of the spectral response observed over the wavenumber range studied. It is not intended as a physically meaningful model. The model coefficients listed in Table 8.2 cannot be used to extrapolate the limits of certification without incurring significant error. Extrapolation of the model outside the certification limits of 200 cm^{-1} and 3500 cm^{-1} is not a supported use of this SRM.

This SRM is not intended for use as a standard for the determination of absolute spectral irradiance or radiance.

Luminescence Spectrum on the Wavelength Scale: The equation describing the mean luminescence spectrum of the glass SRM is given in Equation 8.1, where ($\Delta\upsilon$) is the wavenumber in units of Raman shift (cm^{-1}). For correction of spectra where the x-axis is in wavelength with units of nanometers, the same model coefficients can be used to calculate ISRM(λ) through the following coordinate transformation as Equations 8.4 and 8.5:

$$\begin{aligned}\mathrm{ISRM}(\lambda) = [10^7/\lambda] \times [&A_0 + A_1 \times Z^1 + A_2 \times Z^2 + A_3 \times Z^3 \\ &+ A_4 \times Z^4 + A_5 \times Z^5]\end{aligned} \tag{8.4}$$

where:

$$Z = 10^7 \times [(1.0/\lambda_L) - (1.0/\lambda)] \tag{8.5}$$

and λ_L is the wavelength of the laser in nanometers and λ is the measurement wavelength in nanometers. The prefactor of 10^7 in the first term of Equation 8.4 is needed only if it is desired to preserve the numerical value of spectral areas computed relative to the two x-axis coordinate systems. This definition is from reference (2), with NIST basic application of intensity standard.

8.2 Wavenumber Standard Reference Materials for Raman

Several standard materials are recommended for alignment of the Raman wavenumber axis. These include those listed in Table 8.3. A virtual spectrum of a silicon wafer is shown in Figure 8.1 with a sharp peak at 520.7 cm^{-1}.

Other wavenumber standards include: naphthalene, 1,4 bis (2-methylstyryl) benzene (BMB), sulfur, 50/50 (v/v) toluene/acetonitrile, 4-acetamidophenol, benzonitrile, and polystyrene.[19] An infrared spectrum of polystyrene is given in Figure 8.2, with a table of wavenumber peaks (Tables 8.4 and 8.5).

Table 8.3 Reference materials used for Raman wavenumber calibration.

Material type	Peak position (cm^{-1})	Peak uncertainty (cm^{-1})
Silicon[15]	520.7	±0.5
O_2 (Air)	1555.0	±1.0
N_2 (Air)	2332.0	±1.0
Calcite ($CaCO_3$)[16]	156	—
	281	±1.0
	713	—
	1085	±2.0
Cyclohexane[17]	284.10	±0.78
	426.30	±0.41
	801.30	±0.96
	1028.30	±0.45
	1157.60	±0.45
	1266.34	±0.94
	1444.40	±0.3
	2664.40	±0.42
	2852.90	±0.32
	2923.80	±0.36
	2938.30	±0.51
Indene[18]	730.4	±0.5
	1018.3	±0.5
	1205.6	±0.5
	1552.7	±0.5
	1610.2	±0.5
	2892.2	±1.0
	3054.7	±1.0

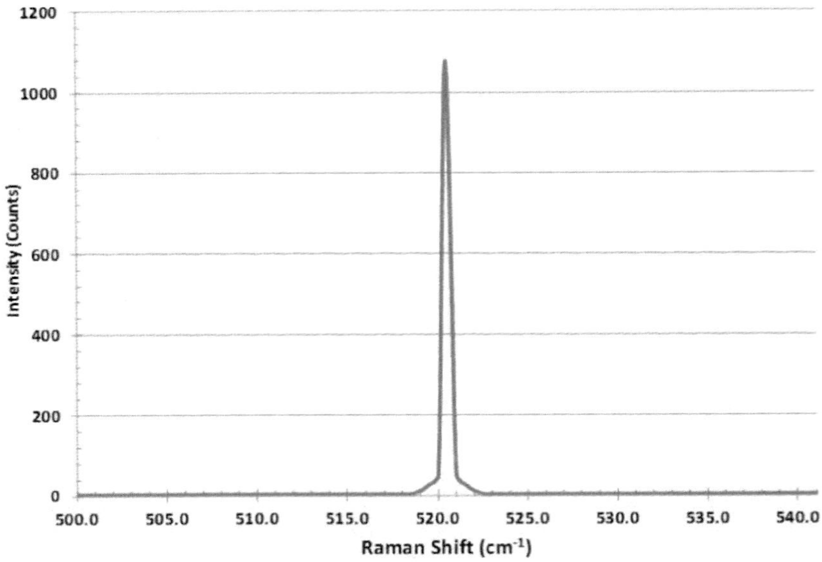

Fig. 8.1 Virtual silicon wafer spectrum with sharp peak at 520.7 cm^{-1} (wavenumbers).

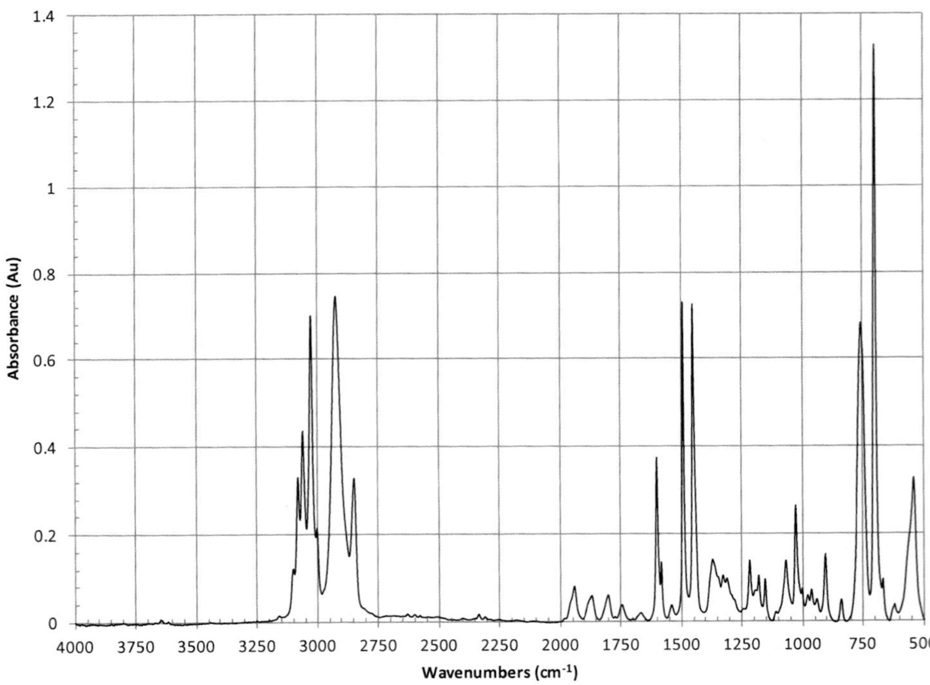

Fig. 8.2 Standard polystyrene film converted to absorbance from transmittance spectrum with film holder and 4 cm^{-1} resolution. 64 co-added scans per spectrum, Nicolet Model 510 FT-IR Spectrometer.

Table 8.4 Standard polystyrene film peak table from sample in Figure 8.2. Peak selection made using a five-point cubic spline, peak interpolation method on data with 4 cm^{-1} resolution.[20–26]

Peak position (cm^{-1})	%T	NIST certified peak position*
3100.9744	75.6287	
3081.6898	46.8141	3082.19
3058.5482	36.6493	3060.03
3025.7642	19.9316	3026.42
3000.6942	61.1944	3001.40
2923.5555	17.9556	
2848.3453	47.0112	2850.13
1941.9659	83.2527	
1870.6127	87.4307	
1803.1163	87.1904	
1747.1908	91.7715	
1600.6273	42.2176	1601.35
1583.2711	73.4623	1583.13
1540.8449	91.9904	
1492.6332	18.6207	
1452.1354	18.9335	
1371.1398	72.4310	
1328.7135	78.6375	
1311.3573	79.9875	
1218.7909	72.8249	
1195.6493	85.1615	
1182.1501	78.8698	
1155.1515	80.3401	1154.64
1068.3705	72.8427	1069.20
1027.8727	54.4572	1028.35
1002.8027	84.5943	
979.6611	87.5156	
964.2333	84.9182	
943.0202	89.2676	
906.3793	70.3824	906.82
840.8115	89.5645	842.08
755.9589	20.7826	
698.1049	4.7028	
667.2495	80.4758	
620.9663	91.9022	
539.9707	47.0627	545.48

*NIST measurements are at made using 0.5 cm^{-1} resolution setting, with center of mass peak picking algorithm. Data reproduced with permission from NIST.

Table 8.5 Peak positions for polystyrene spectra at different resolution settings (from Figure 8.2), using a five-point cubic spline, peak interpolation method.[20–26]

2 cm^{-1}	Peak height	4 cm^{-1}	Peak height	8 cm^{-1}	Peak height	16 cm^{-1}	Peak height	32 cm^{-1}	Peak height
2741.922	0.955	2741.922	0.903	2741.922	0.853	2740.833	0.720	2737.791	0.635
2617.731	0.642	2622.283	0.659	2621.633	0.654	2623.737	0.635	2622.482	0.611
2539.056	0.681	2539.706	0.674	2541.006	0.658	2536.565	0.652	NA	NA
2475.335	1.948	2474.684	2.396	2463.631	2.064	2468.909	1.917	2469.901	1.599
2351.794	1.108	2354.395	1.054	2353.745	1.029	2355.716	0.998	2362.745	0.933
2309.530	0.808	2309.530	0.765	2307.580	0.760	2308.227	0.712	2310.332	0.643
2169.734	0.912	2170.385	0.907	2169.734	0.882	2167.711	0.813	2163.575	0.732
2144.376	0.858	2144.376	0.937	2145.026	0.858	2146.894	0.727	NA	NA
1765.301	0.213	1764.651	0.202	1765.301	0.211	1763.729	0.198	1761.740	0.192
1678.823	0.693	1679.473	0.682	1681.424	0.695	1680.460	0.665	1682.537	0.624

8.2.1 *The Meaning of the NIST Uncertainty Number for SRMs*

NIST uncertainty is a measure of uncertainty unlike the typical manufacturer report as either precision or accuracy. For example, NIST uncertainty has been calculated from Equation 8.6.[27]

$$U = \sqrt{A^2 + B^2 + C^2} \tag{8.6}$$

where A = twice the largest standard deviation of measurement of multiple emission lines (measurands) versus the spectrometer measured values over a period of 31 days (or some other designated period); B = twice the standard deviation of the uncertainty in the peak selection method used; and C is the maximum variation in the standard due to temperature changes over a specified range and conditions. (Note: The number 2 (twice) is referred to as the k value also known as the *coverage factor*, see reference).[28]

For example, if in testing a wavenumber standard, the reported error for multiple wavelengths using emission lines for the NIST spectrometer was ±0.285 nm for all the data points measured over 31 days; the NIST actual reported uncertainty for that SRM is equal to Equation 8.7.

$$U = \sqrt{0.57^2 + 0.41^2 + 0.15^2} = 0.718\,\text{nm} \tag{8.7}$$

This value might be rounded up to ±1 nm considering the added uncertainty of the measurement spectrometer measurements and material stability over the lifetime of an SRM.[27]

A more detailed description of uncertainty computations and rationale with respect to wavelength/wavenumber SRMs is found in several references.[28–30] There are significant complexities to the question of uncertainty in reference materials and so the reader is referred directly to these three citations. One may also obtain more detailed information on the history of the international metrical committees responsible for defining the technical details for uncertainty associated with measurement results; the reader is referred to Appendices C and D in reference 28 for historical details.

Table 8.6 Spectral response (as $D^* = (cm \cdot Hz^{1/2})/W$) for Raman and some infrared detectors (in microns).

Microns	CCD	InAs (-77C)	InGaAs, typ (27C)	InGaAs, ext (27C)	InGaAs IR (-30C)	InGaAs IR (-50C)	Ge (-77C)	Ge (-196C)	Ge:Au (-196C)	InSb (-196C)	PbTe (-196C)	PbSe (-196C)
0.00	2.00E+13	5.00E+10	3.00E+12	3.00E+10	3.00E+10	5.00E+10	1.50E+13	4.50E+13				
1.00	3.75E+13	1.00E+11	4.80E+12	4.00E+10	4.00E+10	7.00E+10	2.50E+13	5.50E+13				
1.50	2.50E+12	1.50E+11	5.50E+12	5.00E+10	1.45E+11	2.80E+11	3.75E+13	9.50E+13				
2.00		2.00E+11	5.00E+10	6.50E+10	3.10E+11	7.00E+11	1.00E+13	7.50E+12		4.50E+10	6.00E+08	7.00E+08
2.50		2.50E+11		4.50E+10	1.80E+11	2.50E+11	2.50E+12	2.50E+12		5.50E+10	8.00E+08	9.00E+08
3.00		3.10E+11		6.00E+09	5.00E+10	7.00E+10	1.00E+12	1.00E+12		6.00E+10	1.50E+09	2.00E+09
4.00		8.00E+10					3.50E+11	3.50E+11	3.60E+09	9.00E+10	2.50E+09	3.00E+09
4.50							3.00E+11	3.00E+11	3.80E+09	1.00E+11	1.00E+09	3.50E+09
5.00							2.00E+11	2.00E+11	3.50E+09	9.50E+10		4.00E+09
5.50							9.00E+10	9.00E+10	3.00E+09	3.00E+10		2.20E+09
6.00							8.00E+10	8.00E+10	2.70E+09			1.50E+09
6.50							6.00E+10	6.00E+10	2.00E+09			7.00E+08
7.00							5.80E+10	5.80E+10	1.50E+09			
7.50							5.50E+10	5.50E+10	1.00E+09			
8.00							5.00E+10	5.00E+10	7.50E+08			
8.50							4.80E+10	4.80E+10	4.00E+08			
9.00							4.70E+10	4.70E+10	1.00E+08			
9.50							4.60E+10	4.60E+10				
10.00							4.50E+10	4.50E+10				
10.50							4.40E+10	4.40E+10				
11.00							4.30E+10	4.30E+10				

(Continued)

Table 8.6 (*Continued*)

Microns	CCD	InAs (-77C)	InGaAs, typ (27C)	InGaAs, ext (27C)	InGaAs IR (-30C)	InGaAs IR (-50C)	Ge (-77C)	Ge (-196C)	Ge:Au (-196C)	InSb (-196C)	PbTe (-196C)	PbSe (-196C)
11.50							4.20E+10	4.20E+10				
12.00							4.10E+10	4.10E+10				
12.50							4.00E+10	4.00E+10				
13.00							4.00E+10	4.00E+10				
13.50							4.00E+10	4.00E+10				
14.00							4.00E+10	4.00E+10				
14.50							4.00E+10	4.00E+10				
15.00							4.00E+10	4.00E+10				
15.50							4.00E+10	4.00E+10				
16.00							4.00E+10	4.00E+10				
16.50							4.00E+10	4.00E+10				
17.00							4.00E+10	4.00E+10				
17.50							4.00E+10	4.00E+10				
18.00							4.00E+10	4.00E+10				
18.50							4.00E+10	4.00E+10				
19.00							4.00E+10	4.00E+10				
19.50							4.00E+10	4.00E+10				
20.00							4.10E+10	4.10E+10				

(*Continued*)

8.3 Raman Materials Data

The following tables and figures illustrate the properties of detectors, sources and optical materials updated to modern material specifications. Raman detector response curves are given in Tables 8.6 and 8.7 and Figures 8.3 and 8.4. Raman laser excitation wavelengths are shown in Table 8.8 and Figure 8.5. Optical material transmittance data is given in Tables 8.9 and 8.10 and Figures 8.6 through 8.8.[31–35]

Table 8.7 Raman detector sensitivity ranges.

Detector sensitivity ranges (μm)	Start	End
Silicon (Si)	0.3	1.1
Indium Gallium Arsenide (InGaAs), typical	0.9	1.7
Indium Gallium Arsenide (InGaAs), extended	0.65	2.4
Indium Gallium Arsenide (InGaAs), IR extended	0.65	2.7
Indium Antimonide (InSb)	1.8	6.8
Lead Telluride (PbTe)	1.5	4.5
Deuterated Triglycine Sulfate (DTGS/KBr)	0.83	25
Mercury Cadmium Telluride (MCT)	1	17
Silicon Photodiode Detectors	0.19	1.1
Photomultipler Tube (PMT), S1 Type Photocathode	0.3	1.2
Photomultipler Tube, Cesium Galium Arsenide (CsGaAs) Photocathode	0.3	0.93
Photomultipler Tube, Cesium Indium Galium Arsenide (CsInGaAs) Photocathode	0.3	1
Photomultipler Tube, Cesium Antiminide CsSb Photocathode	0.25	0.75
Photomultipler Tube, Multialkali (Na-K-Sb-Cs) Photocathode	0.3	0.93
Photomultipler Tube, CsI Photocathode	0.2	0.9
Photomultipler Tube, S1 Type Photocathode	0.3	1.2
Silicon (Si) CCD	0.2	1.1
Indium Gallium Arsenide (InGaAs) CCD	0.8	1.7
Ultraviolet range	0.19	0.36
Visible Range	0.36	0.78
Near Infrared Range	0.78	2.5
Infrared Range	2.5	25
Raman Shift Range	2.5	25

8.3.1 *Raman Detector Response Curves*

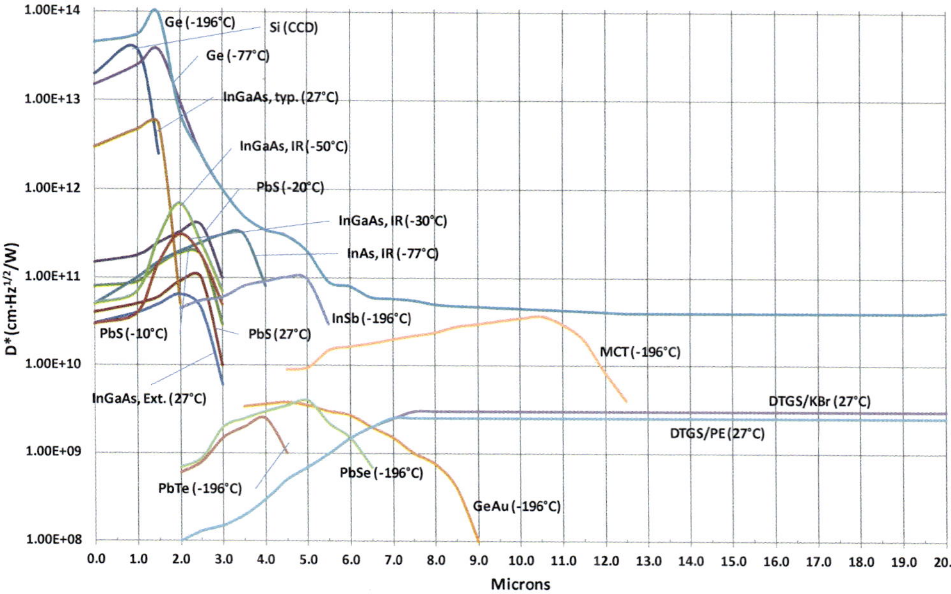

Fig. 8.3 Spectral response (as $D^* = (cm \cdot Hz^{1/2})/W$) for Raman and some infrared detectors (in microns).

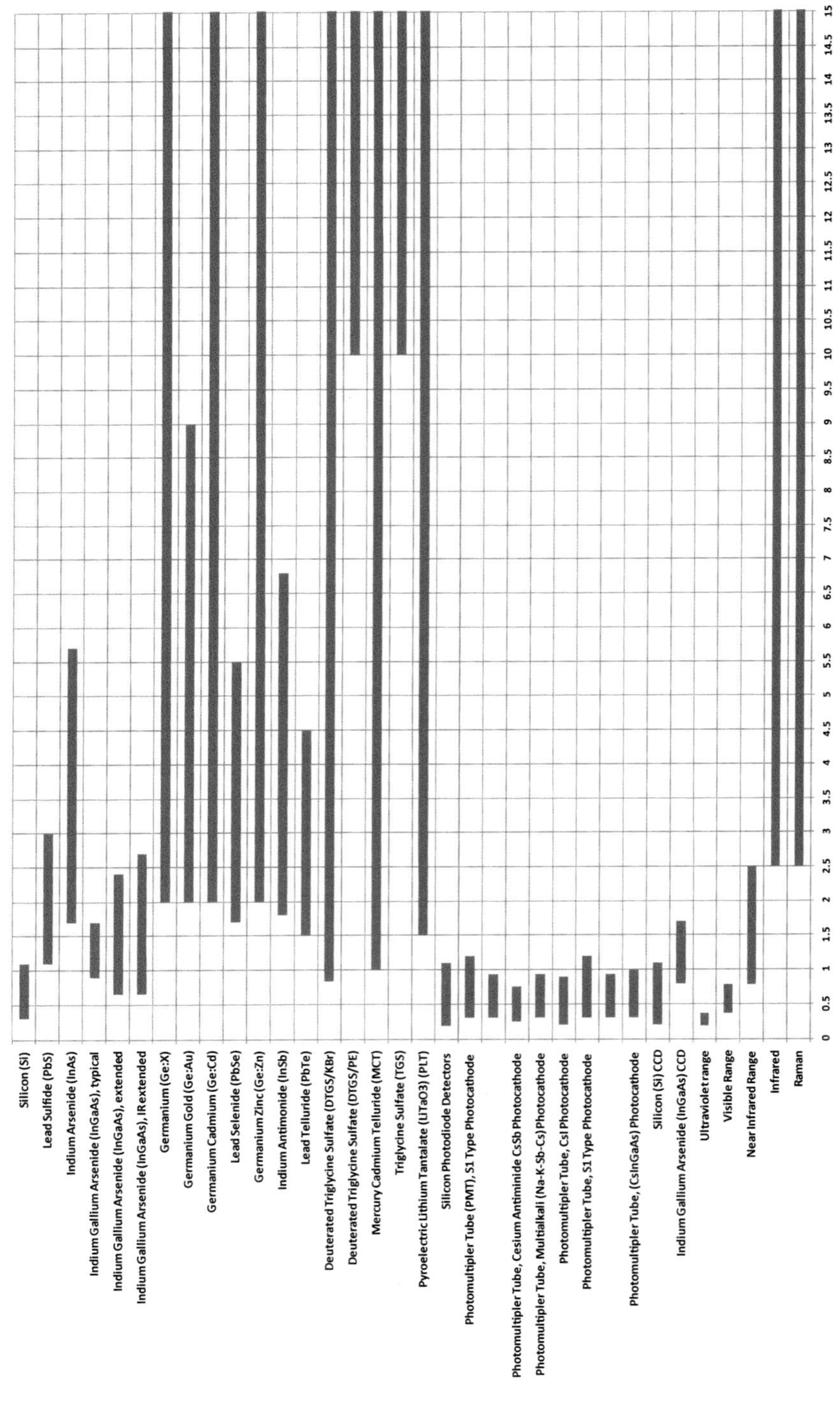

Fig. 8.4 Raman detector sensitivity ranges (in microns).

8.3.2 *Raman Laser Excitation Ranges*

Table 8.8 Raman laser excitation sources with start and end wavelengths (in nm).

Laser emission ranges (nm)	Start (nm)	End (nm)
Ultraviolet (Laser Diode), 244 nm	243.5	244.5
Ultraviolet (Laser Diode), 257 nm	256.5	257.5
Ultraviolet (Laser Diode), 325 nm	324.5	325.5
Ultraviolet (Laser Diode), 364 nm	363.5	364.5
Visible (Laser diode), 455 nm	454.5	455.5
Visible (Laser diode), 473 nm	472.5	473.5
Visible (Laser diode), 488 nm	487.5	488.5
Visible (Laser diode), 514 nm	513.5	514.5
Visible (Laser diode), 532 nm	531.5	532.5
Helium-neon laser (He:Ne), 632.8 nm	632.7	632.9
Visible (Laser diode), 660 nm	659.5	660.5
Visible (Laser diode), 785 nm	784.5	785.5
Near infrared (VCSEL), 850 nm	829.5	830.5
Near infrared (Laser diode), 980 nm	979.5	980.5
NIR: Neodymium yttrium aluminum garnet (Nd:YAG) laser, 1064 nm	1063.9	1064.1

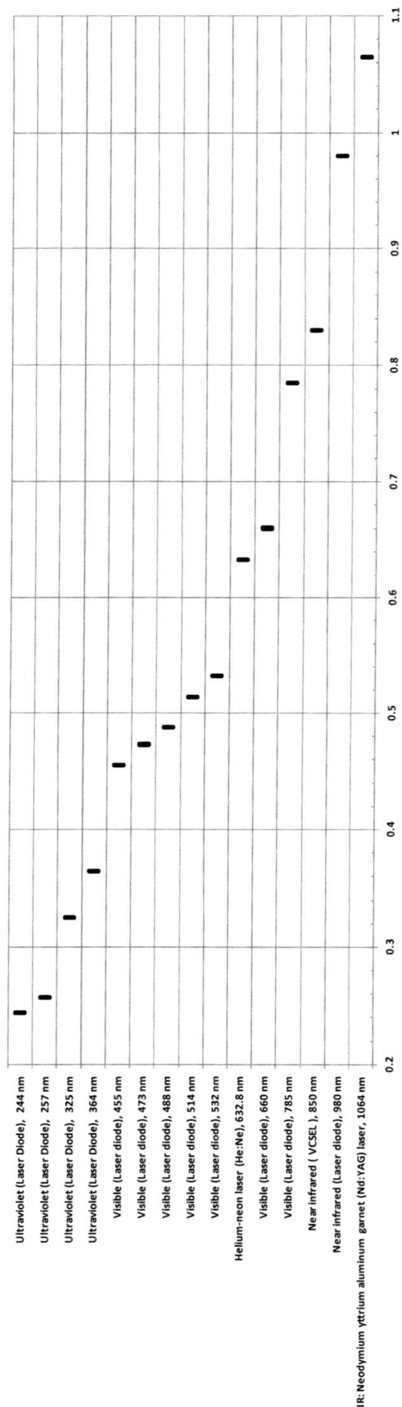

Fig. 8.5 Figure showing most common Raman laser excitation wavelengths.

8.3.3 *Raman Optical Material Transmittance Ranges*

Table 8.9 Optical materials for Raman sampling and optics.

Optical material spectral ranges	Start	End	Difference
Methacrylate, Polymer	0.25	1.1	0.85
Silica, UV Grade Fused (SiO_2)	0.2	2.5	2.30
Silica, Synthetic Fused (SiO_2)	0.23	2.6	2.37
Quartz, Crystalline (SiO_2)	0.24	2.6	2.36
Quartz, extremely low O-H	0.19	2.7	2.51
Glass, Flint (SF 10)	0.38	2.35	1.97
Glass, Flint (SF 8)	0.355	2.35	2.00
Glass (BK 7)	0.315	2.35	2.04
Glass, "Optical Crown"	0.32	2.3	1.98
Glass, "Borosilicate Crown"	0.36	2.35	1.99
Glass, Pyrex®	0.31	2.5	2.19
Glass, Zerodur® Schott	0.4	2.85	2.45
Glass, Tempax®	0.36	2.35	1.99
Sapphire (Aluminum Oxide) (Al_2O_3) 5 mm is 50 %T at 5.1 microns	0.15	5.1	4.95
Sodium Chloride (NaCl)	0.25	16	15.75
Suprasil 300	0.19	3.6	3.41
Diamond (C)	0.22	4	3.78
Silica, Fused as pure SiO_2 (IR Quality)	0.2	4	3.80
Quartz, Anhydrous	0.2	2.75	2.55
Zirconium Fluoride (ZrF)	0.9	4.76	3.86
Chalcogenide (NIR-IR measurements)	2.22	11.1	8.88
Cesium Iodide (CsI)	0.3	50	49.70
Lead Sulfide (PbS)	1.1	3	1.90
Potassium Bromide (KBr)	0.25	26	25.75
Potassium Chloride (KCl)	0.25	20	19.75
Thallium Bromide-Iodide (KRS-5)	0.6	40	39.40
Germanium (Ge)	1.1	30	28.90
Arsenic Sulfide (As_2S_3)	0.6	15	14.40
Magnesium Fluoride (MgF_2)	0.6	9.5	8.90
Zinc Selenide (ZnSe) (IRTRAN-4)	0.6	26	25.40
Barium Fluoride (BaF_2)	0.2	13	12.80
Zinc Sulfide (ZnS) (Cleartran or IRTRAN-2)	0.6	15	14.40
Calcium Fluoride (CaF_2) (IRTRAN-3)	0.2	9	8.80
Silver Bromide (AgBr)	0.5	35	34.50
Polyethylene, Polymer (High Density)	16	333	317.00
Glass, Silica, Fused "Water Free" (SiO_2); 10 mm is 82%T at 3.15 microns	0.15	3.15	3.00

(Continued)

Table 8.9 (*Continued*)

Optical material spectral ranges	Start	End	Difference
Glass, Corning Optical Glasses as HPFS® 7978 and 7979 50%T/cm	0.15	4.35	4.20
Glass, Corning Optical Glasses as HPFS® 7980; 50%T/cm	0.2	2.6	2.40
Glass, Corning Optical Glasses as HPFS® 865X and 797X 50%T/cm	0.15	2.6	2.45
Glass, ESCO Optical Glasses as S1-UVA and S1-UVB Working Range (Band 1)	0.17	2.2	2.03
Glass, ESCO Optical Glasses as S1-UVA and S1-UVB Working Range (Band 2)	2.9	3.6	0.70
Glass, ESCO Optical Glasses as I2-IR Working Range	0.25	3.6	3.35
Glass, ESCO Optical Glasses as G1 and A1 Working Range	0.27	2.5	2.23
Glass, ESCO Optical Glasses as BK7 Working Range	0.35	2	1.65
Ultraviolet range	0.19	0.36	0.17
Visible Range	0.36	0.78	0.42
Near Infrared Range	0.78	2.5	1.72
Infrared	2.5	25	22.50
Raman	2.5	25	22.50

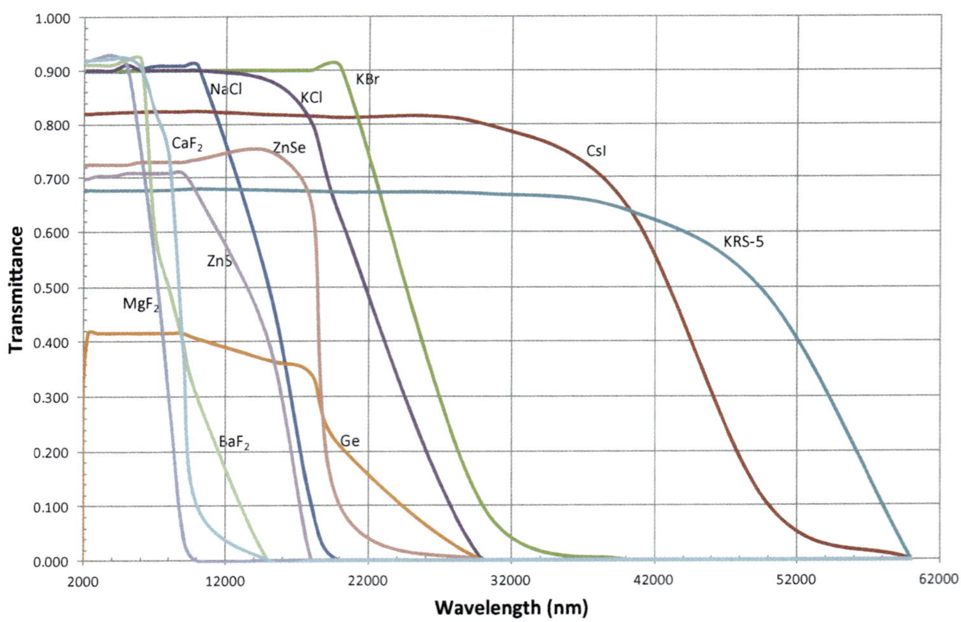

Fig. 8.6 Optical materials transmittance data (in T with respect to nm).

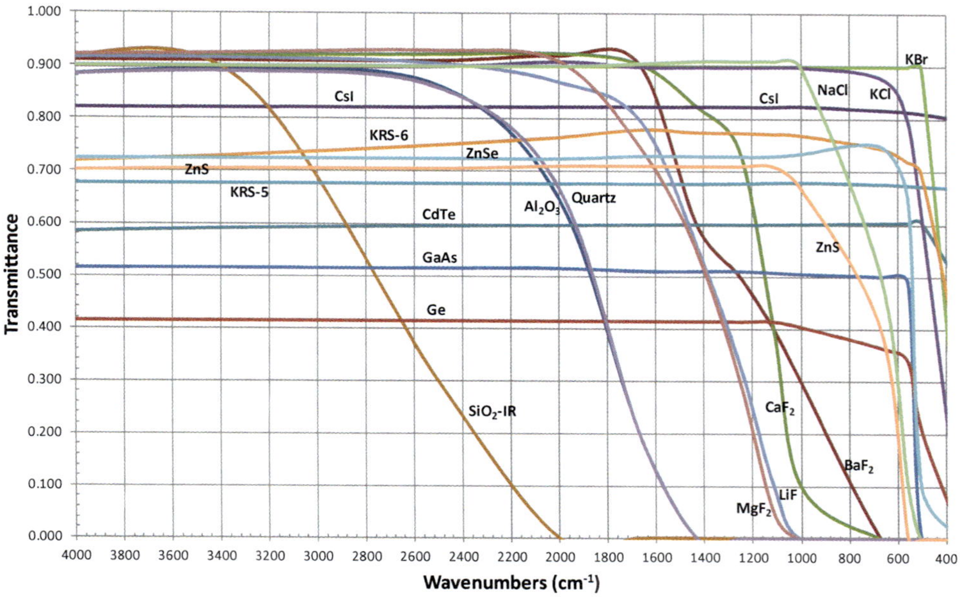

Fig. 8.7 Optical materials transmittance data (in T with respect to wavenumbers, cm^{-1}).

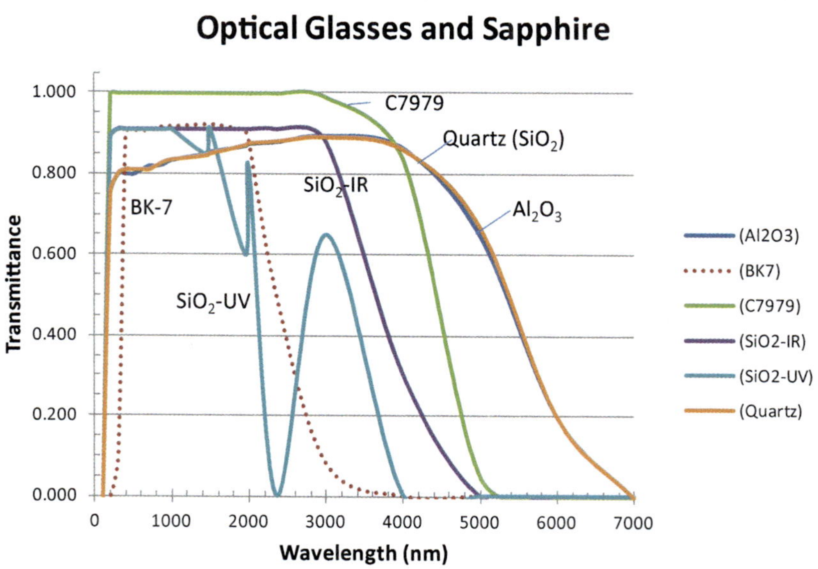

Fig. 8.8 Optical glasses (i.e., BK7, Corning 7979, IR glass, UV glass, Quartz) and sapphire (Al_2O_3) transmittance data versus wavelength (in nanometers) useful for Raman.

Table 8.10 Optical glass and sapphire transmittance data versus wavelength (in nanometers) useful for Raman.

Lambda (nm)	(Al2O3)	(BK7)	(C7979)	(SiO2-IR)	(SiO2-UV)	(Quartz)
100	0.000	0.000	0.0000	0.000	0.000	0.000
200	0.750	0.000	0.9970	0.890	0.890	0.750
300	0.800	0.100	0.9999	0.910	0.910	0.800
400	0.800	0.900	1.0000	0.910	0.910	0.810
500	0.800	0.905	1.0000	0.910	0.910	0.810
600	0.810	0.905	1.0000	0.910	0.910	0.810
700	0.820	0.905	1.0000	0.910	0.910	0.810
800	0.820	0.910	1.0000	0.910	0.910	0.820
900	0.825	0.915	1.0000	0.910	0.910	0.830
1000	0.835	0.915	1.0000	0.910	0.910	0.835
1450	0.850	0.920	1.0000	0.910	0.850	0.845
1500	0.855	0.920	1.0000	0.910	0.910	0.850
1950	0.870	0.905	0.9999	0.910	0.600	0.870
2000	0.875	0.880	0.9997	0.910	0.810	0.875
2350	0.880	0.500	1.0000	0.910	0.000	0.880
3000	0.895	0.080	0.9907	0.880	0.650	0.890
4000	0.865	0.000	0.8369	0.300	0.000	0.860
5000	0.650	0.000	0.0500	0.000	0.000	0.670
6000	0.200	0.000	0.0000	0.000	0.000	0.200
7000	0.000	0.000	0.0000	0.000	0.000	0.000

References

1. NIST Raman SRM website: https://www-s.nist.gov/srmors/detail.cfm (Used with Permission)
2. NIST Standard Reference Material (SRM)® 2241 Certificate, Relative Intensity Correction Standard for Raman Spectroscopy: 785 nm Excitation, National Institute of Standards and Technology, Gaithersburg, MD 20899, 6 pages (28 February 2014).
3. NIST Standard Reference Material (SRM)® 2242 Certificate, Relative Intensity Correction Standard for Raman Spectroscopy: 532 nm Excitation, National Institute of Standards and Technology, Gaithersburg, MD 20899, 6 pages (22 October 2013).
4. NIST Standard Reference Material (SRM)® 2244 Certificate, Relative Intensity Correction Standard for Raman Spectroscopy: 1064 nm Excitation, National Institute of Standards and Technology, Gaithersburg, MD 20899, 6 pages (03 December 2009).
5. NIST Standard Reference Material (SRM)® 2245 Certificate, Relative Intensity Correction Standard for Raman Spectroscopy: 633 nm Excitation, National Institute of Standards and Technology, Gaithersburg, MD 20899, 6 pages (27 September 2011).
6. NIST Standard Reference Material (SRM)® 2246 Certificate, Relative Intensity Correction Standard for Raman Spectroscopy: 830 nm Excitation, National Institute of Standards and Technology, Gaithersburg, MD 20899, 7 pages (31 August 2012).
7. ASTM E1840-96(2007) Standard Guide for Raman Shift Standards for Spectrometer Calibration, ASTM International, 100 Barr Harbor Drive, PO Box C700, West Conshohocken, PA, 19428-2959 USA (2007).

8. K.J. Frost and R.L. McCreery, Calibration of Raman Spectrometer Response Function with Luminescence Standards: An Update, *Appl. Spectrosc*, **52**(12) (1998), 1614–1618.
9. W. May, R. Parris, C. Beck II, J. Fassett, R. Greenberg, F. Guenther, G. Kramer, S. Wise, T. Gills, J. Colbert, R. Gettings and B. MacDonald, Definition of Terms and Modes Used at NIST for Value-Assignment of Reference Materials for Chemical Measurements; *NIST Special Publication 260-136* (2000); http://www.nist.gov/srm/publications.cfm (accessed Feb 2014).
10. A.L. Rukhin, Weighted Means Statistics in Interlaboratory Studies, *Metrologia*, **46**, (2009) 323–331.
11. R. Dersimonian and N. Laird, Meta-Analysis in Clinical Trials, *Control. Clin. Trials*, **7**(3) (1986) 177–188.
12. S.D. Horn, R.A. Horn and D.B. Duncan, Estimating Heteroscedastic Variances in Linear-Models, *J. Am. Stat. Assoc.* **70**(350) (1975) 380–385.
13. JCGM 100:2008; *Evaluation of Measurement Data — Guide to the Expression of Uncertainty in Measurement* (GUM 1995 with Minor Corrections); Joint Committee for Guides in Metrology (2008); available at http://www.bipm.org/utils/common/documents/jcgm/JCGM_100_2008_E.pdf (accessed Feb 2014);
14. B.N. Taylor and C.E. Kuyatt, *Guidelines for Evaluating and Expressing the Uncertainty of NIST Measurement Results*; NIST Technical Note 1297; U.S. Government Printing Office: Washington, DC (1994); available at http://www.nist.gov/pml/pubs/index.cfm (accessed Feb 2014).
15. J. Dubessy, M.-C. Caumon, S. Sharma and F. Rull, Instrumentation in Raman spectroscopy, part 2: how to calibrate your spectrometer, 14 June, 2012, *Raman spectroscopy applied to Earth sciences and cultural heritage — M.C. Caumon, International GeoRaman Conference,* http://georaman2014.wustl.edu/
16. ASTM E2529-06(2014), Standard Guide for Testing the Resolution of a Raman Spectrometer, Book of Standards Volume: 03.06, ASTM International, 100 Barr Harbor Drive, PO Box C700, West Conshohocken, PA, 19428-2959 USA (2014).
17. ASTM E1840-96(2014), Standard Guide for Raman Shift Standards for Spectrometer Calibration, Book of Standards Volume: 03.06, ASTM International, 100 Barr Harbor Drive, PO Box C700, West Conshohocken, PA, 19428-2959 USA (2014).
18. J.R. Ferraro and K. Nakamoto, *Introductory Raman Spectroscopy*, 1st edn. (Academic Press, Boston, 1994).
19. The Richard McCreery Research Group provides the greatest detail of information at the following link: http://www.chem.ualberta.ca/~mccreery/ramanmaterials.html
20. NIST SRM website: URL at http://www.nist.gov/srm/index3column.cfm (Used with Permission)
21. NIST SRM URL at https://www-s.nist.gov/srmors/detail.cfm (Used with Permission)
22. D. Gupta, L. Wang, L.M. Hanssen, J.J. Hsia and R.U. Datla, Polystyrene Films for Calibrating the Wavelength Scale of Infrared Spectrophotometers — SRM 1921, *NIST Special Publication 260-122*, National Institute of Standards and Technology, Gaithersburg, MD (April 1995), 25 pages.
23. NIST Standard Reference Material (SRM)® 1921 Certificate, Infrared Transmission Wavelength Standard, National Institute of Standards and Technology, Gaithersburg, MD 20899, 6 pages (7 April 1995).
24. NIST Standard Reference Material (SRM)® 1921b Certificate, Infrared Transmission Wavelength Standard, National Institute of Standards and Technology, Gaithersburg, MD 20899, 7 pages, Certificate Issue Date: 19 February 2014.

25. W.H. Press, S.A. Teukolsky, W.T. Vetterling and B.P. Flannery, *Numerical Recipes 3rd Edition: The Art of Scientific Computing* 3rd edn. (Cambridge University Press, 2007, 120–124).
26. W.H. Press, S.A. Teukolsky, W.T. Vetterling and B.P. Flannery, *Numerical Recipes 3rd Edition: The Art of Scientific Computing* 3rd edn. (Cambridge University Press, 2007, 120–124). Algorithm as implemented into eFTIR software peak picking algorithm.
27. V.R. Weidner, P.Y. Barnes and K.L. Eckerle, A Wavelength Standard for the Near Infrared Based on the Reflectance Of Rare-Earth Oxides, *J. Research NBS*, **91**(5), 243-253 (1986).
28. B.N. Taylor and C.E. Kuyatt, Guidelines for Evaluating and Expressing the Uncertainty of NIST Measurement Results, *National Institute of Standards and Technology (NIST), Technical Note 1297*, 20 pages (1994).
29. NIST Standard Reference Material (SRM)® 2036 Certificate, Near-Infrared Wavelength/Wavenumber Reflection Standard, National Institute of Standards and Technology, Gaithersburg, MD 20899, 7 pages (24 January 2013).
30. S.J. Choquette, D.L. Duewer, L.M. Hanssen and E.A. Early, Standard reference material 2036 near-infrared reflection wavelength standard, *Applied Spectroscopy* **59**(4), 496–504 (2005).
31. Hamamatsu: http://www.hamamatsu.com/resources/pdf/ssd/infrared_techinfo_e.pdf
32. A. Rogalski, Infrared Detectors at the Beginning of the Next Millennium, Opto−Electron. Rev. **9**(2) (2001) 173–187.
33. A. Rogalski, History of infrared detectors, *Opto−Electron. Rev.* **20**(3) (2012) 279–308.
34. E.L. Dereniak and G.D. Boreman, *Infrared detectors and systems* (Wiley, New York, NY, 1996).
35. D.L. Shumaker and J.S. Accetta (Eds), *The Infrared and Electro Optical Systems Handbook* (8 Volume Set) Hardcover (Press Monographs, SPIE-International Society for Optical Engine, January 1993).

Chapter 9

Glossary of Raman Spectroscopy Terms

Contents

Raman — Rudimentary Glossary 251
References 255

Raman — Rudimentary Glossary

(Note: More details for each topic are found within Volume 3 (Chapter 9), the index, the glossary, and the various other volumes of this text set).[1–5]

ANALYTE — The particular material or quality to be determined in an analytical measurement.

APODIZATION OR APODIZATION FUNCTION — A function used to remove the "ripple" effect on either side of spectral peaks or bands following the application of the Fourier transform to an interferogram. These ripples are minimized by applying a triangular function or Happ-Genzel (smoothed triangular function) to the spectral data to produce smaller ripples and slightly broadened peak widths.

AUTOSCALING FOR SPECTRAL DATA SET — Each intensity data point for each wavenumber for a set of spectra is mean centered (i.e., centered) and then scaled by dividing each point by the standard deviation of the intensities at each data point. This scaling may overemphasize noisy data such that important spectral information is diminished or obscured.

BOXCAR APODIZATION — Is the apodization function applied to FT transformed interferogram spectra that maximizes both peak height and resolution (band halfwidth) as compared to other apodization functions. It is the apodization function most often used for high resolution spectroscopy. This is a square waveform apodization function as contrasted to triangular or Happ-Genzel (smoothed triangular function) apodization functions.

CALIBRATION — Determining the optical response (Raman intensity, in counts) of some analytical method to known amounts of pure analyte. This may be a univariate or multivariate method for quantitative spectroscopy.

CALIBRATION CURVE — The results of a calibration when graphed, usually Cartesian coordinates, e.g., concentration (in moles/liter or grams/liter) versus Raman intensity.

CENTERBURST — An interferogram has a centerburst, also referred to as the ZPD or 'zero-path-difference' location. When the path of the fixed and moving mirror are identical in the interferometer, the interferogram is at maximum interference; referred to as the centerburst.

CONCENTRATION — The amount of analyte in a given volume of solution, e.g., moles per liter.

CUVET (A.K.A. CUVETTE) — Transparent receptacle in which sample solutions are introduced into the light path of spectrometers. These are usually cylindrical or rectangular for clear liquid transmittance measurements. Raman spectrometers may use borosilicate glass containers due to the low Raman scattering of the O-H group.

DETECTOR — Device used to detect the intensity of the radiation reflected from, or transmitted through, a sample. There is detailed information for Raman intensity detectors within this set of volumes.

DOUBLE-SIDED INTERFEROGRAM — An interferogram has a 'center-burst', also referred to as the ZPD or 'zero-path-difference' location. When the path of the fixed and moving mirror are identical in the interferometer, the interferogram is at maximum interference. The centerburst may be placed anywhere along the recorded data for the interferogram. When it is placed in the center of the interferogram it is termed a double-sided interferogram (also known as an asymmetric interferogram).

FREQUENCY — The number of times per unit that the magnitude of an electromagnetic wave goes from maximum to minimum then back to maximum amplitude. The unit for the number of waves per second is Hertz (Hz).

INTENSITY CORRECTION — A correction of the Y-axis (intensity) for Raman spectra such that a standard lamp and or a standard lamp and standard reference sample are used in combination with a polynomial fit across the entire wavenumber axis in order to standardize the intensity measurements of any Raman spectrometer. It is analogous to a photometric correction accomplished for absorption spectroscopy. See Chapters 6 and 8 for further explanation.

INTERFEROGRAM — A wave pattern created using an interferometer where waves are superimposed and later interpreted to compute a pattern of amplitude versus frequency for a measurement of light interaction with materials. This interferometer pattern is interpreted using a Fourier transform to compute a Raman intensity versus Raman shift spectrum.

LIMIT OF DETECTION — Lowest amount of analyte that may be seen above noise level of instrument for those measurement conditions.

LINEAR RESPONSE — Straight line response between concentration and measured intensity signal. In context, this means that for a doubling of the analyte concentration, the detectable and measureable signal intensity will double. This allows a linear calibration model or equation to be used to determine the concentration of analyte through a range of intensity levels.

LINEARITY — The intensity response of a Raman spectrometer changing in a predictable and linear fashion with increasing analyte concentration.

NOISE — Any signal generated by the detector that is not directly responding to the Raman energy scattered at the measured wavenumber.

PARETO SCALING FOR A SPECTRAL DATA SET — For this scaling method, each intensity data point for each wavenumber for a set of spectra is mean centered (i.e., centered) and then scaled by dividing each point by the square root of the standard deviation of the intensities at each data point. This scaling method diminished noisy data and allows more important data to be emphasized.

PATHLENGTH — The distance the light passes through the sample in its holder. In practical terms, the inside dimension of the sample holding device, or in reflection measurement mode, the length of the light passing through the sample holding device. In Raman measurements the physical pathlength is considered infinite, as the Raman intensity is determined by the matrix chemistry and its scattering properties relative to the excitation laser.

REFERENCE — A material with known characteristics, such as peaks of identified wavenumber positions, or photometric values of Stokes scattered energy in intensity or counts as per specific wavenumber.

REFERENCE STANDARD — Reference material wherein the wavelength or intensity properties are documented or certified. This reference standard is used to calibrate the basic performance of a spectrometer. Government issued reference standards are often termed SRMs or Standard Reference Materials.

RETARDATION — Is defined as the optical path difference between the two beams in the interferometer. Higher retardation gives better resolution, increases the number of data points, and results in larger data computer files (i.e., greater number of data points).

SCANNING — The process where the wavenumber range of the system is viewed in serial order, usually from highest to lowest wavenumber i.e., highest frequency. This usually occurs when the grating is rotated about its axis, or an interferometer is sequenced for scanning an interferogram and converting to a spectrum.

SENSITIVITY — See "Limits of Detection".

SIGNAL — The output of the detector due to its response to energy emerging from the sample in the sample holder. It consists of an analog signal converted to a digital signal and then further converted to units such as counts, depending upon the measurement geometry and computational methods.

SIGNAL-TO-NOISE RATIO — The numerical ratio of the total signal to the noise of the instrument as root mean squared (RMS) deviation of repeated signal measurements. The RMS is the standard deviation of the Stokes scattered intensity measurements at each wavenumber.

SINGLE-SIDED INTERFEROGRAM — An interferogram has a 'center-burst', also referred to as the ZPD or 'zero-path-difference' location. When the path of the fixed and moving mirror are identical in the interferometer, the interferogram is at maximum interference. The centerburst may be placed anywhere along the recorded data for the interferogram. When it is placed near the start of the interferogram it is termed a single-sided interferogram (also known as a symmetric interferogram).

SOURCE — Also known as the excitation laser for Raman spectrometers. This is the origin of the light used in the spectrometer. Multiple types of lasers are used for Raman measurements.

SPECTRUM — Series of wavenumbers of radiation, belonging to a specific portion of the electromagnetic spectrum, e.g., the Raman shift spectrum, where the Stokes scattering of a sample measurement is recorded with respect to wavenumber position.

WAVENUMBER CORRECTION — The process of aligning the X-axis or wavenumber axis for a Raman spectrometer. It is accomplished by measuring a standard reference material with carefully known wavelength or wavenumber peak positions and making adjustments with the instrument to align to these reference standards. See Chapters 6 and 8 for further explanation.

WAVENUMBER (CM^{-1}) — For Raman spectroscopy, the wavenumber represents the number of waves (cycles) occurring per centimeter designated as cm^{-1} for the Raman shift corresponding to the molecular vibrational frequency for measured materials. It is equivalent to 1×10^7/wavelength in nanometers, e.g., 2500 nm is equivalent to 4000 cm^{-1}.

ZERO FILLING — Is a process of adding a number of data points consisting of amplitude values located between corresponding interferogram data points to increase the number of data points. This action is completed prior to applying the FFT to the interferogram in order to produce a spectrum. Zero-filling adds interpolated points that lie between the real interferogram data points. The resultant data is smoother in appearance and more suitable for application of the FFT algorithm (proper number of data points). The appearance and scattering intensity accuracy of the spectrum is improved, but there is no improvement in the linewidth resolution of the spectrum.

ZPD — An interferogram has a centerburst, also referred to as the ZPD or 'zero-path-difference' location. When the path of the fixed and moving mirror are identical within the interferometer, the interferogram is at maximum interference; there is zero path difference between the light traveling through the two channels of the Michelson interferometer.

References

1. J.M. Chalmers and P.R. Griffiths, *Handbook of Vibrational Spectroscopy* 1st edn. (John Wiley & Sons, 2002).
2. Roy Cain, personal communications, *A User's Guide to Fourier Transform Infrared Spectrometers*, (ASTM, March 2006). Roy was an active leader in ASTM activities and was a recipient of the ASTM Fellow, ASTM Award of Merit in (1999).
3. ASTM E2529-06 Standard Guide for Testing the Resolution of a Raman Spectrometer, ASTM International, 100 Barr Harbor Drive, PO Box C700, West Conshohocken, PA, 19428–2959 USA (2006).
4. ASTM E1683-02(2007) Standard Practice for Testing the Performance of Scanning Raman Spectrometers, ASTM International, 100 Barr Harbor Drive, PO Box C700, West Conshohocken, PA, 19428–2959 USA (2007).
5. ASTM E1840-96(2007) Standard Guide for Raman Shift Standards for Spectrometer Calibration, ASTM International, 100 Barr Harbor Drive, PO Box C700, West Conshohocken, PA, 19428–2959 USA (2007).

Volume 5 Index

A

A=B=C systems, 39–40
4-Acetamidophenol (active ingredients of Tylenol) (Aldrich A730-2), 25
Acetylenes (-C≡ C-), 38
Advanced Raman measurement techniques, 95
Agricultural sciences, 183
Aliphatic methyl C-H stretching band, 32
Aliphatic methylene C-H stretching band, 33
Alkane band intensities: Raman vs. MIR, 27
Alkanes (paraffins), 31, 183
Alkene C=C=C stretching, 185
Alkenes (olefins), 34
Alkenes, 175, 183, 185
Amide I band, 175
Amide II band, 175
Amides (R-C=ONH-X), 43
Amides, 184
Amines (R-NH-X), 48, 184
Amorphous materials (organic), continuous sheets, and polymeric films, 109
Amperes per watt (A/W), 83
Anhydrides (R-C=O-C=O-R'), 45
Animal sciences, 185
Apodization function, 251
Applications of Raman spectroscopy, 7, 22
Ar, argon, emission lines, xxxiii–xxxiv
Area conversions, xv
Ar-H, 175
Aromatic C=C stretching, 185
Aromatic C-H stretching, 175
Aromatic systems, 39, 183
Array based Raman spectrometer, 75, 79
Association of Official Analytical Chemists (AOAC) International, 221
ASTM E1683-02(2014)e1, 25
ASTM E1840-96(2014), 25
ASTM E2529-06(2014), 25
ASTM E2911-13, 25
ASTM International Standards, 221
Attenuation losses over time/wear, 93–94
Avogadro constant, xiii

B

Background correction modeling, 169
Band pass (bandpass) and resolution, 21, 90
Basic applications of Raman spectroscopy, 22
Beer's law description, xxvii–xxix
Benzonitrile (Baker B883-07), 25
Beverages, 207
Biotechnology, 192
1,4 Bis (2-methylstyryl) benzene (BMB) (Aldrich #25,740-0), 25
Boltzmann constant, xiii
Boxcar apodization, 251
Branched chain hydrocarbons, 33

C

c, speed of light, xiii
C=C stretch frequency with respect to carbon size, 36
C=C stretch vibrations table, 35
C=C stretching, 175, 185
C=O carbonyl stretching functional group band, 179, 181, 184
C=O stretching, 175
C=S stretching, 175, 185
C≡C stretching, 175, 185
C≡N stretching, 175
Calcite, 95
Calibrated irradiance white-light source spectrum, 167

Calibration and basic maintenance, 24
Calibration basics, 24, 162
Carbonyl compounds (>C=O), 42–43, 175
Carbonyl group, special cases, 43
CARS (Coherent Anti-Stokes Raman Spectroscopy), 18, 57, 59, 66, 112
Centerburst, 252
C-H bending, 179
C-H out-of plane bending vibrations, 38
C-H stretch of alkenes, 34–35
C-H stretching, 179–180
CH_2 asymmetric stretching, 175
CH_2 symmetric stretching, 175
CH_3 asymmetric stretching, 175
CH_3 bending, 175
CH_3 symmetric stretching, 175
Characteristics of Raman measurements, 174–175
Chemical analysis, 27
Chemical production, 205
Clinical chemistry, 209
cm^{-1} conversions, xxvii
C-O stretching, 175, 179–181
Coefficients of the certified polynomial, intensity calibration, 231
Common symbols used in matrix notation for regression methods, 128
Comparing sample and reference spectra, 153
Components of a Raman spectrometer, 11
Compton and Raman effects, 52
Compton effect, 52
Computing the derivative for spectroscopy, 139
Computing the probability that two sample spectra are statistically identical, 153
Concentration conversions, xix
Contents of volume 5, vi, xi
Continuous sheets and polymeric films, 109
Corrected intensity spectrum equation, 166
Corrected Raman spectrum equation, 71
Corrected Raman spectrum, 123
Correction vector ratio spectrum, 168
Correlation between vectors (spectra), 156
Correlation spectral matching (or dot product), 151
Cosign similarity for spectral matching, 156
Cosmetics, 205
Coupling of vibrations, 30
Critical values for the t distribution, 155, 157
C-S stretching, 175, 185
Cumulated double bond systems (A=B=C), 39
Cyclic groups, 33
Cyclohexane (Mallinckrodt analytical reagents), 25
Cyclohexane, 95

D

D* for Raman detectors, 84, 238–239, 241
D*, detectivity equation, 83
Data file formats and extensions, 123–124
Data formatting, 123
Data processing for Raman spectroscopy, 126
Density conversions, xvii
Derivatives in spectroscopy, 139
Detector response curves, 241
Detector responsivity (R), 83
Detector: performance and sensitivity, 83
Detectors for spectrophotometers, xxx
Differences between infrared and Raman measurement techniques, 9
Dispersive monochromator-based Raman, 12–13
Dispersive Raman microscope optical configuration, 76
Dispersive Raman spectrometer with array spectrograph, 75
Dividing the Raman spectrum, 29
Dot product spectral matching, 151
Double monochromator Raman spectrometer, 74
Double-sided interferogram, 252
Dynamic range, 22

E

Early work in Raman spectroscopy, 7
Earth sciences and mineralogy, 205
Electricity and magnetism conversions, xix
Electromagnetic spectrum summary, xxiv–xxvii, 4
Electronic components used in spectrometry, 89
Emission sources, peak positions, xxxiii–xxxiv
EMR, 4
Emulsions, and slurries, 105
Energy, work, quantity of heat conversions, xviii
Entrance and exit pupils (apertures), 90
Environmental sciences, 206
Esters (R-CO-OR'), 44, 184
Ethers (C-O-C), 44, 184
Ethers and alcohols, 175

Euclidean distance, 151
European Pharmacopoeia Commission, 220
eV, electron volts conversions, xxvii
Explosives, and propellants, 210
Extensions to the S-G derivative method, 145

F

Faraday constant, xiii
Fermi resonance (or second order coupling), 30
Fiber optic Raman, 17
Fingerprint frequencies, 29
Fisher's z' transformation, 155
Fluid flow conversions, xix
Fluorescence Raman background, 82
Fluorescence, 87
Food and beverages, 207
Forensic science, 208
Fourier transform Raman spectrometer (interferometer) configuration, 77
Frequencies of alkanes (paraffins), 31
Frequency conversions, xxvii
Fuels research, 211
Full Width at Half Maximum (FWHM), 91
Fundamental grating equation, 163–164

G

Gas emission peak positions, xxxiii–xxxiv
Gas phase analysis, 208
Gas sample measurements, 102
Gases and vapors, 102
Gem-dimethyl groups, 33
General concepts of calibration, 24
Glass sampling vials for Raman measurements, 103
Glossary, 251
Granules or pellets, 110
Group frequencies of alkenes (olefins), 34
Group frequencies, 26

H

H bonding, 175
Helium:neon gas laser illustration, 80
Helium-neon laser (He:Ne), 632.8 nm, 79–80
Hg, mercury emission lines, xxxiii–xxxiv
History and applications of Raman, 174
Hit Quality Index (HQI), 156
H-O-H bands associated with Raman water spectrum, 192
Hydrogen bonding (X-H.......Y), 45
Hz, Hertz, conversions, xxvii

I

Illustration of a Raman spectrum showing Raman shift, 10
Illustration of aliphatic methyl C-H stretching band, 32
Illustration of aliphatic methylene C-H stretching band, 33
Illustration of calcite spectral resolution test band, 163
Illustration of energy states involved in infrared, Raman, and fluorescence phenomena, 11
Illustration of fluorescence interference, 16
Illustration of stimulated Raman measurements, 58
Illustration of the conversion of an interferogram to a Raman spectrum, 77
Illustration of the conversion of an interferogram to a Raman spectrum with background reference correction, 78
Illustration of the major signal phenomena for Raman, 55
Illustration that polynomials best fit may be used to approximate a derivative curve, 140–141
Illustrations of quantum energy transitions, 54
Indene, 95
Individual wavenumber step-scan, 17
Influence of resolution on SNR, 88
Infrared spectroscopy summary, xxiii
Infrared versus Raman measurement techniques, 9
Instrument physics, 209
Instrument problems, 26
Instrumentation for measuring Raman spectra, 70
Instruments (optical configurations), 8
Intensity calibration (simplified method), 165
Intensity calibration (NIST method), 230
Intensity correction vector equation, 165
Intensity correction, 229–230, 252
Interferogram, 253
Interferometer instruments, 73
International Organization for Standardization (ISO) Standards, 221
Interpretive Raman spectroscopy, 26
Intractable materials, 110
Iso-groups, 33

J

Jablonski type diagram, 54
Jones unit, 83

K

KBr or KCl matrix, 107
KBr pellet cell, 108
Ketones, 184
Kinetics, 17
Kr, krypton emission lines, xxxiii–xxxiv

L

Laboratory dispersive monochromator-based Raman spectrometer as a microscope design, 13
Laboratory dispersive monochromator-based Raman spectrometer design, 12
Laboratory Michelson interferometer-based Raman spectrometer microscope, 14
Lamp sources for spectrophotometers, xxix
Laser attenuation, when passing through an optical fiber, 80
Laser excitation ranges, 14, 76, 79, 243
Lasers used in Raman pharmaceutical applications, 221
Length conversions, xiv
Limitations of the Savitzky-Golay (S-G) method, 143
Linear and non-linear A=B=C systems, 39–40
Linear response, 253
Liquid sampling, 103–104
Liquid waveguide, 110
Liquid/slurry vial measurements, 105
Loadings matrix (V), 130
Locations of C-H out-of plane bending vibrations, 38
Low levels of analyte in solution, 108
Lumps, granules, or pellets, 110

M

Mahalanobis distance measurements, 150
Mass conversions, xvi
Medicine and clinical chemistry, 209
Mercury emission lines, xxxiii–xxxiv
Methods used for Raman data processing, 126
Methyl C-H stretching band, 32
Methyl group with attached atoms (X-CH_3), 34
Methylene C-H stretching band, 33
Microscope design, 13
Mid-IR summary, xxiii
Military, explosives, and propellants, 210
Mineralogy, 205
Molar gas constant, xiii
Molecular and electronic spectroscopy unit conversion, 5
Monochromator instruments, 72
Multiple regression, 130
Multivariate Linear Regression (MLR), 127
Multivariate regression, 127

N

N=N (azo-) stretching, 175
N_2, 95
n-alkanes asymmetric and symmetric C-H stretching for methyl (CH_3) group bands, 177
n-alkanes asymmetric and symmetric C-H stretching for methylene (CH_2) group bands, 178
n-alkanes with typical functional groups and band locations, 176
n-alkanes, 31
Naphthalene (Mallinckodt #94848), 25
Natural gas, and fuels research, 211
Ne, neon, emission lines, xxxiii–xxxiv
Near infrared (laser diode), 980 nm, 79
Near infrared (VCSEL), 850 nm, 79
Near infrared spectroscopy summary, xxii–xxiii
Neodymium yttrium aluminum garnet (Nd:YAG) laser, 1064 nm
Neon, emission lines, xxxiii–xxxiv
NEP, 83
N-H asymmetric stretching, 175
N-H bending, 175
N-H group frequencies, 47
N-H scissoring-bending, 175
N-H stretching functional group bands, 182
N-H stretching, 175
N-H, 45
NIR laser, 79
NIST SRM 2241, 230
NIST SRM 2242, 230
NIST SRM 2243, 230
NIST SRM 2244, 230
NIST SRM 2245, 230
NIST SRM 2246, 230
NIST SRMs for Raman spectrometer calibration, 230
NIST uncertainty number for SRMs, 237
Nitriles (R-C≡N), 38

Normal hydrocarbons (n-alkanes), 31

O

O_2, 95
O-H and N-H, 45
O-H stretching, 45–46, 175, 179–180, 182
Olefinic C=C stretch, 35
Olefinic C-H bend, 37
Olefinic C-H stretch, 35
Olefins, 34
Optical and electronic designs for Raman spectrometers, 72
Optical material transmittance ranges, 245
Optical materials for Raman sampling and optics, 245–248
Optical materials for spectrophotometers, xxxi–xxxii
Outline of volume 5, v
Out-of-plane bending vibrations for benzene, 41
Overview of volume 5, v

P

P=O stretching, 185
Paraffins, 31
Pareto scaling, 253
Partial Least Squares — Singular Value Decomposition (PLS-SVD), 133
Partial Least Squares Regression (PLSR), 134
Pastes, emulsions, and slurries, 105
Pathlength, 24, 253
PCR calculations, 132
Peak positions, 92
Peak positions, emission sources, xxxiii–xxxiv
Pellets, 110
Peroxides (-C-O-O-C-), 45
Petroleum, natural gas, and fuels research, 211
Pharmaceutical industry, 211
Planck constant, xiii
Plane angle conversions, xviii
Plant sciences, paper and pulp, 212
PLS - SVD, 133–136
PLS calculations, 133–136
Polarization modulation, 17
Polymer science, 215
Polymeric films, 109
Polynomial background correction, 168
Polynomial correction for x-axis, 165
Polystyrene (Aldrich #18,243-5), 25
Polystyrene film peak table, 235–236
Polystyrene film, 234
Powders, 107
Practical aspects for Raman spectral measurements, 22
Precision test, wavenumbers, 158
Predicted values, 129
Presentation formats, 122
Pressure conversions, xvii
Principal Component(s) Analysis (PCA), 129
Principal Component(s) Regression (PCR), 131
Process control applications, 215
Propellants, 210
Properties of spectrophotometers, 90

Q

Qualitative applications, 17, 23
Quantitative analysis, 16–17, 23, 70, 220
Quantitative and qualitative applications, 23
Quantitative applications, 23
Quantum efficiency (Q), detector, 83–84

R

r to z' table, 155
r to z' transformation, 153
Radioactivity conversion, xx
Raman active molecular bonds, 28
Raman and infrared band frequency locations for atom attached to methyl group (X-CH3), 34
Raman band positions and strengths for major functional groups, 175
Raman band spectra-structure tables and figures, 175–191, 193–204, 217–219
Raman configuration for liquid sample measurements, 104
Raman detector sensitivity ranges, 240–242
Raman detectors (as photon counting types), 85
Raman effect, 53
Raman equations for quantitative analysis, 16–21
Raman functional group band positions tables, 186–191, 193–204
Raman fundamental group frequency and fingerprint frequency locations, 176
Raman instrument comparison and evaluation methods, 94

Raman intensity (in counts) spectrum for C-H and C=C functional group bands, 178
Raman laser excitation sources, 14, 79, 243–244
Raman materials data, 240
Raman measurement techniques, 18, 56
Raman measurement theory, 70
Raman relationship for quantitative analysis, 70
Raman, rudimentary glossary, 251
Raman shift for three different laser sources, 10
Raman spectral measurements, practical aspects, 22
Raman spectral region, definition, 7
Raman spectral resolution test(s), 157
Raman spectra-structure correlation charts, 175–191, 193–204, 217–219
Raman spectrometer calibration basics, 162
Raman spectrometer components, 76
Raman spectrometer detectors, 15
Raman spectrometer measurement modes, 17, 57
Raman spectroscopic instruments (optical configurations), 8
Raman spectroscopy measurements, 16
Raman spectroscopy summary, xxiii–xxiv
Raman theory, 53
Raman versus infrared measurement techniques, 9
Raman wavenumber calibration reference materials, 233
Raman wavenumber calibration, 95
Rate conversions, xvi
Rayleigh scattered intensity, 55
Reference materials, vi
Regression vector, 129
Relative alkane band intensities: Raman vs. MIR, 27
Resolution and bandpass, 90
Responsivity, detector, 83
Retardation, 254
Ring solids rotating cell, 109
Rotating liquid/slurry cell, 106
RRS/RR (Resonance Raman Spectroscopy) measurements, 60
RRS/RR (Resonance Raman Spectroscopy), 18, 57, 66, 112–113

S

S=O stretching, 185
Sample preparation and handling, 25
Sample type versus sampling method, 101
Sample/solvent problem, 26
Sampling method, 101
Savitzky-Golay (S-G) method, 143
Scattered fraction of Raman compared to source, 8
Scores matrix (T), 130
Second order coupling, 30
Second-order polynomial equation, 164
Selection of measurement technique and pathlength, 24
Semiconductors, electronics, 216
SERRS (Surface-Enhanced Resonance Raman Spectroscopy), 18, 58, 112, 114
SERRS (Surface-Enhanced Resonance Raman Spectroscopy) measurements, 62
SERS (Surface-Enhanced Raman Spectroscopy), 18, 58, 66, 112–113
SERS (Surface-Enhanced Raman Spectroscopy) measurements, 61
S-H (thiol-) stretching, 175, 185
Signal-to-noise ratio (SNR) in a Raman spectrometer, 21, 85–86
Silicon wafer spectrum, 234
Silicon, 95
Single reflection monochromator Raman spectrometer, 72
Single transmission monochromator Raman spectrometer, 73
Single-sided interferogram, 254
Singular Value Decomposition (SVD), 135
Slurries, 105
SNR and resolution, 88
SNR in a Raman spectrometer, 85
Solids, 106
Solid-state flashlamp pump laser illustration, 80
Solid-state P-I-N (p-i-n) junction semiconductor diode laser illustration, 81
Solid-state VCSEL (Vertical-Cavity Surface-Emitting Laser) pump laser illustration, 81
Solutions, 108
Solvent grades, 25
SORS (Spatially Offset Raman Spectroscopy) (offset position), 116
SORS (Spatially Offset Raman Spectroscopy) (zero position), 115
SORS (Spatially Offset Raman Spectroscopy) measurements, 64

SORS (Spatially Offset Raman Spectroscopy), 19, 60, 66, 116
Sources for spectrophotometers, xxix
Special cases of the carbonyl group, 43
Special Raman sampling techniques, 111
Specific surface conversions, xvi
Spectral data manipulation, 122
Spectral matching (correlation or dot product), 151
Spectral range, 121
Spectral resolution factors, 162
Spectral resolution illustrated, 160–161
Spectral resolution, 162
Spectral search algorithms and terminology, 150
Spectral searches (comparisons), 149
Spectra-structure correlation charts, 175–191, 193–204, 217–219
spectrophotometers, properties, 90
Spectroscopic regions of interest for chemical analysis, 27
Spectroscopy theory, 4
Spectrum acquisition parameters, 119
Spectrum acquisition, presentation format and data manipulation procedures, 119
Speed of light, c, xiii
SPPERS (Surface Plasmon Polariton (SPP) Enhanced Raman Spectroscopy), 19, 61, 66, 116–117
SRM 2241, 230
SRM 2242, 230
SRM 2243, 230
SRM 2244, 230
SRM 2245, 230
SRM 2246, 230
SRMs for Raman intensity correction, 229
SRS (Spontaneous Raman Spectroscopy) measurements, 18, 58, 63, 66, 113–114
SRS/SPSRS (Stimulated Raman Spectroscopy) measurements, 59, 63, 114–115
S-S stretching, 175
Standard approaches to spectral searches, 149
Standard Reference Materials (SRMs) for Raman intensity correction, 229
Standard reference methods and practices, 220
Standard scan, 17
Statistics for comparing spectra, 153
Statistics of spectral searches (comparisons), 149
Stefan-Boltzmann constant, xiii
Step-scan phase-resolved, 17
Step-scan space-resolved, 17
Step-scan time-resolved, 17
Stimulated Raman measurements, 58
Stimulated Raman spectroscopy (SRS), 111
Stimulated Raman, 18, 56, 66, 111
Stokes shift spectrum, 5
Sulfur (Aldrich #41,498-0), 25
Surface analysis, 216
SVD of data matrix, 131, 135, 137–138

T

Teflon liquid waveguide, 110
Temperature conversions, xvii
Terminology and symbols, vii
TERS (Tip-Enhanced Raman Spectroscopy) (bottom illumination), 117
TERS (Tip-Enhanced Raman Spectroscopy), 19, 60, 65–66, 116–117
Tertiary butyl groups, 33
Textiles, 217
The data matrix (A), 130
Theoretical limit of D* for a detector system, 84
Thiol (S-H) stretching, 175
Third-order polynomial equation, 164
50/50 (v/v) Toluene/acetonitrile (Mallinckrodt analytical reagents), 25
Transmission Raman, 17
Triple bonds (-C≡), 38
Triple monochromator Raman spectrometer, 75
Troubleshooting, 26
Two-dimensional correlation, 17
Typical settings for liquids and solids measurements using dispersive Raman spectroscopy, 120
Typical settings for liquids and solids measurements using FT-Raman spectroscopy, 120

U

Ultraviolet (laser diode), 244 nm, 79
Ultraviolet (laser diode), 257 nm, 79
Ultraviolet (laser diode), 325 nm, 79
Ultraviolet (laser diode), 364 nm, 79
Ultraviolet spectroscopy summary, xxi
Uncertainty number for SRMs, 237
Unit conversion, 5–6
Unit conversions, electronic spectroscopy, xxiv–xxvii

Unit conversions, molecular spectroscopy, xxiv–xxvii
Unit conversions: wavelength, wavenumber, frequency, eV, Hz, xxvii
United States Pharmacopeia (USP) Methods, 220

V

Vibrational states in benzene, 41
Viscous materials, 105
Visible (laser diode), 455 nm, 79
Visible (laser diode), 473 nm, 79
Visible (laser diode), 488 nm, 79
Visible (laser diode), 514 nm, 79
Visible (laser diode), 532 nm, 79
Visible (laser diode), 660 nm, 79
Visible (laser diode), 785 nm, 79
Visible spectroscopy summary, xxi–xxii
Volatile materials, 104
Volume conversions, xv

W

Waveguide Raman, 17
Wavelength conversions, xiv, xxvii
Wavenumber calibration, 95
Wavenumber correction, 254
Wavenumber precision test, 158
Wavenumber repeatability test, 158
Wavenumber SRMs for Raman, 233
Wavenumber/wavelength calibration, 163
Wavenumbers conversions, xxvii

X

Xe, xenon, emission lines, xxxiii–xxxiv
X-H functional groups (O-H and N-H), 45

Z

Z observed statistic, 155

Master Index for All 5 Volumes (Abbreviated)

A

A Standard Illuminant, A, ***Vol. 2:*** 157

Abrasion measurement techniques, ***Vol. 2:*** 78; ***Vol. 4:*** 135

Absorbance spectrum formula, ***Vol. 2:*** 45–46

Absorbance/response accuracy test, ***Vol. 1:*** 128; ***Vol. 2:*** 103–104, 133; ***Vol. 3:*** 231

Absorbance/response repeatability test, ***Vol. 1:*** 129; ***Vol. 2:*** 104, 133; ***Vol. 3:*** 232

Absorption bands, ***Vol. 1:*** 144–146, 158–159; ***Vol. 2:*** 147–156; ***Vol. 3:*** 279, 292–346, 348–351; ***Vol. 4:*** 233–248, 274–297; ***Vol. 5:*** 175–191, 193–204, 217–219

4-Acetamidophenol (active ingredients of Tylenol) (Aldrich A730-2), ***Vol. 5:*** 25

Acetone, ***Vol. 2:*** 14, 20

Acetonitrile, ***Vol. 2:*** 20

Acetylenes (-C≡C-), ***Vol. 1:*** 83; ***Vol. 4:*** 32; ***Vol. 5:*** 38

Acid phosphatase, ***Vol. 2:*** 151

Acidic diphenylcarbazide method for chromium, ***Vol. 2:*** 148

Advanced FTIR measurement techniques, ***Vol. 4:*** 88

Advanced Raman measurement techniques, ***Vol. 5:*** 95

Advances in infrared spectroscopy, ***Vol. 4:*** 274

Agricultural analysis, ***Vol. 3:*** 3, 283; ***Vol. 5:*** 183

Alcohols (R-OH), ***Vol. 1:*** 83; ***Vol. 2:*** 149; ***Vol. 4:*** 233, 238

Aldehydes (R-C=O–H), ***Vol. 1:*** 83

Aldolase, ***Vol. 2:*** 149

Aliphatic C-H stretching spectrum, ***Vol. 4:*** 237; ***Vol. 5:*** 32–33

Alkaline phosphatase, ***Vol. 2:*** 149, 151

Alkanes (paraffins), ***Vol. 3:*** 3, 281, 315–316, 331–332; ***Vol. 4:*** 25, 235–236, 241; ***Vol. 5:*** 27, 31, 183

Alkenes (>C=C<), ***Vol. 1:*** 83; ***Vol. 2:*** 3, 21, 147, 155; ***Vol. 3:*** 215, 316–317, 331–333; ***Vol. 4:*** 29, 233, 241; ***Vol. 5:*** 34, 175, 183, 185

Alkynes (C≡C), ***Vol. 3:*** 316–318, 332–334; ***Vol. 4:*** 233, 241; ***Vol. 5:*** 175, 183, 185

Alpha-amylase, ***Vol. 2:*** 149

Alpha-chymotrypsin, ***Vol. 2:*** 149

Alpha-hydroxybutyrate dehydrogenase, ***Vol. 2:*** 150

Aluminon method for beryllium, ***Vol. 2:*** 148

Aluminum, ***Vol. 2:*** 148, 201–202, 207; ***Vol. 3:*** 416, 419, 432

Amide I band, ***Vol. 4:*** 233, 240; ***Vol. 5:*** 175

Amide II band, ***Vol. 4:*** 233, 240; ***Vol. 5:*** 175

Amides (R-C=ONH-X), ***Vol. 1:*** 83, 144, 146, 158–159; ***Vol. 3:*** 3, 324, 326, 339, 342; ***Vol. 4:*** 37, 175, 233, 240–241; ***Vol. 5:*** 43, 175, 184, 233, 240

Amines (R-NH-X), ***Vol. 1:*** 83, 144, 146, 158–159; ***Vol. 3:*** 3, 278, 327, 343, 348; ***Vol. 4:*** 42, 240, 293; ***Vol. 5:*** 48, 184, 189, 202,

Amino acid polymers, ***Vol. 3:*** 3

4-Aminoantipyrine and potassium ferricyanide method for phenol, ***Vol. 2:*** 148

1-Aminobutane (primary amine) spectrum, ***Vol. 4:*** 240

Ammonia nitrogen, ***Vol. 2:*** 148

Amorphous materials (organic), continuous sheets, and polymeric films, ***Vol. 4:*** 104; ***Vol. 5:*** 109

Amorphous materials, ***Vol. 1:*** 87; ***Vol. 2:*** 74

Amperes per watt (A/W), ***Vol. 5:*** 83

Analysis of variance testing for flow cells, ***Vol. 2:*** 122
Analysis of variance testing for flow sampling systems, ***Vol. 2:*** 127–128
Anharmonic oscillator, ***Vol. 3:*** 43–48
Anhydrides (R-C=O -C=O-R'), ***Vol. 4:*** 39; ***Vol. 5:*** 45
Animal sciences, ***Vol. 4:*** 253; ***Vol. 5:*** 185
Antibonding, ***Vol. 2:*** 3
AOAC 2000.1 method, ***Vol. 4:*** 249
AOAC international methods, ***Vol. 3:*** 349–352; ***Vol. 4:*** 249, 302
AOCS Cd 14d-99 method, ***Vol. 4:*** 249
Apodization function, ***Vol. 4:*** 361–362; ***Vol. 5:*** 251
Appearance applications, ***Vol. 2:*** 15
Aromatic (C=C), ***Vol. 2:*** 3, 21, 147, 155; ***Vol. 3:*** 3; ***Vol. 4:*** 34, 233, 236, 241; ***Vol. 5:*** 39, 175, 183, 185
Arsenic, ***Vol. 2:*** 148
Arylsulfatase, ***Vol. 2:*** 149
Ascorbic acid method for phosphorus, ***Vol. 2:*** 148
Assigning band locations, ***Vol. 3:*** 53
Association of Official Analytical Chemists (AOAC) international methods, ***Vol. 4:*** 302; ***Vol. 5:*** 221
ASTM D6122-10, ***Vol. 3:*** 352
ASTM E1252-98 (2013)e1, ***Vol. 3:*** 353
ASTM E131, ***Vol. 1:*** 46; ***Vol. 3:*** 352
ASTM E1614-94 (2013), ***Vol. 3:*** 354
ASTM E1655-05 (2012), ***Vol. 3:*** 354
ASTM E168-06, ***Vol. 3:*** 352
ASTM E1767, ***Vol. 2:*** 213
ASTM E1790-04 (2010), ***Vol. 3:*** 354
ASTM E1791-96 (2008)e1, ***Vol. 3:*** 354
ASTM E1840-96 (2014), ***Vol. 5:*** 25
ASTM E1865-97 (2013), ***Vol. 3:*** 354
ASTM E1866-97 (2013), ***Vol. 3:*** 355
ASTM E1944-98 (2013), ***Vol. 3:*** 355
ASTM E2105-00 (2010), ***Vol. 3:*** 355
ASTM E2106-00 (2011), ***Vol. 3:*** 356
ASTM E2310-04 (2009), ***Vol. 3:*** 356
ASTM E2529-06 (2014), ***Vol. 5:*** 25
ASTM E2617-10, ***Vol. 3:*** 356
ASTM E2642-09, ***Vol. 3:*** 356
ASTM E2719-09, ***Vol. 3:*** 356
ASTM E275-01, ***Vol. 3:*** 352
ASTM E2911-13, ***Vol. 5:*** 25
ASTM E308, ***Vol. 2:*** 146
ASTM E334-01 (2013), ***Vol. 3:*** 353
ASTM E387-04 (2009), ***Vol. 3:*** 353
ASTM E903, ***Vol. 2:*** 180–181
ASTM E903, ***Vol. 2:*** 195
ASTM E932-89 (1997), ***Vol. 3:*** 353
ASTM G-173-03, ***Vol. 2:*** 194-195
ASTM International standards, ***Vol. 3:*** 352–356
ASTM International standards, ***Vol. 4:*** 302
ASTM International standards, ***Vol. 5:*** 221
ASTM International, ***Vol. 3:*** 27
ASTM standard methods, ***Vol. 1:*** 140
ASTM E1421-99 (2009), ***Vol. 3:*** 353
ASTM E1683-02 (2014)e1, ***Vol. 5:*** 25
ASTM E1982-98 (2013), ***Vol. 3:*** 355
ASTM E2056-04 (2010), ***Vol. 3:*** 355
Asymmetric stretching of O-C–H, ***Vol. 3:*** 36
ATR (attenuated total reflectance), ***Vol. 4:*** 46, 116
Autoscaling, ***Vol. 1:*** 119; ***Vol. 2:*** 93; ***Vol. 3:*** 203; ***Vol. 4:*** 206, 208
Aventurine, green, ***Vol. 2:*** 152
Axinite, ***Vol. 2:*** 152
Azo group (R-N=N-R'), ***Vol. 1:*** 83

B

B, international blue, ***Vol. 2:*** 147
Band pass (bandpass) and resolution, ***Vol. 1:*** 10, 69; ***Vol. 2:*** 10–12, 56–58; ***Vol. 3:*** 19–21, 116; ***Vol. 4:*** 84; ***Vol. 5:*** 21, 90
Band shifts, UV, ***Vol. 1:*** 160
Bandpass filters, ***Vol. 2:*** 51–52
Baseline correction algorithms, ***Vol. 1:*** 103, 110; ***Vol. 2:*** 90; ***Vol. 3:*** 179–180
Basic BCRA (British Ceramic Research Series II) achromatic tiles spectra, ***Vol. 2:*** 230–231, 244–252
Basic CERAM/BCRA (British Ceramic Research series II) chromatic tiles spectra, ***Vol. 2:*** 244–245
Basic red orange color standards spectra, ***Vol. 2:*** 242–243
BCRA (British Ceramic Research Series II) achromatic tiles spectra, ***Vol. 2:*** 246–247
BCRA (British Ceramic Research series II) chromatic tiles spectra, ***Vol. 2:*** 244–245
BCRA (British Ceramic Research) G-GS9-02c standards spectra, ***Vol. 2:*** 248–252
Beer's law description, ***Vol. 1:*** xxi–xxii; ***Vols. 2–5:*** xxvii–xxix
Benzene vapor standard, ***Vol. 1:*** 200

Benzene, ***Vol. 4:*** 36
Benzonitrile (Baker B883-07), ***Vol. 5:*** 25
Beryllium, ***Vol. 2:*** 148
Beta-carotene (vitamin A), ***Vol. 2:*** 147
Beta-galactosidase, ***Vol. 2:*** 150
Beta-glucosidase, ***Vol. 2:*** 150
Beta-glucuronidase, ***Vol. 2:*** 150
Beverages, ***Vol. 4:*** 259; ***Vol. 5:*** 207
Bias tests, ***Vol. 2:*** 134; ***Vol. 3:*** 182, 446
Biotechnology, ***Vol. 4:*** 253; ***Vol. 5:*** 192
1,4 Bis (2-methylstyryl) benzene (BMB) (Aldrich #25,740-0), ***Vol. 5:*** 25
BK 7 glass, ***Vol. 2:*** 17
Black body radiator, ***Vol. 2:*** 24–28
Black optical surfaces, absorbance, ***Vol. 3:*** 405–411
Black photometric standards spectra, ***Vol. 2:*** 230
Blue, sky, ***Vol. 2:*** 147, 156
Blue, ***Vol. 2:*** 16, 147, 156
Blue/green, ***Vol. 2:*** 16, 147, 156
Borosilicate crown glass, ***Vol. 2:*** 17
Boxcar apodization, ***Vol. 4:*** 362; ***Vol. 5:*** 251
Boxcar smoothing, ***Vol. 1:*** 119; ***Vol. 2:*** 94; ***Vol. 3:*** 205
Bright line emission spectrum, ***Vol. 2:*** 11; ***Vol. 3:*** 20; ***Vol. 4:*** 28; ***Vol. 5:*** 33

C

C=C alkenes, ***Vol. 4:*** 233; ***Vol. 5:*** 35–36, 175, 185
C=O carbonyl stretching functional group band, ***Vol. 4:*** 233; ***Vol. 5:*** 175, 179, 181, 184
C=S stretching, ***Vol. 5:*** 175, 185
C≡C stretching, ***Vol. 4:*** 233; ***Vol. 5:*** 175, 185
C≡N stretching, ***Vol. 4:*** 233; ***Vol. 5:*** 175
Cadmium reduction method for nitrate nitrogen, ***Vol. 2:*** 148
Calcite, ***Vol. 5:*** 95
Calibrated irradiance white-light source spectrum, ***Vol. 5:*** 167
Calibration emission lamp standards, ***Vol. 1:*** 125
Calibration set, ***Vol. 3:*** 438
Calibration transfer, ***Vol. 3:*** 185; ***Vol. 4:*** 159
Carbohydrates, ***Vol. 3:*** 3
Carbonyl (>C=O), band locations, ***Vol. 1:*** 144, 146, 159; ***Vol. 3:*** 3, 318–320, 323–324, 336, 338–339, 350; ***Vol. 4:*** 36–37, 233, 237, 241; ***Vol. 5:*** 42–43, 175
Carboxylic acid, ***Vol. 1:*** 83, 144, 159; ***Vol. 3:*** 323, 338; ***Vol. 4:*** 239, 241, 287; ***Vol. 5:*** 184, 188, 198
Carotenes band, ***Vol. 2:*** 147
CARS (Coherent Anti-Stokes Raman Spectroscopy), ***Vol. 5:*** 18, 57, 59, 66, 112
Catalase (liberation of H_2O_2), ***Vol. 2:*** 149
CD analysis for protein structure, ***Vol. 1:*** 155–156
Cellulose acetate butyrate, ***Vol. 4:*** 234
Cellulose, ***Vol. 3:*** 3
Centerburst, ***Vol. 4:*** 362; ***Vol. 5:*** 252
Centroid (center of mass), ***Vol. 2:*** 113
CERAM/BCRA (British Ceramic Research series II) chromatic tiles spectra, ***Vol. 2:*** 244–245
Ceramic spectra, ***Vol. 2:*** 230–233, 266; ***Vol. 3:*** 433–434
C-F stretch band locations, ***Vol. 4:*** 242
CFS-12-02c spectra, ***Vol. 2:*** 258
Chemical absorptions chart for visible region, ***Vol. 2:*** 155
Chemical applications, ***Vol. 2:*** 15; ***Vol. 3:*** 284
Chi-square, ***Vol. 3:*** 263–264
Chloramine-T method for cyanide, ***Vol. 2:*** 148
Chloride, ***Vol. 2:*** 149
Chlorine reaction method for thiocyanate, ***Vol. 2:*** 148–149
Chloroform, ***Vol. 2:*** 20
Chlorophyll a band, ***Vol. 2:*** 147
Chlorophyll b band, ***Vol. 2:*** 147
Cholinesterases, ***Vol. 2:*** 149
Chromic-sulfuric acid solution,
Chromic-sulfuric acid wash, ***Vol. 1:*** 18; ***Vol. 2:*** 19
Chromium, ***Vol. 2:*** 148
Chromophores for visible spectroscopy, ***Vol. 2:*** 20, 147–152, 205–212
Chromophores, UV, ***Vol. 1:*** 19, 144–145, 159
CIE (International Commission on Illumination), ***Vol. 2:*** 146
CIE 1931 Standard 10° Observer, ***Vol. 2:*** 158
CIE 1931 Standard 2° Observer, ***Vol. 2:*** 161
CIE 1931 supplementary standard (2°) Observer, ***Vol. 2:*** 171–173
CIE 1964 $\bar{x}_{10}$, $\bar{y}_{10}$, $\bar{z}_{10}$ color coordinates, ***Vol. 2:*** 161
CIE 1964 Supplementary Standard (10°) Observer, ***Vol. 2:*** 164

1976 CIE UCS (Uniform Color Space) Diagram, ***Vol. 2:*** 178
CIE 1976 UCS Diagram, ***Vol. 2:*** 178
CIE color designations, ***Vol. 2:*** 147
CIE color system, ***Vol. 2:*** 146
Circular dichroism (CD), ***Vol. 1:*** 155
Circumsolar or spectral irradiance, ***Vol. 2:*** 192
Cleaning of cuvets/cuvettes, ***Vol. 2:*** 19
Cleaning sample cup windows, ***Vol. 3:*** 28
Clear liquids, ***Vol. 2:*** 17
Clear solids (optical materials), ***Vol. 2:*** 17
Clinical chemistry, ***Vol. 4:*** 267; ***Vol. 5:*** 209
Cloudiness, sample, ***Vol. 2:*** 70
Cluster/discriminant analysis, ***Vol. 4:*** 157
C-N stretch band locations, ***Vol. 4:*** 242
C-O stretching spectrum, ***Vol. 4:*** 237
C-O stretching, ***Vol. 4:*** 233; ***Vol. 5:*** 175, 179–181
Cobalt solution, ***Vol. 1:*** 17
Coefficient of variation, ***Vol. 2:*** 138
Coefficients of the certified polynomial, intensity calibration, ***Vol. 5:*** 231
$\bar{x}_{10}, \bar{y}_{10}, \bar{z}_{10}$ Color coordinates, ***Vol. 2:*** 158, 161, 163
Color applications, ***Vol. 2:*** 15
Color coordinate computations, ***Vol. 2:*** 157
Color coordinate map, ***Vol. 2:*** 164
Color coordinate values, suite of calculations, illuminant A, ***Vol. 2:*** 179–180
Color coordinate values, suite of calculations, illuminant D65, ***Vol. 2:*** 179–180
Color coordinate values, ***Vol. 2:*** 146, 157
$\bar{x}, \bar{y}, \bar{z}$ color coordinates, ***Vol. 2:*** 157
$\bar{x}_2, \bar{y}_2, \bar{z}_2$ color coordinates, ***Vol. 2:*** 157, 161, 163
Color definitions, ***Vol. 2:*** 2–3, 156
Color designations, ***Vol. 2:*** 16
Color difference (ΔE) using L*a*b* coordinates, ***Vol. 2:*** 176
Color difference using Hunter lab coordinates, ***Vol. 2:*** 177–178
Color difference using L*u*v* color space coordinates, ***Vol. 2:*** 178
Color filter set CFS-12-02c spectra, ***Vol. 2:*** 258
Color measurements, ***Vol. 2:*** 146, 156–157
Color reactions, ***Vol. 2:*** 147–152
Color theory, ***Vol. 2:*** 146
Color wavelengths, ***Vol. 2:*** 16
Colorimetric methods of analysis, ***Vol. 2:*** 147–149
Colorimetry methods, ***Vol. 2:*** 147–149
Common symbols used in matrix notation for regression methods, ***Vol. 5:*** 128
Compressed pellets, ***Vol. 4:*** 111
Compression cell: diamond anvil cell, ***Vol. 4:*** 125
Compton and Raman effects, ***Vol. 5:*** 52
Computation of the regression line, ***Vol. 2:*** 138
Computing the derivative for spectroscopy, ***Vol. 5:*** 139
Computing the probability that two sample spectra are statistically identical, ***Vol. 5:*** 153
Computing the spectral deviation due to high order stray light, ***Vol. 3:*** 126
Conformity index, ***Vol. 2:*** 119; ***Vol. 3:*** 255, 259
Construction of optical fibers, ***Vol. 3:*** 161
Constructive interference as additive integer waves, ***Vol. 4:*** 74
Continuous sheet samples, ***Vol. 2:*** 74; ***Vol. 4:*** 104; ***Vol. 5:*** 109
Continuum regression (CR), ***Vol. 4:*** 161
Contrast, ***Vol. 2:*** 36–37; ***Vol. 3:*** 81
Copper, ***Vol. 2:*** 148
Corrected Raman intensity spectrum equation, ***Vol. 5:*** 166
Corrected Raman spectrum equation, ***Vol. 5:*** 71, 123
Correction vector ratio spectrum, ***Vol. 5:*** 168
Cosmetics, ***Vol. 5:*** 205
Coupling of vibrations, ***Vol. 3:*** 51; ***Vol. 4:*** 23; ***Vol. 5:*** 30
Creatine kinase, ***Vol. 2:*** 149
Critical values for the t distribution, ***Vol. 5:*** 155, 157
Cross-correlation, ***Vol. 3:*** 244
Crystalline quartz (SiO_2), ***Vol. 2:*** 17
C-S stretch band locations, ***Vol. 4:*** 242; ***Vol. 5:*** 175, 185
C-Si stretch band locations, ***Vol. 4:*** 242
Cu(II) ion, ***Vol. 2:*** 147
Cumulated double bond systems (A=B=C), ***Vol. 4:*** 33; ***Vol. 5:*** 39
Cut off filter, ***Vol. 2:*** 14, 51, 53
Cut off wavelengths, solvents, ***Vol. 1:*** 18, 146–147
Cut on filter, ***Vol. 2:*** 51–52
C-X stretch fundamental vibrations, ***Vol. 4:*** 242
Cyanide, ***Vol. 2:*** 148
Cyclic groups, ***Vol. 4:*** 28; ***Vol. 5:*** 33

Cyclohexane (Mallinckrodt analytical reagents), ***Vol. 5:*** 25
Cyclohexane, ***Vol. 2:*** 20; ***Vol. 5:*** 95

D

D65 Standard Illuminant, ***Vol. 2:*** 157, 161
D-amino acid oxidase, ***Vol. 2:*** 149
Dark corrected spectra, ***Vol. 1:*** 52
Dark red, ***Vol. 2:*** 16, 147, 156
Dark signal, ***Vol. 2:*** 45–46, 48
dB (decibels), ***Vol. 1:*** 72
DC deuterium lamp (185–375 nm), ***Vol. 2:*** 6, 49
Deep pink photometric standards spectra, ***Vol. 2:*** 230–232
Delrin (black polymer) spectrum, ***Vol. 2:*** 262
De-noising, ***Vol. 4:*** 163
Deoxyribonuclease I, ***Vol. 2:*** 150
Depolymerization method, ***Vol. 2:*** 150
Derivative and smoothing applications in spectroscopy, ***Vol. 2:*** 90, 96; ***Vol. 3:*** 211; ***Vol. 5:*** 139
Derivatives, comparing line shapes and resolution, ***Vol. 3:*** 215
Derivatives, comparing magnitude of ordinates scales, ***Vol. 3:*** 215
Derivatives, Savitzky-Golay derivative function for different conditions, ***Vol. 3:*** 218–222
Destructive interference as additive half wavelength multiples of waves, ***Vol. 4:*** 74
Detrend, ***Vol. 3:*** 261
Deuterium lamp, ***Vol. 1:*** 125; ***Vol. 2:*** 6
Dialysis, ***Vol. 4:*** 134
Diamine oxidase, ***Vol. 2:*** 150
Diaminobenzidine method for selenium, ***Vol. 2:*** 148
Diamond (C), ***Vol. 2:*** 17, 76
Diamond anvil cell, ***Vol. 4:*** 125
Diamond ATR, ***Vol. 4:*** 126
Diamond-based accessories, ***Vol. 4:*** 125
Diazotized sulfanilic acid method for nitrite nitrogen, ***Vol. 2:*** 148
Dichloromethane (amylene stabilized) as a wavelength standard, ***Vol. 3:*** 402–404
Didymium oxide, ***Vol. 1:*** 17, 125, 174–175; ***Vol. 2:*** 19; ***Vol. 3:*** 392–393
Diffraction, ***Vol. 1:*** 29; ***Vol. 2:*** 8, 9, 24, 30, 51–54; ***Vol. 3:*** 16, 61
0°/45° Diffuse reflection measurement geometry, ***Vol. 2:*** 161
Diffuse gold photometric reflectance standard, ***Vol. 3:*** 379–380; ***Vol. 4:*** 338–340
Diffuse reflectance (DRIFTS), ***Vol. 4:*** 120
Diffuse reflectance and scatter correction, ***Vol. 4:*** 171
Diffuse reflectance measurement of films, ***Vol. 2:*** 74
Digital filtering, ***Vol. 4:*** 170
Digital light processing (DLP) spectrometer, ***Vol. 3:*** 15
Digital micromirror device (DMD) spectrometer, ***Vol. 3:*** 15
Dimethyl ketone carbonyl (>C=O) stretching, spectrum, ***Vol. 4:*** 237
Diode array detector spectrophotometer, ***Vol. 2:*** 46, 48
Dipole model, ***Vol. 3:*** 41, 440
Direct absorption methods, ***Vol. 2:*** 147–152
Direct normal irradiance, ***Vol. 2:*** 192
Direction vector and oscillation vector of light, ***Vol. 2:*** 196, 201–202
Discrete photometers, ***Vol. 1:*** 50
Discriminant analysis, PCA based, ***Vol. 3:*** 245; ***Vol. 4:*** 157
Dispersion orders from a diffraction grating, ***Vol. 1:*** 9, 64–65; ***Vol. 2:*** 8–9, 12; ***Vol. 3:*** 16
Dispersive infrared spectrometers, ASTM, ***Vol. 3:*** 353
Dispersive monochromator-based Raman, ***Vol. 5:*** 12–13
Dispersive Raman microscope optical configuration, ***Vol. 5:*** 76
Dispersive Raman spectrometer with array spectrograph, ***Vol. 5:*** 75
Dispersive spectrometers, ***Vol. 1:*** 31
Distillation, ***Vol. 4:*** 133
Dithizone method for cadmium, ***Vol. 2:*** 148
Dithizone method for silver, ***Vol. 2:*** 149
Dithizone method for vanadium, ***Vol. 2:*** 149
DLP spectrometer, ***Vol. 3:*** 15
DMD spectrometer, ***Vol. 3:*** 15
Dot product spectral matching, ***Vol. 5:*** 151
Double beam, ***Vol. 3:*** 94, 441
Double bond systems (A=B=C), ***Vol. 4:*** 33
Double monochromator, ***Vol. 1:*** 5
Double-beam measurements, ***Vol. 2:*** 48
Double-sided interferogram, ***Vol. 4:*** 362
Drift Tests, ***Vol. 1:*** 133–134
DRIFTS, ***Vol. 4:*** 120

Dual monochromator scanning fluorometer, ***Vol. 1:*** 56
Dynamic range and stray radiant energy, ***Vol. 1:*** 75
Dynamic range versus R and T, ***Vol. 1:*** 90

E

Earth sciences and mineralogy, ***Vol. 4:*** 254; ***Vol. 5:*** 205
Edge filters, ***Vol. 2:*** 51
Elastic scattering, ***Vol. 2:*** 29
Elastomers (rubbers), ***Vol. 3:*** 23
Electronic components used in spectrometry, ***Vol. 5:*** 89
Electronic field vector, ***Vol. 2:*** 196
Emission lines, gases, ***Vol. 1:*** xxvii–xxviii; ***Vols. 2–5:*** xxxiii–xxxiv
Emission spectrometers, ***Vol. 4:*** 57
Emissivity ranges, infrared sources (in microns), 350–351
Emissivity, ***Vol. 1:*** 30
Emitter, infrared, ***Vol. 4:*** 251
EMR (electromagnetic radiation), ***Vol. 1:*** 2
Emulsions, and slurries, ***Vol. 5:*** 105
EN 14500, ***Vol. 2:*** 181
EN 410, ***Vol. 2:*** 180
Energy spectrum test, ***Vol. 4:*** 213–214
Energy unit conversions, ***Vol. 3:*** 5
Entrance and exit pupils (apertures), ***Vol. 5:*** 90
Entrance and exit pupils, ***Vol. 1:*** 69
Environmental sciences, ***Vol. 4:*** 255; ***Vol. 5:*** 206
Enzymatic methods, ***Vol. 2:*** 149–151; ***Vol. 2:*** 149–151
Eriochrome cyanine R method for aluminum, ***Vol. 2:*** 148
Ester carbonyl (>C=O) stretching spectrum,
Esters (R-C=O–OR'), ***Vol. 1:*** 83; ***Vol. 4:*** 39, 239; ***Vol. 5:*** 44, 184
Etendue, ***Vol. 1:*** 73, 121
Ethanoic Acid (carboxylic acid) spectrum, ***Vol. 4:*** 239
Ethanol, ***Vol. 2:*** 20
Ethers (R-O-R'), ***Vol. 1:*** 83; ***Vol. 4:*** 38; ***Vol. 5:*** 44, 184
Ethers and alcohols, ***Vol. 4:*** 233; 241; ***Vol. 5:*** 175
Euclidean distance, ***Vol. 3:*** 240, 255
European Pharmacopoeia (EP), ***Vol. 2:*** 13; ***Vol. 5:*** 220
Evaporation, ***Vol. 4:*** 133
Example of reporting photometric accuracy, ***Vol. 4:*** 324
Experimental design plots, ***Vol. 3:*** 200
Expert calibration systems, ***Vol. 4:*** 157
Explosives, and propellants, ***Vol. 4:*** 268; ***Vol. 5:*** 210
Exponential fitted baseline correction, ***Vol. 1:*** 107
Extensions to the S-G derivative method, ***Vol. 5:*** 145
External or specular reflectance, ***Vol. 4:*** 119
Extinction coefficient, absorptivity, ***Vol. 2:*** 9–10, 280
Extraterrestrial radiation, ***Vol. 2:*** 192
Extreme ultraviolet, ***Vol. 1:*** 82
Eye function, ***Vol. 2:*** 156

F

Failure of signal averaging test, ***Vol. 1:*** 134
Far Ultraviolet, FUV, ***Vol. 1:*** 82
Fat in meat, AOAC, ***Vol. 3:*** 351
Fats, ***Vol. 2:*** 70
Feed and forage products applications, ***Vol. 3:*** 357
Fermi resonance (or second order coupling), ***Vol. 5:*** 30
Fermi resonance, ***Vol. 3:*** 52, 442
Ferric thiocyanate method for chloride, ***Vol. 2:*** 149
Fiber (acid detergent) in forages, AOAC, ***Vol. 3:*** 350
Fiber bundle, ***Vol. 2:*** 78
Fiber optic Raman, ***Vol. 5:*** 17
Fiber optics, ***Vol. 2:*** 78; ***Vol. 3:*** 160, 354
Field of view (FOV), ***Vol. 2:*** 158
Film holder for diffuse reflectance, ***Vol. 1:*** 37–38, 88–89; ***Vol. 2:*** 74
Film thickness measurement, ***Vol. 1:*** 37–38; ***Vol. 4:*** 46
Filter 10-1255 spectrum, ***Vol. 2:*** 234
Filter 20-1255 spectrum, ***Vol. 2:*** 234
Filter 30-1255 spectrum, ***Vol. 2:*** 234
Filter based photometer, ***Vol. 2:*** 8
Filter wheel, ***Vol. 1:*** 8
Fine chemicals and chemical production, ***Vol. 4:*** 256
Fingerprint analogy, ***Vol. 1:*** 5
Fingerprint frequencies, ***Vol. 2:*** 5; ***Vol. 4:*** 22, 234; ***Vol. 5:*** 29

First and second derivative nomenclature, **Vol. 2:** 96
First derivative zero cross-over, **Vol. 1:** 137; **Vol. 2:** 113
First through fourth derivative nomenclature, **Vol. 3:** 211
Fisher's z' transformation, **Vol. 5:** 155
Fixed mirror, **Vol. 1:** 7
Flat baseline correction (two-point baseline correction), **Vol. 1:** 104; **Vol. 2:** 90; **Vol. 3:** 180
Flint glass (SF 10), **Vol. 2:** 17
Flint glass (SF 8), **Vol. 2:** 17
Flour and ground grain products applications, **Vol. 3:** 357
Fluorescence emission, **Vol. 1:** 80
Fluorescence instrument calibration, ASTM, **Vol. 3:** 356
Fluorescence lifetime, **Vol. 1:** 42
Fluorescence Raman background, **Vol. 5:** 82
Fluorescence, **Vol. 1:** 38; **Vol. 5:** 87
Fluorescent and Raman standards, **Vol. 1:** 203
Fluoride, **Vol. 2:** 148
Fluorilon 10% reflectance (R10) spectrum, **Vol. 2:** 262
Fluorilon 99% reflectance (R99) spectrum, **Vol. 2:** 262
Fluorilon and carbon black mixture, 10% R, **Vol. 3:** 372–373
Fluorilon and carbon black mixture, 50% R, **Vol. 3:** 373–374
Fluorilon FW99-02 reflectance, **Vol. 1:** 186–191
Fluorilon photometric standards, compared, **Vol. 3:** 379
Fluorilon, 99% R, **Vol. 2:** 42, 262; **Vol. 3:** 375–378
Fluorometer scanning functions, **Vol. 1:** 41
Fluorometer, monochromator type, **Vol. 1:** 40
Fluorometers, **Vol. 1:** 54
Fluorophore, excitation and emission, **Vol. 1:** 40
Food and beverages, **Vol. 4:** 259; **Vol. 5:** 207
Food and foodstuff products applications, **Vol. 3:** 357
Forage analysis, AOAC, **Vol. 3:** 350–351
Forage products applications, **Vol. 3:** 357
Forensic science, **Vol. 4:** 261; **Vol. 5:** 208
Fourier transform, **Vol. 4:** 67, 75
Fourier-domain (FFT) smoothing, **Vol. 1:** 119; **Vol. 2:** 95; **Vol. 3:** 205
FOV, field of view, **Vol. 2:** 158
Fovea, of eye, **Vol. 2:** 156
FTIR spectrometer basics, **Vol. 4:** 66
Fuels research, **Vol. 4:** 268; **Vol. 5:** 211
Full width at half maximum (FWHM), **Vol. 2:** 11, 51–52, 57–58
Fundamental bands, **Vol. 2:** 69
Fundamental grating equation, **Vol. 5:** 163–164
F-value, F-statistic, **Vol. 3:** 441
FWHM (Full Width at Half Maximum), **Vol. 1:** 71

G

G, international green, **Vol. 2:** 147
Gallic acid method for vanadium, **Vol. 2:** 149
Gamma-glutamyl transferase, **Vol. 2:** 150
Gas analysis, ASTM, **Vol. 3:** 354–355
Gas cells, **Vol. 1:** 90; **Vol. 4:** 109
Gas chromatography-infrared (GC-IR), **Vol. 4:** 128
Gas emission lines, **Vol. 1:** xxvii–xxviii; **Vol. 2–5:** xxxiii–xxxiv
Gas phase analysis, **Vol. 4:** 262; **Vol. 5:** 208
Gas phase water spectrum, **Vol. 3:** 454
Gas sample measurements, **Vol. 5:** 102
Gases and vapors, **Vol. 4:** 98; **Vol. 5:** 102
GC-IR, **Vol. 4:** 128
Gem-dimethyl groups, **Vol. 4:** 28; **Vol. 5:** 33
Gemstones, colors, **Vol. 2:** 151–152
General imaging, **Vol. 4:** 252
General sampling considerations, **Vol. 4:** 96, 99
Glan-Taylor prism, **Vol. 2:** 204
Glass (BK 7), **Vol. 2:** 76
Glass absorbance data versus wavelength (in nanometers), **Vol. 4:** 358–359
Glass sampling vial for Raman measurements, **Vol. 5:** 103
Glass transmittance data versus wavelength (in nanometers), **Vol. 4:** 358
Glass tungsten halogen lamp, **Vol. 2:** 49
Glass, "borosilicate crown", **Vol. 2:** 76
Glass, "optical crown", **Vol. 2:** 76
Glass, Corning optical glasses as HPFS®7978 and 7979, **Vol. 2:** 76
Glass, Corning optical glasses as HPFS®7980, **Vol. 2:** 76
Glass, Corning optical glasses as HPFS®865X and 797X, **Vol. 2:** 76
Glass, ESCO optical glasses as BK7, **Vol. 2:** 76

Glass, ESCO optical glasses as G1 and A1, ***Vol. 2:*** 76
Glass, ESCO optical glasses as I2-IR, ***Vol. 2:*** 76
Glass, ESCO optical glasses as S1-UVA and S1-UVB, ***Vol. 2:*** 76
Glass, Flint (SF 10), ***Vol. 2:*** 76
Glass, Flint (SF 8), ***Vol. 2:*** 76
Glass, Pyrex®, ***Vol. 2:*** 76
Glass, silica, 76 fused "Water Free" (SiO_2), ***Vol. 2:*** 76
Glass, Tempax®, ***Vol. 2:*** 76
Glass, ***Vol. 1:*** 16; ***Vol. 2:*** 70, 76
Glass, Zerodur® Schott, ***Vol. 2:*** 76
Global distance (GD), ***Vol. 3:*** 243
Global tilts, for spectral irradiance, ***Vol. 2:*** 192
Globar (silicon carbide), ***Vol. 4:*** 251
Glucose based polymers, ***Vol. 3:*** 3
Glucose oxidase, ***Vol. 2:*** 150
Glucose phosphate isomerase, ***Vol. 2:*** 150
Glucose-6-phosphate dehydrogenase, ***Vol. 2:*** 150
Glutamate pyruvate transaminase, ***Vol. 2:*** 150
Glutamate-oxaloacetate transaminase, ***Vol. 2:*** 150
Gold photometric reflectance standard, ***Vol. 3:*** 379–380; ***Vol. 4:*** 338–340
Good laboratory practice (GLP) guidelines, ***Vol. 1:*** 96
Good laboratory technique, ***Vol. 2:*** 21
Grain products applications, ***Vol. 3:*** 357
Granules or pellets, ***Vol. 5:*** 110
Granules, ***Vol. 2:*** 75
Grating angle equation, ***Vol. 1:*** 63
Grating efficiency curve, ***Vol. 1:*** 65
Grating equations, ***Vol. 2:*** 53–54
Grating resolution equation, ***Vol. 1:*** 63
Grating, intensity distribution, ***Vol. 1:*** 63
Grazing angle infrared reflectance, ***Vol. 4:*** 49
Green, ***Vol. 2:*** 16, 147, 156
Grey photometric standards spectra, ***Vol. 2:*** 229, 231–232, 236–238, 244–249
Grinding of samples, ***Vol. 2:*** 70
Group frequencies, ***Vol. 3:*** 50; ***Vol. 4:*** 20; ***Vol. 5:*** 26
Group theory, ***Vol. 3:*** 43

H

H bonding, ***Vol. 5:*** 175
h, metric-hue angle, ***Vol. 2:*** 176
ΔH, Metric-hue angle difference, ***Vol. 2:*** 176–177
Harmonic oscillator, ***Vol. 3:*** 18, 46
Haze in sample, ***Vol. 2:*** 70
Helium-neon laser (He:Ne), 632.8 nm, ***Vol. 5:*** 79–80
8°/Hemispherical spectral reflectance factor geometry, ***Vol. 2:*** 276
Heteropoly blue method for silica, ***Vol. 2:*** 148
Hexokinase, ***Vol. 2:*** 150
High optical density, ***Vol. 2:*** 17
High resolution measurement mode, ***Vol. 4:*** 89
Highly absorbing samples, ***Vol. 2:*** 17
High-pass filter, ***Vol. 2:*** 51–52
Hit Quality Index (HQI), ***Vol. 5:*** 156
H-O-H bands associated with Raman water spectrum, ***Vol. 5:*** 192
Holmium oxide, ***Vol. 1:*** 17, 171–173; ***Vol. 2:*** 19; ***Vol. 3:*** 393–394, 400–401
Homogenized samples, ***Vol. 2:*** 72
Hooke's law, ***Vol. 3:*** 41
HPLC grade solvents, ***Vol. 2:*** 20; ***Vol. 3:*** 28
Human color perception, ***Vol. 2:*** 156
Hunter 1948 Lab Color Coordinate Space, ***Vol. 2:*** 177
Hunter, Richard, S., ***Vol. 2:*** 177
Hydrogen bonding (X-H.......Y), ***Vol. 3:*** 45; ***Vol. 4:*** 40; ***Vol. 5:*** 45
Hydrogen Lyman-α, ***Vol. 1:*** 82
Hydrophilic molecules, ***Vol. 2:*** 70

I

I, lamp current, ***Vol. 2:*** 29
I, light emerging from the sample, ***Vol. 2:*** 10
I_0, incident light, ***Vol. 2:*** 10
Ideal harmonic oscillator, ***Vol. 3:*** 41, 46
Ideal line shape, ***Vol. 2:*** 11
Illuminance and contrast, ***Vol. 3:*** 81
Illuminant D65, ***Vol. 2:*** 167–168
Illumination, ***Vol. 3:*** 81
Imaging applications, ***Vol. 3:*** 287
Imaging, ***Vol. 4:*** 219
Indene, ***Vol. 5:*** 95
Indigo, ***Vol. 2:*** 16, 147, 156
Indium gallium arsenide detectors or arrays (InGaAs), ***Vol. 2:*** 9, 50, 68
Individual wavenumber step-scan measurement mode, ***Vol. 4:*** 89; ***Vol. 5:*** 17
Influence of resolution on SNR, ***Vol. 5:*** 88
Infrasil®, ***Vol. 1:*** 16; ***Vol. 2:*** 17, 70

Instrument design studies, ***Vol. 4:*** 262
Instrument line shape (ILS) test, ***Vol. 3:*** 237
Instrument optical quality performance tests, ***Vol. 3:*** 229
Instrument/calibration performance test, ***Vol. 2:*** 133
Integrating sphere solar measurements, ***Vol. 2:*** 181
Integrating sphere, ***Vol. 1:*** 32; ***Vol. 2:*** 33, 34
Intensity correction (NIST method), ***Vol. 5:*** 165, 229–230, 252
Interactance, ***Vol. 2:*** 16
Interference filter based photometer, ***Vol. 3:*** 14
Interferogram into spectrum, ***Vol. 4:*** 72
interferogram ZPD (zero path difference), ***Vol. 4:*** 66
Interferogram, double-sided, ***Vol. 4:*** 362
Interferogram, single-sided, continuous source, ***Vol. 4:*** 70
Interferogram, single-sided, discrete laser source, ***Vol. 4:*** 70
Interferogram, single-sided, ***Vol. 4:*** 364
Internal reference correction, ***Vol. 2:*** 45–46
Internal reflectance as ATR (attenuated total reflectance), ***Vol. 4:*** 46, 116
Internal umbrella (symmetrical) bending, ***Vol. 3:*** 38
International color (blue), ***Vol. 2:*** 147
International color (green), ***Vol. 2:*** 147
International color (red), ***Vol. 2:*** 147
International Organization for Standardization (ISO) standards, ***Vol. 4:*** 298; ***Vol. 5:*** 221
Interpolation, ***Vol. 2:*** 113
Interpretation, spectral, ***Vol. 3:*** 192
Interpretive spectroscopy, ***Vol. 3:*** 50, 192; ***Vol. 4:*** 20; ***Vol. 5:*** 26
Iodide, ***Vol. 2:*** 148
Ion exchange, ***Vol. 4:*** 134
Ionic materials, ***Vol. 2:*** 70
IPAC GmbH color standards spectra, ***Vol. 2:*** 250–252
IR expert calibration systems, ***Vol. 4:*** 157
IR Spectra-Structure Band Positions Charts, ***Vol. 4:*** 233–248, 274–297
Irradiance, ***Vol. 1:*** 151
IRRAS (infrared reflection absorption spectroscopy) measurement mode, ***Vol. 4:*** 90
ISO 11151-1:2000, ***Vol. 3:*** 348
ISO 12099:2010, ***Vol. 3:*** 348
ISO 14132-5:2008, ***Vol. 3:*** 349
ISO 15063:2011, ***Vol. 3:*** 349
ISO 21543:2006 (IDF 201: 2006), ***Vol. 3:*** 347
ISO 9050, ***Vol. 2:*** 180
ISO NIR standards, ***Vol. 3:*** 347
ISO standards, infrared, ***Vol. 4:*** 298
ISO/TS 10867:2010, ***Vol. 3:*** 348
ISO/TS 10868:2011, ***Vol. 3:*** 348
ISO12099 plotting graphics, ***Vol. 3:*** 270
ISO-21348, ***Vol. 1:*** 82
Isocitrate dehydrogenase, ***Vol. 2:*** 150
Iso-groups, ***Vol. 4:*** 28; ***Vol. 5:*** 33
Isooctane, ***Vol. 2:*** 20

J

Jablonski type diagram, ***Vol. 5:*** 54
Jacquinot or J-stop, ***Vol. 1:*** 7; ***Vol. 2:*** 11; ***Vol. 4:*** 68–69
JCAMP-DX (.dx), ***Vol. 2:*** 82
Jones unit, ***Vol. 1:*** 60; ***Vol. 2:*** 51; ***Vol. 5:*** 83

K

KBr or KCl matrix, ***Vol. 5:*** 107–108
Ketones (R-C=O -R'), ***Vol. 1:*** 83; ***Vol. 5:*** 184
Kinetics measurement mode, ***Vol. 4:*** 90; ***Vol. 5:*** 17
K-nearest neighbor (KNN), ***Vol. 3:*** 246
Kramers-Kronig correction, ***Vol. 3:*** 181
Kubelka-Munk transform, ***Vol. 1:*** 120
Kyanite, ***Vol. 2:*** 152

L

L*, lightness variable, ***Vol. 2:*** 176
L*a*b* color space coordinates, ***Vol. 2:*** 174–176
L*a*b* spherical color space, ***Vol. 2:*** 175
L*C*h Color Space, ***Vol. 2:*** 176
L*u*v* Color Space, ***Vol. 2:*** 178
L, x, R color (CIE Standard Observer), ***Vol. 2:*** 147
Lactate dehydrogenase, ***Vol. 2:*** 149–150
Lambertian surface, ***Vol. 2:*** 33, 77
L-amino acid oxidase, ***Vol. 2:*** 149
Lamp life, ***Vol. 2:*** 28–29
Lamp sources, ***Vols. 2–5:*** xxix
L-Arginase, ***Vol. 2:*** 149
Laser attenuation, when passing through an optical fiber, ***Vol. 5:*** 80
Laser diode, ***Vol. 2:*** 46
Laser excitation ranges, ***Vol. 5:*** 14, 76, 79, 243

LC-IR, ***Vol. 4:*** 128
Least squares fit, ***Vol. 3:*** 445
Leucine amino peptidase, ***Vol. 2:*** 150
Leuco crystal violet method for iodide, ***Vol. 2:*** 148
Level 0 testing, ***Vol. 4:*** 213–214
Level 1 testing, ***Vol. 4:*** 214
Level zero and level one tests, (FT-MIR), ASTM, ***Vol. 3:*** 353, 355
Lifetime of emission sources, ***Vol. 2:*** 28–29
Light equations, ***Vol. 3:*** 5
Lignin polymers, ***Vol. 2:*** 149; ***Vol. 3:*** 3
Limit of detection, ***Vol. 2:*** 280
Limitations of the Savitzky-Golay (S-G) method, ***Vol. 5:*** 143
Line shape, ***Vol. 2:*** 11
Linear and non-linear A=B=C systems, ***Vol. 5:*** 39–40
Linear regression on curved data, ***Vol. 1:*** 116
Linear regression on linear data, ***Vol. 1:*** 114
Linear regression, ***Vol. 2:*** 136–138
Linear response, ***Vol. 2:*** 10; ***Vol. 5:*** 253
Linearity checks, ***Vol. 1:*** 17; ***Vol. 2:*** 19, 280; ***Vol. 3:*** 27
Lipase, ***Vol. 2:*** 150
Lipophilic molecules, ***Vol. 2:*** 70
Liquid chromatography-infrared (LC/IR), ***Vol. 3:*** 356; ***Vol. 4:*** 128
Liquid sampling, liquid waveguide, ***Vol. 5:*** 110
Liquid sampling, ***Vol. 1:*** 84; ***Vol. 2:*** 70; ***Vol. 3:*** 137; ***Vol. 4:*** 99; ***Vol. 5:*** 103–105
Lithium carbonate, ***Vol. 2:*** 14
Loadings matrix (V), ***Vol. 5:*** 130
Local distance (LD), ***Vol. 3:*** 243
Local mode theory (Morse oscillator), ***Vol. 3:*** 49
Lorentzian curve, ***Vol. 2:*** 11
Low levels of analyte in solution, ***Vol. 5:*** 108
Low resolution measurement mode, ***Vol. 4:*** 89
Low-OH glass, ***Vol. 2:*** 70
Low-OH silica fibers, ***Vol. 2:*** 78
Low-pass filter, ***Vol. 2:*** 51
Luminance, ***Vol. 2:*** 37; ***Vol. 3:*** 81
Luminescence, ***Vol. 2:*** 32–33; ***Vol. 3:*** 81
Luminometers, ***Vol. 3:*** 64
Lumps, granules, or pellets, ***Vol. 2:*** 75; ***Vol. 4:*** 105; ***Vol. 5:*** 110

M

M, y, G color (CIE Standard Observer), ***Vol. 2:*** 147
Mahalanobis distance measurements, ***Vol. 2:*** 117; ***Vol. 3:*** 240, 256; ***Vol. 5:*** 150
Mapping, ***Vol. 4:*** 253
MASC, ***Vol. 1:*** 122
Matching index, ***Vol. 2:*** 117; ***Vol. 3:*** 243, 257
Matte red photometric standards spectra, ***Vol. 2:*** 230–232
Matte white photometric standards spectra, ***Vol. 2:*** 230–232
MBR (minimum blue reflectance) reflectance wavelength reference standards spectra, ***Vol. 2:*** 236–240
MBR blue-green color standards reflectance data, ***Vol. 2:*** 238–239
0°/<5° Measurement geometry, ***Vol. 2:*** 162
0°/135° Measurement geometry, ***Vol. 2:*** 162
0°/180° Measurement geometry, ***Vol. 2:*** 277
0°/45° Measurement geometry, ***Vol. 2:*** 160
0°/45° Measurement geometry, ***Vol. 2:*** 277
22.5°/22.5° Measurement geometry, ***Vol. 2:*** 160
45°/0° Measurement geometry, ***Vol. 2:*** 158
d/8° Measurement geometry, ***Vol. 2:*** 159
Medical applications, ***Vol. 3:*** 283
Medicine and clinical chemistry, ***Vol. 4:*** 267; ***Vol. 5:*** 209
Melamine, ***Vol. 5:*** 207
Metal and ceramic surfaces, absorbance, ***Vol. 3:*** 432
Metal on quartz filter, ***Vol. 1:*** 17, 182–183, 185
Metal surfaces spectra, ***Vol. 2:*** 261, 265
Metalized film etalon, ***Vol. 3:*** 383
Methacrylate, ***Vol. 2:*** 17
Methyl methacrylate, ***Vol. 2:*** 17, 70
Metric chroma (C*), ***Vol. 2:*** 176
Metric-hue angle difference, ***Vol. 2:*** (ΔH), 176–177
Metric-hue angle, (h), ***Vol. 2:*** 176
Military, explosives, and propellants, ***Vol. 4:*** 268; ***Vol. 5:*** 210
Milliliter capacity for visible cuvets/cuvettes, ***Vol. 2:*** 18
Mineralogy, ***Vol. 5:*** 205
MIR Spectra-Structure Band Positions Charts, ***Vol. 4:*** 243–248
Mirror reflectance, ***Vol. 2:*** 8

ml capacity for visible cuvets/cuvettes, ***Vol. 2:*** 18
Mobile phase, ***Vol. 4:*** 100
Moisture measurement, ***Vol. 3:*** 3, 351; ***Vol. 4:*** 268
Molybdosilicate method for silica, ***Vol. 2:*** 148
Monoamine oxidase, ***Vol. 2:*** 150
Morse oscillator model, ***Vol. 3:*** 49
Moving-averaged-segment convolution, ***Vol. 1:*** 122
Mulls, ***Vol. 4:*** 113
Multiple regression, ***Vol. 5:*** 127–130
Multiplicative scatter (or signal) correction (MSC), ***Vol. 3:*** 260
Multivariate Linear Regression (MLR), ***Vol. 5:*** 127
Multivariate mathematics, ***Vol. 3:*** 447
Multivariate regression, ***Vol. 5:*** 127
Multi-way data methods, ***Vol. 4:*** 172–173
mzData (.XML), ***Vol. 2:*** 82

N

N, N-diethyl-p-phenylenediamine (DPD) method for chlorine, ***Vol. 2:*** 148
N=N (azo-) stretching, ***Vol. 5:*** 175
N_2, ***Vol. 5:*** 95
NADH and NADPH, ***Vol. 2:*** 149
n-Alkanes, spectra, ***Vol. 4:*** 25, 235–236; ***Vol. 5:*** 31, 176–178
Naphthalene (Mallinckodt #94848), ***Vol. 5:*** 25
Natural gas, and fuels research, ***Vol. 3:*** 357; ***Vol. 4:*** 268; ***Vol. 5:*** 211
Near ultraviolet, NUV, ***Vol. 1:*** 82
Neocuproine method for copper, ***Vol. 2:*** 148
Neodymium yttrium aluminum garnet (Nd:YAG) laser, ***Vol. 5:*** 1064 nm
Nesslerization method for ammonia or organic nitrogen, ***Vol. 2:*** 148
n−Hexane, ***Vol. 2:*** 20
Nickel solution, ***Vol. 1:*** 17
Nicotinic acid standard, ***Vol. 1:*** 181, 184
NIST SRM 1920a spectra, ***Vol. 3:*** 388–391
NIST SRM 1921b, ***Vol. 4:*** 331
NIST SRM 2036 spectrum, ***Vol. 2:*** 228–229
NIST SRM 2241, ***Vol. 5:*** 230
NIST SRM 2242, ***Vol. 5:*** 230
NIST SRM 2243, ***Vol. 5:*** 230
NIST SRM 2244, ***Vol. 5:*** 230
NIST SRM 2245, ***Vol. 5:*** 230
NIST SRM 2246, ***Vol. 5:*** 230
NIST SRM 930D/930d, ***Vol. 2:*** 217–219, 234–235
NIST SRM-2036 reflectance wavelength standard, ***Vol. 3:*** 394–395
NIST SRM-2065 reflectance wavelength standard, ***Vol. 3:*** 396–397
NIST SRMs for Raman spectrometer calibration, ***Vol. 5:*** 230
NIST SRMs, ***Vol. 1:*** 126; ***Vol. 2:*** 217; ***Vol. 3:*** 388–391, 394–395; ***Vol. 4:*** 331; ***Vol. 5:*** 230
NIST uncertainty number for SRMs, ***Vol. 3:*** 395; ***Vol. 4:*** 340; ***Vol. 5:*** 237
NIST wavenumber Standard Reference Material for infrared, ***Vol. 4:*** 331
NIST, National Institute of Standards and Technology, ***Vol. 2:*** 18, 217
Nitrate nitrogen, ***Vol. 2:*** 148
Nitriles (C≡N),
Nitriles (R-C≡N), ***Vol. 1:*** 83; ***Vol. 4:*** 33, 233; ***Vol. 5:*** 38
N-Methyl acetamide (secondary amide) for N-H stretching spectrum, ***Vol. 4:*** 240
N-mode principal component analysis algorithm, ***Vol. 4:*** 174
Noise compensation in modeling, ***Vol. 4:*** 168
Noise equivalent power (NEP), ***Vol. 2:*** 50–51
Noise estimates, chemometrics, ***Vol. 4:*** 162
Normal mode theory (The ideal harmonic oscillator), ***Vol. 3:*** 41–46

O

O_2, ***Vol. 5:*** 95
Observer angles, ***Vol. 2:*** 156
Oil-based mulls, ***Vol. 4:*** 113
Oils, ***Vol. 2:*** 70
Opal glass standards spectra, ***Vol. 2:*** 233–236
Opaque filter method, ASTM, ***Vol. 3:*** 353
Opaque liquids, ***Vol. 2:*** 17
Opaque solids, ***Vol. 2:*** 17
Optical black surfaces, absorbance, ***Vol. 3:*** 405–411
Optical crown glass, ***Vol. 2:*** 17
Optical glass and sapphire data versus wavelength (in nanometers), ***Vol. 1:*** 91, 206–210; ***Vol. 2:*** 260–265, 270–275; ***Vol. 3:*** 405; ***Vol. 4:*** 358–359; ***Vol. 5:*** 245–248
Optical glass materials spectra, ***Vol. 2:*** 264
Optimizing the integrating sphere as a detector, ***Vol. 3:*** 157

Orange photometric standards spectra, ***Vol. 2:*** 230–232, 240–243, 252–253
Orange, ***Vol. 2:*** 16, 147, 156
Orange/red, ***Vol. 2:*** 16, 147, 156
Order sorting filter, ***Vol. 3:*** 447
Organic nitrogen, ***Vol. 2:*** 148
Oscillator, ***Vol. 3:*** 43
Outlier detection algorithms, ***Vol. 3:*** 240, 447
Out-of-plane bending vibrations for benzene., ***Vol. 5:*** 41

P

Paper and pulp, ***Vol. 4:*** 270
Parafac/gram algorithm, ***Vol. 4:*** 174
Pareto scaling, ***Vol. 5:*** 253
Partial evaporation or distillation, ***Vol. 4:*** 133
Partial Least Squares — Singular Value Decomposition (PLS-SVD), ***Vol. 5:*** 133
Partial Least Squares Regression (PLSR), ***Vol. 5:*** 134
PAS (photoacoustic spectroscopy) measurement mode, ***Vol. 4:*** 90, 123
Pathlength correction, ***Vol. 1:*** 120
Pathlength measurement, cuvette, ***Vol. 1:*** 85
Pathlength, ***Vol. 1:*** 85, 120; ***Vol. 2:*** 9–10, 16, 280; NIR, ***Vol. 3:*** 6; ***Vol. 5:*** 24, 253
PCA plots, ***Vol. 3:*** 188
PCR calculations, ***Vol. 5:*** 132
Pepsin, ***Vol. 2:*** 150
Peroxidase, ***Vol. 2:*** 150
Peroxides (-C-O-O-C-), ***Vol. 4:*** 39; ***Vol. 5:*** 45
Petroleum, natural gas, and fuels research, ***Vol. 4:*** 268, 284; ***Vol. 5:*** 211
PFS (specular factor) mirror reflectance standard, ***Vol. 3:*** 381–382
Pharmaceutical company applications, ***Vol. 2:*** 15; ***Vol. 3:*** 283; ***Vol. 4:*** 269; ***Vol. 5:*** 211
Phenate method for ammonia nitrogen, ***Vol. 2:*** 148
Phenol, ***Vol. 2:*** 148
Phosphorescence, ***Vol. 1:*** 40, 42, 80
Phosphorus, ***Vol. 2:*** 148
Photoacoustic spectroscopy (PAS) measurement mode, ***Vol. 4:*** 90, 123
Photometric accuracy, ***Vol. 4:*** 324
Photomultipler tube, (CsI) photocathode, ***Vol. 2:*** 50
Photomultipler tube, (CsInGaAs) photocathode, ***Vol. 2:*** 50
Photomultipler tube, cesium antimonide (CsSb) photocathode, ***Vol. 2:*** 50
Photomultipler tube, cesium gallium arsenide (CsGaAs) photocathode, ***Vol. 2:*** 50
Photomultipler tube, multialkali (Na-K-Sb-Cs) photocathode, ***Vol. 2:*** 50
Photomultipler tube, S1 type photocathode, ***Vol. 2:*** 50
Photomultiplier tube (PMT), ***Vol. 2:*** 8, 9, 50, 68
Photomultiplier tube detectors, ***Vol. 2:*** 8, 9, 50, 68
Photon (photo) detectors, ***Vol. 2:*** 50
Piperazine in drugs, AOAC, ***Vol. 3:*** 352
Plant sciences, paper and pulp, ***Vol. 4:*** 270; ***Vol. 5:*** 212
Plotting data for visual assessment, ***Vol. 1:*** 111
PLS and PLS — SVD, ***Vol. 5:*** 133–136
p-Nitroalinine method, ***Vol. 2:*** 150
Polar materials, ***Vol. 2:*** 70
Polarization modulation, ***Vol. 4:*** 89; ***Vol. 5:*** 17
Polarized light, ***Vol. 1:*** 67, 148; ***Vol. 2:*** 54–55, 196; ***Vol. 3:*** 80, 114
Poly(methylmethacrylate), ***Vol. 1:*** 16
Polymer science, ***Vol. 4:*** 270; ***Vol. 5:*** 215
Polymeric films, ***Vol. 4:*** 104; ***Vol. 5:*** 109
Polymers and rubbers (elastomers), ***Vol. 3:*** 23
Polystyrene (Aldrich #18,243-5), ***Vol. 5:*** 25
Polystyrene (crystalline form), ***Vol. 3:*** 384–387; ***Vol. 4:*** 331–334; ***Vol. 5:*** 234–236
Polystyrene subtraction test, ***Vol. 4:*** 214–216
Polystyrene, 1 mm thickness, NIR, ***Vol. 4:*** 335–337
Post-dispersive spectrometer, ***Vol. 3:*** 448
Potassium acid phthalate, ***Vol. 1:*** 17
Potassium bromide, ***Vol. 2:*** 14
Potassium chloride, ***Vol. 2:*** 14
Potassium chromate standard, ***Vol. 1:*** 180
Potassium dichromate standard, ***Vol. 1:*** 180–181
Potassium dichromate, ***Vol. 1:*** 17; ***Vol. 2:*** 19
Potassium ferromanganate, ***Vol. 2:*** 14
Potassium iodide, ***Vol. 2:*** 14
Potential energy curve, bond, ***Vol. 3:*** 41
p-polarized light, ***Vol. 2:*** 196–204
Predicted versus actual plot, ***Vol. 2:*** 91; ***Vol. 3:*** 192
Pre-dispersive spectrometer, ***Vol. 3:*** 448
Primary amine spectrum, ***Vol. 4:*** 240
Principal Component(s) Analysis (PCA), ***Vol. 3:*** 188; ***Vol. 5:*** 129

Principal Component(s) Regression (PCR), ***Vol. 5:*** 131
Principal components scores-based discriminant analysis, ***Vol. 3:*** 245
Prism, light propagation, ***Vol. 1:*** 66–67; ***Vol. 3:*** 109
Process monitoring applications, ***Vol. 3:*** 285
2-Propanol spectrum, ***Vol. 4:*** 238
Propellants, ***Vol. 4:*** 268; ***Vol. 5:*** 210
Protein (crude) in forages, AOAC, ***Vol. 3:*** 350
Protein (crude) in wheat, AOAC, ***Vol. 3:*** 350
Protein in meat, AOAC, ***Vol. 3:*** 351
Protein structure analysis, ***Vol. 1:*** 155–156
Pulp, ***Vol. 4:*** 270
Pyrex, ***Vol. 2:*** 17
Pyrolysis-based measurements, ***Vol. 4:*** 135
Pyruvate kinase, ***Vol. 2:*** 151

Q

Qualitative analysis and conformity index, ***Vol. 1:*** 14; ***Vol. 2:*** 15, 115; ***Vol. 3:*** 7–8, 25, 255, 280, 352–354; ***Vol. 4:*** 90; ***Vol. 5:*** 16–17, 23, 70, 220
Quantitative applications, ***Vol. 1:*** 14; ***Vol. 3:*** 7–8, 24–25, 280; ***Vol. 4:*** 90; ***Vol. 5:*** 23
Quartz, crystalline (SiO_2), ***Vol. 2:*** 76
Quartz, extremely low O-H, ***Vol. 2:*** 17, 76
Quartz, ***Vol. 2:*** 17, 70, 76
Quinine sulfate (sulphate) solution, ***Vol. 1:*** 204

R

R (international red), ***Vol. 2:*** 147
r to z' table, ***Vol. 5:*** 155
r to z' transformation, ***Vol. 5:*** 153
Raman effect, ***Vol. 5:*** 53
Raman equations for quantitative analysis, ***Vol. 5:*** 16–21, 70
Raman functional group band positions tables, ***Vol. 5:*** 186–191, 193–204
Raman laser excitation sources, ***Vol. 5:*** 14, 79, 243–244
Raman spectra-structure correlation charts, ***Vol. 5:*** 175–191, 193–204, 217–219
Raman theory, ***Vol. 5:*** 53
Rare earth oxide similar to NIST SRM 1920a spectra, ***Vol. 3:*** 388–391
Rare earth oxide wavelength standard, ***Vol. 1:*** 178–179
Red 16, 147, ***Vol. 2:*** 156
Red, dark, ***Vol. 2:*** 147
0% Reflectance, double beam solar measurements, ***Vol. 2:*** 183
0% Reflectance, single beam solar measurements, ***Vol. 2:*** 183
100% Reflectance, double beam solar measurements, ***Vol. 2:*** 184
100% Reflectance, single beam solar measurements, ***Vol. 2:*** 184
Reflectance factor, ASTM, ***Vol. 3:*** 354
Reflectance measurements, polarized light, ***Vol. 1:*** 148; ***Vol. 3:*** 80
Reflectance standard, 10% R, ***Vol. 3:*** 372–373
Reflectance standard, 50% R, ***Vol. 3:*** 373–374
Reflectance standard, 99% R, ***Vol. 3:*** 375–378
Refractive index of film, ***Vol. 1:*** 38
Regression coefficient, ***Vol. 3:*** 449
Regression method symbols, ***Vol. 5:*** 128
Regression vector, ***Vol. 5:*** 129
Regression, ***Vol. 2:*** 136–138
Residual versus actual plot, ***Vol. 2:*** 91; ***Vol. 3:*** 195
Resolution standards, ***Vol. 1:*** 200–201
Resolution, ***Vol. 1:*** 10, 69–72, 135; ***Vol. 2:*** 12–13, 110; ***Vol. 3:*** 19–21; ***Vol. 4:*** 76, 216
RGB (red-green-blue) color standards spectra, ***Vol. 2:*** 157–158, 163, 258–260, 262–263
RRS/RR (Resonance Raman Spectroscopy) measurements, ***Vol. 5:*** 18, 57, 60, 66, 112–113
R-squared, ***Vol. 3:*** 448
Russian opal standards spectra, ***Vol. 2:*** 233–236

S

S, z, B color (CIE Standard Observer), ***Vol. 2:*** 147
Samarium oxide wavelength standard, ***Vol. 1:*** 125, 177
Samarium perchlorate, ***Vol. 1:*** 17; ***Vol. 2:*** 19
0°/<5° Sample measurement geometry, ***Vol. 1:*** 150
0°/45° Sample measurement geometry, ***Vol. 1:*** 149
22.5°/22.5° Sample measurement geometry, ***Vol. 1:*** 149
Sample/solvent problem, ***Vol. 1:*** 19; ***Vol. 2:*** 21
Sapphire (aluminum oxide) (Al_2O_3), ***Vol. 2:*** 17, 76; ***Vol. 4:*** 358–359
Savitzky-Golay derivative function, ***Vol. 1:*** 123; ***Vol. 2:*** 98; ***Vol. 3:*** 206; ***Vol. 5:*** 143

SBW (spectral bandwidth), ***Vol. 1:*** 71
Scaling methods for spectroscopic data, ***Vol. 1:*** 118; ***Vol. 2:*** 92; ***Vol. 3:*** 203
Scatter correction algorithms, ***Vol. 3:*** 260; ***Vol. 4:*** 171
Scores matrix (T), ***Vol. 5:*** 130
Search parameter selection, ***Vol. 4:*** 175
Search software features, ***Vol. 4:*** 203
Searching algorithms, ***Vol. 4:*** 159
Selection rule, ***Vol. 3:*** 47
Selectivity, ***Vol. 2:*** 44
Selenium, ***Vol. 2:*** 148
Self-supporting films, ***Vol. 4:*** 114
Semiconductors, electronics, ***Vol. 4:*** 272; ***Vol. 5:*** 216
Sensitivity, ***Vol. 2:*** 44, 281
SEPt comparison test using Chi-Square, ***Vol. 3:*** 263
SERRS (Surface-Enhanced Resonance Raman Spectroscopy), ***Vol. 5:*** 18, 58, 61–62, 66, 112–114
S-H (thiol-) stretching, ***Vol. 4:*** 242; ***Vol. 5:*** 175, 185
Short wavelength NIR, ***Vol. 2:*** 16
Shortwave near-ir spectral region (SW-NIR), ***Vol. 1:*** 15; ***Vol. 3:*** 2–4
Significance test for validation bias, ***Vol. 2:*** 135
Silica, fused as pure SiO_2 (IR Quality), ***Vol. 2:*** 76
Silica, synthetic fused (SiO_2), ***Vol. 2:*** 76
Silica, UV grade fused (SiO_2), ***Vol. 2:*** 76
Silicate, ***Vol. 2:*** 70
Silicon (Si) photodiode detector, ***Vol. 2:*** CCD, 50
Silicon wafer spectrum, ***Vol. 5:*** 234
Silicon-based photodiode arrays (PDAs), ***Vol. 2:*** 9, 68
Silver diethyldithiocarbamate method for arsenic, ***Vol. 2:*** 148
Silver, ***Vol. 2:*** 148–149
Simple linear regression analysis, ***Vol. 2:*** 136
Simple smoothing function, ***Vol. 1:*** 123
SiO spectrum on aluminum, p-polarized, ***Vol. 2:*** 202
SiO spectrum on aluminum, s-polarized, ***Vol. 2:*** 201
Si-O stretch band locations, ***Vol. 4:*** 242
Size exclusion chromatography-infrared (SEC/IR), ASTM, ***Vol. 3:*** 356
Skew, ***Vol. 3:*** 451
Sky blue, ***Vol. 2:*** 16, 147, 156
Slit profile, for monochromator, ***Vol. 2:*** 11, 57, 281
SMILES format, ***Vol. 3:*** 51
Snell's law, ***Vol. 1:*** 66–67
Sodium chloride (NaCl), ***Vol. 2:*** 17, 76
Sodium nitrate, ***Vol. 2:*** 14
Solar measurement, ***Vol. 2:*** 179–195
Solid-state flashlamp pump laser illustration, ***Vol. 5:*** 80
Solid-state P-I-N (p-i-n) junction semiconductor diode laser illustration, ***Vol. 5:*** 81
Solid-state VCSEL (Vertical-Cavity Surface-Emitting Laser) pump laser illustration, ***Vol. 5:*** 81
Solvent cutoff/cut-off wavelengths, ***Vol. 2:*** 20, 281
Solvent grades, ***Vol. 2:*** 20
Solvent removal, ***Vol. 4:***132
Sorbitol dehydrogenase, ***Vol. 2:*** 151
SORS (Spatially Offset Raman Spectroscopy) (offset position), ***Vol. 5:*** 19, 60, 64, 66, 115–116
SPADNS method for fluoride, ***Vol. 2:*** 148
Spectra-Calc (Archival) data format (fixed point .spc), ***Vol. 2:*** 82
$\bar{x}_{10}, \bar{y}_{10}, \bar{z}_{10}$ Spectral tristimulus values, ***Vol. 2:*** 169
$\bar{x}_2, \bar{y}_2, \bar{z}_2$ Spectral tristimulus values, ***Vol. 2:*** 171–173
Spectral search algorithms and terminology, ***Vol. 3:*** 356; ***Vol. 4:*** 159, 175–200; ***Vol. 5:*** 149–150
Spectral subtraction test, ***Vol. 4:*** 214
Spectral tristimulus values, ***Vol. 2:*** 164
Spectral variance due to cuvette pathlength test, ***Vol. 2:*** 125
Spectral variance due to temperature test, ***Vol. 2:*** 124
Spectral variance test, ***Vol. 2:*** 123
Spectralon^{TM}, ***Vol. 2:*** 42
Spectra-structure correlation charts, ***Vol. 1:*** 144–146, 158–159; ***Vol. 2:*** 147–156; ***Vol. 3:*** 279, 292–346, 348–351; ***Vol. 4:*** 233–248, 274–297; ***Vol. 5:*** 175–191, 193–204, 217–219
Spectrosil®, ***Vol. 1:*** 16; ***Vol. 2:*** 17, 70
0°/180° Specular transmission measurement geometry, ***Vol. 2:*** 162

0°/180° Specular transmission sample measurement geometry, ***Vol. 2:*** 282
Specular reflectance geometry, ***Vol. 2:*** 201
Sphere error, ***Vol. 2:*** 34
Sphere, integrating, ***Vol. 1:*** 32, 34
s-polarized light direction vector, ***Vol. 2:*** 201, 203, 204
s-polarized light, ***Vol. 2:*** 196–203
SPPERS (Surface Plasmon Polariton (SPP) Enhanced Raman Spectroscopy), ***Vol. 5:*** 19, 61, 66, 116–117
SRM 1920a spectra, ***Vol. 3:*** 388–391
SRM 1921b, ***Vol. 4:*** 331
SRM 1935a, ***Vol. 1:*** 126
SRM 2031, ***Vol. 1:*** 17; ***Vol. 2:*** 19
SRM 2031b, ***Vol. 1:*** 126
SRM 2032, ***Vol. 1:*** 126
SRM 2034, ***Vol. 1:*** 126
SRM 2035a, ***Vol. 1:*** 125–126
SRM 2036 spectrum, ***Vol. 2:*** 228–229
SRM 2241, ***Vol. 5:*** 230
SRM 2242, ***Vol. 5:*** 230
SRM 2243, ***Vol. 5:*** 230
SRM 2244, ***Vol. 5:*** 230
SRM 2245, ***Vol. 5:*** 230
SRM 2246, ***Vol. 5:*** 230
SRM 84, ***Vol. 1:*** 17; ***Vol. 2:*** 19
SRM 930d, Filter 10-1255 spectrum, ***Vol. 2:*** 234–235
SRM 930d, Filter 20-1255 spectrum, ***Vol. 2:*** 234–235
SRM 930d, Filter 30-1255 spectrum, ***Vol. 2:*** 234–235
SRM 930D/930d, ***Vol. 1:*** 17; ***Vol. 2:*** 19, 217–219, 234–235
SRM 931, ***Vol. 1:*** 17; ***Vol. 2:*** 19
SRM 931g, ***Vol. 1:*** 126
SRM 935, ***Vol. 1:*** 17; ***Vol. 2:*** 19
SRM 935a, ***Vol. 1:*** 126
SRM, Standard Reference Material (NIST), ***Vol. 2:*** 19
SRMs for Raman intensity correction, ***Vol. 5:*** 229
SRMs, NIST uncertainty number for, ***Vol. 3:*** 395
SRMs, ***Vol. 2:*** 217
SRS (Spontaneous Raman Spectroscopy) measurements, ***Vol. 5:*** 18, 58, 63, 66, 113–114
SRS/SPSRS (Stimulated Raman Spectroscopy) measurements, ***Vol. 5:*** 59, 63, 114–115
S-S stretching, ***Vol. 5:*** 175
Stability test, ***Vol. 4:*** 214
10° Standard Observer, ***Vol. 2:*** 146, 156, 157, 158
2° Standard Observer, ***Vol. 2:*** 146, 156, 157, 158
4° Standard Observer, ***Vol. 2:*** 146
Standard error of cross validation, ***Vol. 3:*** 452
Standard error of estimate, ***Vol. 3:*** 452
Standard error of prediction, ***Vol. 3:*** 452
Standard error of validation (SEV) test, ***Vol. 2:*** 134
Standard illuminant A, ***Vol. 2:*** 167–168
Standard illuminant, D65, ***Vol. 2:*** 157
Standard measurement geometry, ***Vol. 2:*** 276–277
Standard normal variate (SNV), ***Vol. 3:*** 261
Standard observer, ***Vol. 2:*** 146
Standard Reference Materials (SRMs) for photometric accuracy measurements, ***Vol. 4:*** 323
Standard Reference Materials (SRMs) for Raman intensity correction, ***Vol. 5:*** 229
Stannous chloride method for phosphorus, ***Vol. 2:*** 148
Starna green absorbance, ***Vol. 1:*** 185–186
Statistical tests for the significance of bias and slope, ***Vol. 3:*** 181–183
Statistics for comparing spectra, ***Vol. 5:*** 153
Statistics of spectral searches (comparisons), ***Vol. 5:*** 149
Step-scan measurement modes, ***Vol. 4:*** 89; ***Vol. 5:*** 17
Stimulated Raman, ***Vol. 5:*** 18, 56, 58, 66, 111
Stokes shift spectrum, ***Vol. 1:*** 38; ***Vol. 5:*** 5
Stray light standards, ***Vol. 1:*** 191–199
Stray light test, ***Vol. 1:*** 13, 75, 130; ***Vol. 2:*** 105; ***Vol. 3:*** 22, 122, 126
Stray radiant power ratio, ASTM, ***Vol. 3:*** 353
Sulfide, ***Vol. 2:*** 149
Sulfur (Aldrich #41,498-0), ***Vol. 5:*** 25
Sum of squares, ***Vol. 3:*** 452–453
Suprasil 300, ***Vol. 2:*** 17, 270, 272
Surface analysis, ***Vol. 4:*** 272; ***Vol. 5:*** 216
Surrogate mixtures, ASTM, ***Vol. 3:*** 355
SVD of data matrix, ***Vol. 5:*** 131, 135, 137–138
SW-NIR, ***Vol. 3:*** 4
Synthetic fused silica, ***Vol. 2:*** 17

T

T versus A/Au, ***Vol. 2:*** 219–224
Talc as a reference material, NIR, ***Vol. 3:*** 371
Tannin, ***Vol. 2:*** 149
Tanzanite, ***Vol. 2:*** 152
Teaching set, ***Vol. 3:*** 453
Teflon liquid waveguide, ***Vol. 5:*** 110
Tempax®, ***Vol. 2:*** 17
Terminology, ASTM, ***Vol. 3:*** 352
TERS (Tip-Enhanced Raman Spectroscopy), ***Vol. 5:*** 19, 60, 65–66, 116–117
Tertiary butyl groups, ***Vol. 4:*** 28; ***Vol. 5:*** 33
Textiles, ***Vol. 4:*** 274; ***Vol. 5:*** 217
TGA-IR, ***Vol. 4:***129
The data matrix (A), ***Vol. 5:*** 130
The integrating sphere, ***Vol. 3:*** 64
Theoretical limit of D* for a detector system, ***Vol. 3:*** 102
Thermo Galactic data format (.spc), ***Vol. 2:*** 82
Thermogravimetric analysis-infrared (TGA-IR), ***Vol. 3:*** 355; ***Vol. 4:***129
Thiocyanate, ***Vol. 2:*** 149
Thiol (S-H) stretching, ***Vol. 1:*** 83; ***Vol. 5:*** 175
Third-order polynomial equation, ***Vol. 5:*** 164
Threshold Limit Values (TLV), ***Vol. 1:*** 151
Throughput, integrating sphere, ***Vol. 1:*** 32
Tilting filter, ***Vol. 3:*** 453
Timedrive measurement mode, ***Vol. 4:*** 89
Tissue measurement, NIR, ***Vol. 3:*** 6
50/50 (v/v) Toluene/acetonitrile (Mallinckrodt analytical reagents), ***Vol. 5:*** 25
Toluene in hexane resolution standard, ***Vol. 1:*** 201
Trans fat method and band location, ***Vol. 4:*** 249
Transflectance, ***Vol. 2:*** 16
0% Transmittance, double beam solar measurements, ***Vol. 2:*** 186
0% Transmittance, single beam solar measurements, ***Vol. 2:*** 186
0°/180° Transmittance/transmission, ***Vol. 1:*** 148
100% Transmittance, double beam solar measurements, ***Vol. 2:*** 187
100% Transmittance, single beam solar measurements, ***Vol. 2:*** 187
Triangular (weighted smoothing), ***Vol. 1:*** 139
Trimethyl phosphate, ***Vol. 2:*** 20
Triple bonds (-C≡), ***Vol. 4:*** 32; ***Vol. 5:*** 38
Triple monochromator Raman spectrometer, ***Vol. 5:*** 75
Trypsin, ***Vol. 2:*** 151
t-test (or Student's t-statistic), ***Vol. 3:*** 453
Tungsten lamp emission, ***Vol. 2:*** 29
Tungstophosphoric and molybdophosphoric acid method for lignin, ***Vol. 2:*** 149
Turbid samples, ***Vol. 2:*** 17
Turbidity measurement standard, ***Vol. 2:*** 147
Turret-mounted filters, ***Vol. 3:*** 454
Two-dimensional correlation measurement mode, ***Vol. 4:*** 90; ***Vol. 5:*** 17

U

Ultraviolet A, UVA, ***Vol. 1:*** 82
Ultraviolet absorption, ***Vol. 1:*** 5, 80
Ultraviolet B, UVB,, ***Vol. 1:*** 82
Ultraviolet C, UVC, ***Vol. 1:*** 82
Ultraviolet health issues, ***Vol. 1:*** 151
Umbrella (symmetrical) bending, ***Vol. 3:*** 38
Uncertainty number for SRMs, ***Vol. 3:*** 395
Unit conversion, spectroscopy, ***Vol. 1:*** 3, 4
United States Pharmacopeia (USP) methods, ***Vol. 3:*** 347; ***Vol. 4:*** 298; ***Vol. 5:*** 220
Urease, ***Vol. 2:*** 151
USP method <1119>near infrared spectrophotometry, ***Vol. 3:*** 347
UV cut-off wavelengths of solvents, ***Vol. 2:*** 20
UV-A, ***Vol. 1:*** 82
UV-B, ***Vol. 1:*** 82
UV-grade fused silica, ***Vol. 2:*** 17
UV-grade solvent, ***Vol. 1:*** 18

V

Vacuum Ultraviolet, ***Vol. 1:*** 82
Validation of agreement between model and reference, ***Vol. 2:***
Vanadium, ***Vol. 2:*** 149
Vanadomolybdophosphoric acid method for phosphorus, ***Vol. 2:*** 148
Vapor phase measurements, ***Vol. 4:*** 157
VCD (vibrational circular dichroism) measurement mode, ***Vol. 4:*** 90
Violet, ***Vol. 2:*** 16, 147, 154, 156
Visible band locations, ***Vol. 2:*** 213
Visible chromophore band positions, ***Vol. 2:*** 146–152, 213
Visible color designation, ***Vol. 2:*** 16
Visible reference standards, ***Vol. 2:*** 225–260
VLD (vibrational linear dichroism) measurement mode, ***Vol. 4:*** 90

W

Water fluorescent intensity standard, ***Vol. 1:*** 205–206
Water in dried vegetables, AOAC, ***Vol. 3:*** 352
Water vapor spectrum, ***Vol. 3:*** 454
Water, reagent grade, ***Vol. 2:*** 70
Waveguide Raman, ***Vol. 5:*** 17
Wavelength standard WCR-UVVis-02c, ***Vol. 3:*** 397–398
Wavelength/wavenumber measurement standards, ***Vol. 1:*** 171; ***Vol. 2:*** 2, 4–5, 285; ***Vol. 3:*** 383; ***Vol. 4:*** 331; ***Vol. 5:*** 233
Well slide for sampling, ***Vol. 1:*** 86; ***Vol. 2:*** 72
White ceramic comparison spectra, ***Vol. 2:*** 262
Wien's law of displacement, ***Vol. 2:*** 24–27
Wood's anomalies, ***Vol. 1:*** 76

X

X and Y data graphics (plotting), ***Vol. 1:*** 139
X, Y, Z tristimulus values, ***Vol. 2:*** 157–158, 161, 162, 163
X_0, Y_0, Z_0 color coordinates, ***Vol. 2:*** 174
X_{10}, Y_{10}, Z_{10} tristimulus values, ***Vol. 2:*** 165
Xanthine oxidase, ***Vol. 2:*** 151
Xenon arc lamp (180–2500 nm), ***Vol. 2:*** 6
xy chromaticity diagram, ***Vol. 2:*** 174
Xylene, ***Vol. 2:*** 20
xyz chromaticity coordinates, ***Vol. 2:*** 163–164

Y

Yellow photometric standards spectra, ***Vol. 2:*** 230–232, 240–243, 250–252
Yellow, ***Vol. 2:*** 16, 147, 156
Yellow/green, ***Vol. 2:*** 147, 156
Yellow/orange, ***Vol. 2:*** 16, 147, 156
Yellowing chemistry measurements, ***Vol. 1:*** 154

Z

Z dimension (of cuvette), ***Vol. 2:*** 18
Z observed statistic, ***Vol. 5:*** 155
Zero filling, ***Vol. 4:*** 364
Zero optical path difference (ZPD, ZOPD), ***Vol. 4:*** 66–68, 365
Z-test (for dispersion), ***Vol. 3:*** 454